GEOMETRY FORMULAS

Pythagorean Theorem

For a right triangle

$$a^2 + b^2 = c^2$$

Triangle

$$\alpha + \beta + \gamma = 180°$$

$$\text{area} = \tfrac{1}{2}bh$$

$$\text{perimeter} = a + b + c$$

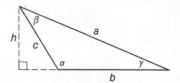

Parallelogram

$$\text{area} = bh$$

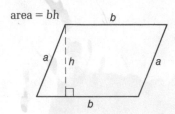

Trapezoid

$$\text{area} = \tfrac{1}{2}h\,(b_1 + b_2)$$

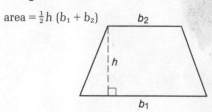

Circle

$$\text{circumference} = 2\pi r$$
$$\text{area} = \pi r^2$$
$$s = r\theta$$
$$\text{area of sector} = \tfrac{1}{2}r^2\theta$$

Sphere

$$\text{surface area} = 4\pi r^2$$
$$\text{volume} = \tfrac{4}{3}\pi r^3$$

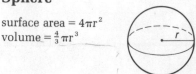

Cone (right circular)

$$\text{lateral surface} = \pi r s$$
$$\text{volume} = \tfrac{1}{3}\pi r^2 h$$

Cylinder (right circular)

$$\text{lateral surface} = 2\pi r h$$
$$\text{volume} = \pi r^2 h$$

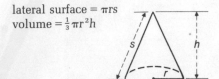

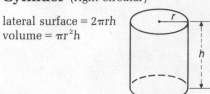

TRIGONOMETRY

Second Edition

TRIGONOMETRY
A Modern Approach

Joseph Elich
UTAH STATE UNIVERSITY

Carletta J. Elich
LOGAN HIGH SCHOOL

Lawrence O. Cannon
UTAH STATE UNIVERSITY

 ADDISON-WESLEY PUBLISHING COMPANY

Reading, Massachusetts • Menlo Park, California
Don Mills, Ontario • Wokingham, England • Amsterdam
Sydney • Singapore • Tokyo • Mexico City
Bogotá • Santiago • San Juan

Jeffrey M. Pepper   Acquisitions Editor
Loretta M. Bailey   Art Editor
bj Art Services   Illustrator
Ann E. DeLacey   Manufacturing Supervisor
Maureen Langer   Text and Cover Designer
Margaret Pinette   Production Editor
Barbara Willette   Copy Editor
Cheryl Wurzbacher   Production Manager

Library of Congress Cataloging in Publication Data
Elich, Joseph, 1918–
   Trigonometry, a modern approach.

   Rev. ed. of: Trigonometry using calculators. 1980.
   Includes index.
   1. Trigonometry   2. Calculating-machines,   I. Elich,
Joseph, 1918–   . Trigonometry using calculators.   II. Elich,
Carletta J., 1935–   .   III. Cannon, Lawrence O.   IV. Title.
QA531.E44   1985   516.2′4′0285   84-9287
ISBN 0-201-10523-3

ABCDEFGHIJ-DO-898765

Preface

In revising the book *Trigonometry Using Calculators*, we felt a need for a title that more adequately describes the contents and philosophy of the book. The title that we liked initially was *Trigonometry: A Modern Approach*, but one reviewer pointed out that "every author since Hipparchus has used that title." After considerable discussion with colleagues, potential users, and publishers, we decided to follow an old and honorable tradition in naming the text.

The primary reason for our choice is that we believe this book really does offer a modern approach to trigonometry. Trigonometry has changed in fundamental ways in recent years, both *in terms of the applications*, which are of greatest importance to users, and *in terms of computational tools*, which affect the way trigonometry is learned and the convenience with which it is applied. This book reflects these changes, emphasizing the significance of the trigonometric functions as usable functions and utilizing part of the incredible technological development of high-speed computing devices as an integral part of the learning process.

Functional Trigonometry and Modern Technology

The most significant applications of trigonometry make use of the six functions, which were originally defined in terms of ratios of sides of triangles but which are now properly considered as functions of real numbers. As the world of science broadens, as more and more disciplines require mathematical treatment, more students need to understand more mathematics. Virtually all cyclic or periodic phenomena are best described by use of trigonometric functions. Students studying engineering and physics have always needed a functional treatment of trigonometry, but the need is now expanding to economics and natural resources, to all areas of business, and to many of the other social sciences.

Use of Calculators

Hand-held scientific calculators have completely changed the way we think about much of the world. They have become *the* tool needed by today's practicing scientists, engineers, and businesspeople. The slide rule, once the trademark of the engineer and a part of the core curriculum of every school in the country, is no longer even part of the vocabulary of engineers. Most engineering students don't know what a slide rule is, but few engineers today could survive for long without a calculator.

The calculator has altered trigonometry in just as fundamental a fashion. As is pointed out in Section 2.2, laboriously computed tables dating from the second century A.D. served all of science until the 1600s when more laborious work provided more detailed tables. The invention of calculus did not change our dependence upon tables. All of the computations performed with these tabular data needed logarithms to make them feasible, and manipulation of logarithms needed either more tables or a mechanical device such as a slide rule. The slide rule was incomparably handier but was severely limited in accuracy. *Now automatic computing devices are as handy as the slide rule and more accurate than the most elaborate tables.* It seems ridiculous to ignore what these remarkable computing tools have made possible.

Trigonometry texts have traditionally devoted a substantial amount of space to logarithms and linear interpolation in connection with extensive tables. Our feeling is that linear interpolation is little needed in a modern trigonometry course. We prefer that students who need it learn specific techniques for the tabular data with which they may have to work. Nonetheless, for those instructors who wish to treat interpolation in trigonometry, we have included a brief discussion in Appendix C. The appendixes include a substantial body of information, including numerous problem sets, and should be used as an important resource.

Looking Ahead to Computers

As technology continues to change, we must recognize that the calculator will not always remain what it is today. We attempt to anticipate some of the possible changes with two features in this book, a discussion of *hand-held computers* and *optional computer problems.*

Computers tend to become smaller, faster, and more powerful as they become more readily available. Several manufacturers are producing hand-held computers with the capabilities of microcomputers, including high-level language programming and even printers. In several footnotes throughout the text we address the use of such computers for needed computation. A brief discussion on their use is included in Appendix A.

Most chapters contain computer problems *for enrichment only*. These are not an essential part of the course, but are available for use when appropriate. See the next section for more information about the computer problems.

Basic Philosophy and Special Features of the Text

A number of features of this revision are significant. Topics have been reorganized and the book has been completely rewritten. A major goal has been to make all of the exposition readable by the student. Obviously every elementary text is intended for the student, but some are conspicuously more successful than others. We believe that student response will confirm that students can truly read and learn from the text.

The text is designed to:

- present and discuss ideas,
- illustrate the ideas by several examples worked in detail,
- give practice with carefully designed problem sets.

To accomplish these aims there are a number of distinctive features that should be noted.

Examples and Exercises

In the discussion of example problems, we explain why particular strategies are useful, and we point out both alternative approaches and pitfalls to be avoided. The problem sets anticipate later work and stimulate related ideas as well as providing the drill necessary to master skills and ideas. Each exercise set includes a variety of problems ranging from simple to challenging. Hints are given for some nonroutine problems.

Introductory Chapter Sections

Key ideas needed for the chapter are included in introductory review sections (with exercises). Instructors may use as much or as little of this material as needed for a specific class.

Chapter Summaries

Important ideas and theorems are summarized with the key formulas from the chapter to help students put their information into a more useful conceptual framework and to serve as a convenient reference.

In the summaries, and throughout the book, students are given an indication of uses and reasons for studying particular topics.

Review Exercises

An integrated set of exercises at the end of each chapter allows students to check mastery of concepts and pull together key ideas.

Optional Computer Problems

As noted above, these enrichment sections are included as a service to students and teachers. We recognize that most trigonometry courses do not have the leisure to permit an instructor to teach programming, nor do we expect programming experience from our students. An increasingly common feature of our schools, however, is that *many of our students are already familiar with computers* and that *many classrooms have computing facilities that are not well-used for the learning of mathematics.* The computer problems allow instructors to encourage students, either by themselves or with a teacher's guidance, to take advantage of a wide variety of problems, to develop computer skills, and more fully to understand the fundamental concepts of the chapter.

Exact and Approximate Calculations

Students are given practice with both *exact* (symbolic) and *approximate* (calculator) numbers and notation. Expressing answers in exact form involves application of definitions and/or basic concepts; giving results in decimal form provides familiarity with numbers as they occur in real-life settings. In this regard, we help students understand the meaning of significant digits in relation to the display shown on their calculators. The fact that the calculator shows eight or ten digits does not automatically invest those figures with meaning. A discussion of significant digits and their relation to the real and ideal worlds is included in Appendix B and in the text where needed.

Degree and Radian Modes

The fact that calculators deal differently with trigonometric functions in radian and degree mode is recognized and explained. For instance, there are two different sine functions as well as two inverse sine functions, depending on the mode. Both are shown to be useful.

Graphing

Graphs are emphasized for their great utility. Many of the crucial features of a function become obvious from a reasonable graph. The trigonometric functions are graphed in Chapter 3 and in Chapter 6,

Contents

6 Graphs of Trigonometric Functions   219

7 Solving Triangles   247

8 Polar Coordinates   294

TRIGONOMETRY

Introductory Concepts

1

The word *trigonometry* suggests a study of measurements related to triangles, but the most important applications of this subject in modern society do not deal with numerical solutions of triangles or the determination of distances that cannot be directly measured. In fact, most applications of trigonometry make no reference to angles at all but use the fact that trigonometric functions are defined on sets of real numbers.

In physics and engineering, one encounters applications in which it is necessary to find some unknown part of a triangle in terms of other given parts. However, the analysis of such unrelated things as alternating circuits, wave motion, weather patterns, tides, musical tones, and economic cycles all involve periodic phenomena; and since trigonometric functions possess periodic properties, it is not surprising that they should occur in a natural way in such applications.

Although it is possible to develop all of trigonometry without any reference to angles or triangles, it is not in our interests to take such an abstract approach in an introductory course.

We shall pursue the conventional geometric presentation in which trigonometric functions are defined on measures of angles, but we shall ultimately see that we are studying nothing but special *real-valued functions defined on sets of real numbers*. This is precisely the setting required for further development of trigonometric functions in calculus and for subsequent applications in a variety of fields.

The main goals of this book are to present:

1. a study of analytic properties of the trigonometric functions in preparation for further study in calculus and beyond.
2. techniques required for solving triangles.

However, before we proceed with this task, we first remind the reader of some facts from geometry that will be needed in our subsequent development.

1.1 Review of Facts From Plane Euclidean Geometry (Optional)

The study of plane geometry implies that we begin with a given plane and that all of the geometric figures considered (lines, rays, angles, triangles, circles, and so on) are subsets of this plane. A *ray* is defined as a half line together with its endpoint; an *angle* is the union of two rays with a common endpoint called the *vertex* of the angle. The two rays are the *sides* of the angle. When the two sides are in opposite directions, we have a *straight angle*. Its measure is 180°. A *right angle* is "half of a straight angle" and has measure 90° (see Fig. 1.1). An *acute angle* is one that has measure less than 90°, while an *obtuse angle* has measure greater than 90°.

We continue with some facts from a prerequisite course in geometry. These will be useful in subsequent portions of the book.

Triangles

If A, B, and C are three noncollinear points, then *triangle ABC* ($\triangle ABC$) is the union of the three line segments $\overline{AB}$, $\overline{AC}$, and $\overline{BC}$. The points A, B, and C are called the *vertices* of the triangle, and the line segments are the *sides* of the triangle.

The three *angles* of the triangle shown in Fig. 1.2 are denoted by $\angle A$, $\angle B$, and $\angle C$ (or merely as angles A, B, and C). For instance, $\angle A$ is the angle consisting of the union of the two rays $\overrightarrow{AB}$ and $\overrightarrow{AC}$. $\angle B$ and $\angle C$ are defined similarly. The size or measure of an angle is frequently denoted by a Greek letter.* In Fig. 1.2 the size of angles at A, B, and C are denoted by α, β, and γ.

The letters a, b, and c denote the *lengths* of the three sides. It is conventional to label the side opposite A as a, and similarly for B, b and C, c. Also, $\overline{AB}$ denotes *the line segment* from A to B while $|\overline{AB}|$ is used for the length of $\overline{AB}$. Thus $|\overline{AB}| = a$.

Note: In mathematical notation it is frequently convenient to let symbols serve multiple duty when the meaning is clear from the

---

* Measures of angles are frequently denoted by Greek letters. The ones most commonly used are α, β, γ, δ, θ, and ϕ.

Greek Alphabet

A	α	Alpha	N	ν	Nu
B	β	Beta	Ξ	ξ	Xi
Γ	γ	Gamma	O	o	Omicron
Δ	δ	Delta	Π	π	Pi
E	ϵ	Epsilon	P	ρ	Rho
Z	ζ	Zeta	Σ	σ	Sigma
H	η	Eta	T	τ	Tau
Θ	θ	Theta	Y	υ	Upsilon
I	ι	Iota	Φ	ϕ	Phi
K	κ	Kappa	X	χ	Chi
Λ	λ	Lambda	Ψ	ψ	Psi
M	μ	Mu	Ω	ω	Omega

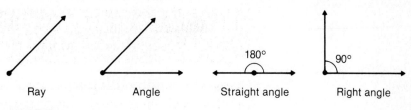

FIGURE 1.1

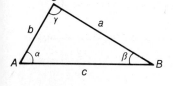

FIGURE 1.2

context of discussion. For instance, A, B, and C may refer to the *vertices*, or they may designate the corresponding *angles*, or even the *size of angles* (when we write A = 45°, for example). Similarly, we shall refer to the *sides* as a, b, and c, or these letters may designate the *lengths of the sides* (when we write a = 5, for instance).

An important property of triangles in Euclidean geometry is that the sum of the angles is 180° for every triangle. This can be written as

$$\angle A + \angle B + \angle C = 180°. \qquad \text{[1.1]}$$

Similar triangles

Two triangles are said to be *similar* if they have the same "shape" but not necessarily the same size. That is, if the corresponding angles of two triangles are equal, then they are similar. In Fig. 1.3 we have two triangles, $\triangle ABC$ and $\triangle DEF$, in which $\angle A = \angle D$, $\angle B = \angle E$, and $\angle C = \angle F$ and so they are similar. A useful and important property of similar triangles is that the ratios of the corresponding lengths of sides are equal. That is, for the similar triangles shown in Fig. 1.3, $a:d = b:e = c:f$, or

$$\frac{a}{d} = \frac{b}{e} = \frac{c}{f}. \qquad \text{[1.2]}$$

FIGURE 1.3

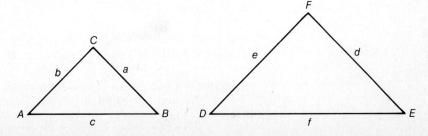

Consequently, from Eq. (1.2) we see that ratios of corresponding sides in each triangle are equal (such as $a/b = d/e$), a fact that allows us to define trigonometric functions for acute angles in Section 2.2.

Isosceles and equilateral triangles

If the lengths of two of the sides of $\triangle ABC$ are equal, then $\triangle ABC$ is called an *isosceles triangle*. The two angles opposite the equal sides are also equal.

If the lengths of all three sides of a triangle are equal, then it is called an *equilateral triangle*. The three angles of an equilateral triangle are also equal; and since their sum is 180°, each must be a 60° angle.

Right triangle

A triangle is said to be a *right triangle* if one of its angles is a right angle. A right triangle with $\angle C = 90°$ is shown in Fig. 1.4. Side c (opposite the right angle) is called the *hypotenuse*, and the other two sides are referred to as *legs* of the right triangle.

Right triangles are characterized by the following theorem.

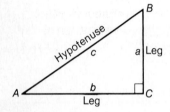

FIGURE 1.4

Pythagorean Theorem

If a, b, and c represent the lengths of the sides of triangle ABC, where c is the longest side, then $\triangle ABC$ is a right triangle if and only if

$$a^2 + b^2 = c^2. \qquad [1.3]$$

This is an extremely important theorem that will be used on numerous occasions throughout this book. The "if" part of the "if and only if" means that if $a^2 + b^2 = c^2$, then $\triangle ABC$ is a right triangle; the "only if" part means that if $\triangle ABC$ is a right triangle, then $a^2 + b^2 = c^2$.

EXAMPLE 1 Suppose $\triangle ABC$ has sides of lengths 3, 4, and 5. Is $\triangle ABC$ a right triangle?

Solution Since $3^2 + 4^2 = 5^2$, then by the "if" part of the Pythagorean Theorem, $\triangle ABC$ is a right triangle. ■

EXAMPLE 2 Suppose $\triangle ABC$ is a right triangle with hypotenuse $c = 337$ cm and one leg $a = 175$ cm. Determine b.

Solution   Applying the Pythagorean Theorem (the "only if" part), we have

$$175^2 + b^2 = 337^2;$$

thus $b = \sqrt{337^2 - 175^2} = 288$ where the evaluation is easily done by use of a calculator. Therefore $b = 288$ cm.   ■

Special cases of right triangles

Here we discuss two special cases of right triangles that are so useful that they should become part of the fundamental vocabulary of every user of trigonometry. The relations are easier to remember if there is a visual association. Each of these special right triangles is *half* of a familiar figure.

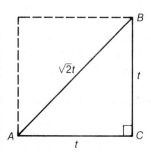

FIGURE 1.5

45°–45° right triangle

We can think of this as "half of a square," as shown in Fig. 1.5. It is an isosceles triangle, and the Pythagorean Theorem gives the relations shown. The ratios of the three sides $a:b:c$ are always $1:1:\sqrt{2}$.

30°–60° right triangle

This is "half of an equilateral triangle," so *the side opposite the 30° angle is always half the length of the hypotenuse* (see Fig. 1.6). The Pythagorean Theorem then shows that the length of the other leg is $\sqrt{3}/2$ times the length of the hypotenuse. The ratio of the sides $a:b:c$ is always $\sqrt{3}/2:1/2:1$; or if the shorter leg has length 1, then the hypotenuse has length 2, and the other leg has length $\sqrt{3}$.

EXAMPLE 3   In $\triangle ABC$, $A = 45°$, $C = 90°$, and $a = 4$. Find b and c.

Solution   Here we have a 45°–45° right triangle, so $b = a = 4$. The Pythagorean Theorem gives

$$c^2 = 4^2 + 4^2 = 32 \qquad \text{and so} \qquad c = \sqrt{32} = 4\sqrt{2}.$$

Thus $b = 4$, and $c = 4\sqrt{2}$.   ■

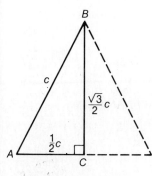

FIGURE 1.6

EXAMPLE 4   In $\triangle ABC$, $A = 30°$, $C = 90°$, and $a = 3$. Find b and c.

Solution   Here we have a 30°–60° right triangle, so the hypotenuse is twice the length of the side opposite the 30° angle. That is, $c = 2a = 2(3) = 6$. Apply the Pythagorean Theorem to find b:

$$b = \sqrt{c^2 - a^2} = \sqrt{6^2 - 3^2} = \sqrt{36 - 9} = \sqrt{27} = 3\sqrt{3}.$$

Therefore, $c = 6$, and $b = 3\sqrt{3}$.   ■

Perimeter and area of a triangle

The *perimeter* P of triangle ABC is the "distance around the triangle," or the sum of the lengths of the sides. That is,

$$P = a + b + c.$$ [1.4]

The area of the region bounded by a triangle is given by the formula

$$\text{Area} = (1/2) \cdot (\text{base}) \cdot (\text{altitude}).$$ [1.5]

For instance, in Fig. 1.7 the area of $\triangle ABC$ is given by

$$\text{Area} = \frac{1}{2} ch \qquad \text{(where } c \text{ is the base)}$$

or $$\text{Area} = \frac{1}{2} ak \qquad \text{(where } a \text{ is the base)}.$$

A similar formula holds if b is taken as the base.

EXAMPLE 5 In a 30°–60° right triangle the perimeter is 24. What are the lengths of the three sides? Give answers in exact form and rounded off to two significant digits.

Solution Let x denote the length of the side opposite the 30° angle. Then the hypotenuse is 2x, and the other leg is $\sqrt{3}x$, as shown in Fig. 1.8. The perimeter is 24, so

$$x + 2x + \sqrt{3}x = 24, \qquad 3x + \sqrt{3}x = 24, \qquad (3 + \sqrt{3})x = 24.$$

Now solve for x:

$$x = \frac{24}{3 + \sqrt{3}} = \frac{24(3 - \sqrt{3})}{(3 + \sqrt{3})(3 - \sqrt{3})} = \frac{24(3 - \sqrt{3})}{6} = 4(3 - \sqrt{3}).$$

In exact form the lengths of the three sides are given by

$$x = 4(3 - \sqrt{3}), \qquad \sqrt{3}x = 4\sqrt{3}(3 - \sqrt{3}), \qquad 2x = 8(3 - \sqrt{3}).$$

FIGURE 1.7

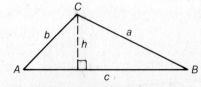
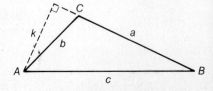

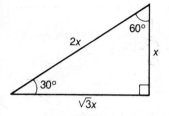

FIGURE 1.8

Using a calculator to evaluate these three lengths and rounding off to two significant digits (see Appendix B), we get

$$x \approx 5.1, \qquad \sqrt{3}x \approx 8.8, \qquad 2x \approx 10.^*\qquad ■$$

Rectangular coordinates

Geometric figures in a plane can be described algebraically (or analytically) by associating in a one-to-one fashion every point in the plane with an *ordered pair* of real numbers. In Fig. 1.9 we show a system of *rectangular coordinates* in which the horizontal number line is labeled as the *x-axis* and the vertical line is the *y-axis*. The point of intersection is called the *origin*.

In addition to the two axes the plane is divided into four quadrants. The diagram illustrates the point-ordered pair correspondence for five points. *A* and *B* are in quadrant I, *C* is in quadrant II, *D* is in quadrant III, and *E* is not in any quadrant.

In general, any point *P* in the plane is associated with an ordered pair (x, y). Frequently, we shall refer to the point *P* as "the point (x, y)." For instance, we shall say "the point $(3, -\sqrt{2})$ is in the fourth quadrant."

FIGURE 1.9

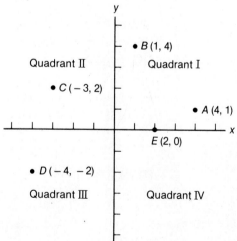

---

*Note that in using exact-form values (radical form) we get genuine exactness. The sum of the three sides is precisely 24:

$$4(3 - \sqrt{3}) + 4\sqrt{3}(3 - \sqrt{3}) + 8(3 - \sqrt{3}) = 12 - 4\sqrt{3} + 12\sqrt{3} - 12 + 24 - 8\sqrt{3} = 24.$$

However, when we use the approximate decimal answers, the sum is

$$5.1 + 8.8 + 10 = 23.9 \approx 24.$$

See Section 1.2 for a discussion of exact form versus approximate decimal form.

Distance between points

Suppose $P(x_1, y_1)$ and $Q(x_2, y_2)$ are two points in the plane as shown in Fig. 1.10 and d is the distance between them. Applying the Pythagorean Theorem to the right triangles shown, we get the following *distance formula* for the distance between P and Q:

$$d = \sqrt{(x_1 - x_2)^2 + (y_1 - y_2)^2}.$$ [1.6]

EXAMPLE 6 Find the distance between points $P(-1, 4)$ and $Q(1, -2)$.

Solution Substitution into Eq. (1.6) gives

$$d = \sqrt{(-1 - 1)^2 + [4 - (-2)]^2} = \sqrt{(-2)^2 + 6^2} = \sqrt{40} = 2\sqrt{10}.$$

Thus the distance between P and Q is $2\sqrt{10}$. ∎

Midpoint of a line segment

Suppose $A(x_1, y_1)$ and $B(x_2, y_2)$ are two given points. The *midpoint* of line segment $\overline{AB}$ is given by

$$M\left(\frac{x_1 + x_2}{2}, \frac{y_1 + y_2}{2}\right).$$

We leave it to the reader to show that $|\overline{AM}| + |\overline{BM}| = |\overline{AB}|$ and $|\overline{AM}| = |\overline{BM}|$.

EXAMPLE 7 Find the coordinates of the midpoint M of the line segment $\overline{AB}$ where $A(-4, 1)$ and $B(2, 5)$.

FIGURE 1.10

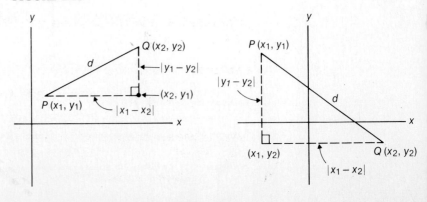

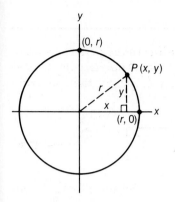

FIGURE 1.11

Solution M is given by $((-4 + 2)/2, (1 + 5)/2)$. Thus, M is the point $(-1, 3)$. ∎

Circles

A *circle* with center at C and radius r is the set of all points P in a plane such that the distance from C to P is equal to r. That is, $|\overline{CP}| = r$.

If the center of the circle is at $(0, 0)$ as shown in Fig. 1.11, then the circle is the set of all points $P(x, y)$ satisfying the equation

$$\boxed{x^2 + y^2 = r^2.}$$ [1.7]

If $r = 1$, the circle is called the *unit circle* and is given by the equation

$$x^2 + y^2 = 1.$$

Circumference of a circle

The *circumference* C of a circle of radius r (or diameter d) is given by

$$\boxed{C = 2\pi r \quad \text{or} \quad C = \pi d,}$$ [1.8]

where $\pi = 3.1415926\ldots$ (a nonterminating and nonrepeating decimal number).

Area of a circle

The *area of the region* bounded by a circle of radius r is given by the formula

$$\boxed{\text{Area} = \pi r^2.}$$ [1.9]

EXAMPLE 8 Find the circumference and area of a circle of radius 2.4 cm.

Solution Substituting 2.4 for r in Eq. (1.8) and Eq. (1.9) gives

$$\text{Circumference} = 2\pi(2.4) = 4.8\pi \text{ cm,}$$
$$\text{Area} = \pi(2.4)^2 = 5.76\pi \text{ cm}^2.$$

We can use a calculator to get decimal approximations (to two significant digits)

$$\text{Circumference} \approx 15 \text{ cm} \quad \text{and} \quad \text{Area} \approx 18 \text{ cm}^2 \quad ∎$$

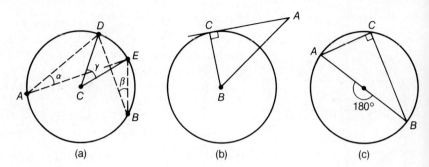

(a)   (b)   (c)

FIGURE 1.12

Other useful properties of circles

1. An angle is called a *central angle* of a circle if its vertex is at the center of the circle and the sides are along radial lines. In Fig. 1.12(a), C is the center of the circle shown, and γ is a central angle that subtends arc $\overset{\frown}{DE}$. If A and B are any points on the circle as shown in the diagram, then *angles α and β are equal.*
   Also, *angle α is half of angle γ*; that is, $\alpha = 1/2\gamma$. These are facts from geometry.

2. A line tangent to a circle is perpendicular to the radial line at the point of tangency. This is shown in Fig. 1.12(b), and we write $\overline{AC} \perp \overline{BC}$.

3. If $\overline{AB}$ is a diameter of a circle and C is a point on the circle as shown in Fig. 1.12(c), then *angle C is a right angle.* Note that this is a consequence of the statement in 1 above where $\gamma = 180°$ and α is angle C.

EXAMPLE 9   In the diagram shown in Fig. 1.13, $\triangle ABC$ is inscribed in a circle of radius 4, $\overline{AB}$ is a diameter, $\angle A = 30°$, and D is the center. Determine the length of $\overline{AC}$ and the angle β.

FIGURE 1.13

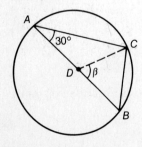

Solution   Since $\overline{AB}$ is a diameter, $\angle C = 90°$. Then $\triangle ABC$ is a 30°–60° right triangle, and so $|\overline{BC}| = \frac{1}{2}|\overline{AB}| = \frac{1}{2}(8) = 4$. Applying the Pythagorean Theorem to right triangle ABC gives $|\overline{AC}| = \sqrt{8^2 - 4^2} = \sqrt{64 - 16} = \sqrt{48} = 4\sqrt{3}$.
   Angle β is a central angle, and by 1 above $\beta = 2(30°) = 60°$. This result can also be seen by noting that $\triangle BCD$ is an equilateral triangle (all three sides have length 4).   ■

EXAMPLE 10   Suppose the lengths of the three sides of a triangle are given by $a = 5$, $b = 3$, and $c = 7$. Find the altitude to side a and the area of the triangle.

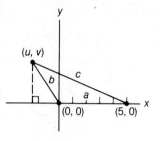

FIGURE 1.14

Solution Let us position the triangle in relation to a system of coordinates as shown in Fig. 1.14, where $(0,0)$ and $(5,0)$ are the endpoints of side a. Suppose (u, v) is the third vertex. Note that v is the altitude to side a. We can find u and v by applying the distance formula to points $(0,0)$ and (u, v) and then to points $(5,0)$ and (u, v).

$$u^2 + v^2 = 3^2 \quad \text{(since } b = 3\text{),} \qquad \text{[1.10]}$$

$$(u - 5)^2 + v^2 = 7^2 \quad \text{(since } c = 7\text{).} \qquad \text{[1.11]}$$

Now expand Eq. (1.11) to get $u^2 - 10u + 25 + v^2 = 49$, and in this result replace the $u^2 + v^2$ by 9 [from Eq. (1.10)]. This gives $-10u + 25 + 9 = 49$, or $u = -\frac{3}{2}$. To find v, substitute $-\frac{3}{2}$ for u in Eq. (1.10) and then solve for v to get $v = \sqrt{9 - (9/4)} = \sqrt{27/4} = 3\sqrt{3}/2$. Thus the altitude to side a is $3\sqrt{3}/2$.

The area of the triangle can now be found by using a as the base:

$$\text{Area} = \frac{1}{2}(5)\left(\frac{3\sqrt{3}}{2}\right) = \frac{15\sqrt{3}}{4}. \quad \blacksquare$$

EXERCISE 1.1

Several problems call for answers in *exact form* or in *approximate decimal form*. See Section 1.2 for a more detailed discussion. Also see Appendix B for a discussion of approximate numbers and significant digits.

1. Draw a triangle with the given measurements. Label it with vertices A, B, C and corresponding angle measures α, β, γ and sides of lengths a, b, c (see Fig. 1.2). Use a protractor and ruler if available; otherwise, a reasonable sketch is sufficient.
   a] $\alpha = 50°$, $\beta = 60°$, and $c = 3$
   b] $\alpha = 30°$, $b = 5$, and $c = 3$

2. In each of the triangles in Problem 1, draw the altitude from vertex C to side $\overline{AB}$ and then estimate its length (to one decimal place if your diagram is sufficiently large and accurately drawn). Use your estimated value of the altitude to get the area of the triangle (approximately).

In Problems 3–10, the two sides (legs) of a *right triangle* are labeled a, b and the hypotenuse is c. Two of the lengths are given. Determine the third. Give answers in simplified exact form (involving square roots where necessary).

3. $a = 5$, $b = 8$ 4. $a = 17$, $c = 33$ 5. $a = \sqrt{3}$, $b = \sqrt{5}$

6. $a = 5$, $b = 12$ 7. $a = 10$, $c = 26$ 8. $a = 2\sqrt{3}$, $b = 3\sqrt{5}$

9. $b = 2730$, $c = 4666$ 10. $a = 24208$, $b = 10575$

11. In a 30°–60° right triangle the side opposite the 30° angle has length 5. Determine
    a] the length of the hypotenuse,
    b] the length of the side opposite the 60° angle.

12. In a 30°–60° right triangle the hypotenuse is 6. Find the lengths of the two legs.

In Problems 13–16, $\triangle ABC$ is labeled as in Fig. 1.2, and $\gamma = 90°$. Give answers in exact form (involving square roots in some cases).

13. If $\alpha = 45°$ and $c = 8$, find β, a, and b.

14. If $\alpha = 45°$ and $a = 4$, determine β, b, and c.

15. If $\beta = 30°$ and $c = 5$, find α, a, and b.

16. If $\alpha = 60°$ and $b = 3$, determine β, a, and c.

In Problems 17–19, give answers in exact form and also as a decimal approximation.

17. Determine the circumference and area of a circle of diameter 3.20.

18. The circumference of a circle is 6.84. Determine its radius.

19. Find the diameter of a circle whose area is 3.48.

In Problems 20 and 21, $\triangle ABC$ is inscribed in a circle of radius 4, and $\overline{AB}$ is a diameter of the circle. Give answers in exact form.

20. If $\angle BAC = 30°$, find the lengths of $\overline{AB}$, $\overline{BC}$, and $\overline{AC}$.

21. If the length of $\overline{AC}$ is 3, determine the lengths of $\overline{AB}$ and $\overline{BC}$.

22. Triangle ABC is inscribed in a circle with center at D. If $\overline{AB}$ is a diameter and $\angle ADC = 48°$, what is the measure of $\angle BAC$?

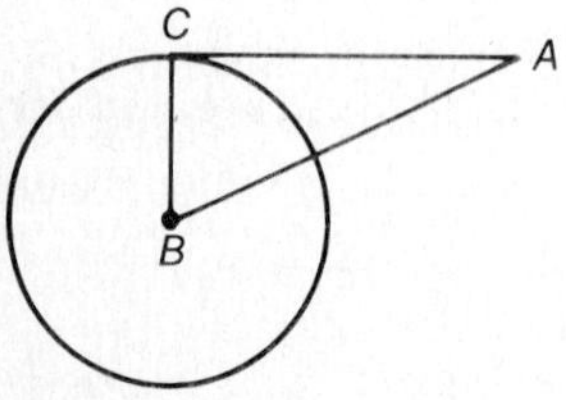

23. The diagram in Fig. 1.15 shows a circle with center at B and diameter 5. The line through A and C is tangent to the circle at C. If the length of $\overline{AB}$ is 6, what is the length of $\overline{AC}$? (Exact and also to two decimal places.)

24. The vertices of $\triangle ABC$ are given by $A(1, 0)$, $B(3, 2)$, $C(4, 1)$. Show that $\triangle ABC$ is a right triangle.* Then determine its perimeter and area.

25. In a right triangle, one angle is 60°, and the perimeter is 5.64. Determine the lengths of the two legs and the hypotenuse (exact and to three significant digits). (See Example 5.)

26. In a right triangle, one angle is 45°, and the perimeter is 16. Determine the lengths of each of the three sides (exact and also to two significant digits).

27. The diagram in Fig. 1.16 shows $\triangle ABC$ inscribed in a circle of radius 6 and center at D. If D is on the line through A and B and $\angle ADC$ is 60°, find the lengths of each of the sides of $\triangle ABC$.

In Problems 28–33, draw a diagram showing the given points P and Q. Then determine the distance between them. Give answers in exact form and also rounded off to two significant digits.

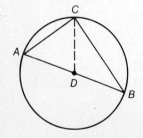

FIGURE 1.15

FIGURE 1.16

---

* Here it is necessary to express lengths as exact numbers (involving square roots) if you want to apply the Pythagorean Theorem. See Example 7 in Section 1.2.

28. $P(-3, 1)$, $Q(2, 4)$ **29.** $P(-2, -2)$, $Q(-3, 4)$

30. $P(3, 1)$, $Q(3, -4)$ **31.** $P(2.5, -3.1)$, $Q(3.3, 2.6)$

32. $P(\sqrt{3}, -1)$, $Q(\sqrt{27}, 2)$ **33.** $P(\sqrt{2}, 3)$, $Q(-\sqrt{8}, 2)$

In Problems 34 and 35, the three vertices A, B, C of a triangle are given. Draw a diagram showing the triangle and label the corresponding sides opposite the vertices as a, b, c. Determine the lengths of the sides; give answers rounded off to two significant digits.

34. $A(-2, 4)$, $B(-3, -1)$, $C(2, 4)$ **35.** $A(2, -3)$, $B(-1, -4)$, $C(3, 5)$

In Problems 36 and 37, the coordinates of three points A, B, and C are given. Determine whether or not $\triangle ABC$ is a right triangle.* Give reasons to support your answers.

36. $A(3, 2)$, $B(-1, -2)$, $C(5, 0)$ **37.** $A(-3, 1)$, $B(0, 4)$, $C(0, -2)$

In Problems 38 and 39, determine whether or not $\triangle ABC$ is isosceles. Give reasons to support your answers.

38. $A(0, 0)$, $B(6, 0)$, $C(3, 4)$ **39.** $A(0, 0)$, $B(0, 5)$, $C(-3, 4)$

In Problems 40 and 41, determine whether or not $\triangle ABC$ is equilateral. Give reasons to support your conclusions.

40. $A(0, 0)$, $B(2, 0)$, $C(1, 1)$ **41.** $A(-1, 0)$, $B(3, 0)$, $C(1, \sqrt{12})$

In Problems 42 and 43, determine whether or not the three given points are on the same line. (Hint: Find distances $|\overline{AB}|$, $|\overline{AC}|$, and $|\overline{BC}|$ and see whether the sum of two of these equals the third.)

42. $A(-2, -1)$, $B(-1, 0)$, $C(1, 2)$ **43.** $A(-2, 3)$, $B(6, -1)$, $C(2, 1)$

In Problems 44 and 45, find the midpoint of the line segment $\overline{AB}$ for the given points A and B.

44. $A(-2, 5)$, $B(4, -3)$ **45.** $A(5, -4)$, $B(2, 3)$

46. In Example 10, three sides of a triangle were given by $a = 5$, $b = 3$, and $c = 7$, and the altitude to side a was determined. Follow a similar technique to find the altitude to side c and then find the area of the triangle. The area should agree with that found in Example 10. (Suggestion: Position the triangle so that side c is on the positive x-axis with endpoints $(0, 0)$ and $(7, 0)$. Also take point (u, v) in the first quadrant so that the endpoints of side b are $(0, 0)$ and (u, v). Then the endpoints of side a are (u, v) and $(7, 0)$.)

47. There are right triangles such that the lengths of the sides are whole numbers and the hypotenuse is 1 greater than one of the legs (say

---

* See the footnote to Problem 24.

$c = b + 1$). For instance, $a = 5$, $b = 12$, and $c = 13$ is such a triangle.

a] Find four other such triangles.

b] Assuming $c = b + 1$, determine the conditions on a and b that will give all such triangles.

(Hint: Apply the Pythagorean Theorem with sides a, b, and $b + 1$ and see how b must be chosen to give whole number lengths.)

48. Follow the instructions of Problem 48 but have the hypotenuse 2 greater than one of the legs (say $c = b + 2$). An example is $a = 8$, $b = 15$, and $c = 17$.

1.2 Measures of Angles: Degrees and Radians

In the preceding section we reviewed angles and angular measure as they occur in a geometry course. We are now interested in extending these notions to include angles whose measures are not limited to measures between 0° and 180°. Also in this section we introduce another unit of measure, the radian. In most of our subsequent discussions this is the unit that will be emphasized, since it is required in the study of calculus and in applications.

We begin by thinking of an angle as being generated by rotating a ray about its endpoint. The initial position of the ray is called the *initial side* of the angle, and the final position is called the *terminal side* of the angle. The point about which rotation takes place (the endpoint of the ray) is called the *vertex* of the angle. Hence an angle will be considered as the union of two rays with a common endpoint *along with a rotation*. The measure of an angle is determined by the amount of rotation. This allows us to have angles with measures greater than 180°. By considering direction of rotation we can also introduce angles with negative measures.

If a ray is rotated about its endpoint from an initial position to a terminal position, then the measure of the resulting angle will be *positive* if rotation is in the *counterclockwise* direction and *negative* if rotation is in the *clockwise* direction. Frequently, we shall say "the angle is positive" to mean "the measure of the angle is positive" and similarly for negative.

We can indicate the measure of an angle (the amount of rotation) with an arrow diagram. In Fig. 1.17, $\angle A$ and $\angle B$ are positive, while $\angle C$ is negative. Also, $\angle B$ involves a rotation of more than one complete revolution.

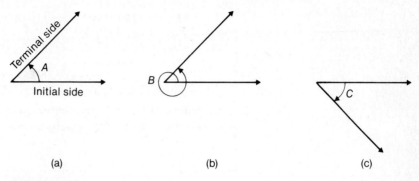

(a)   (b)   (c)

FIGURE 1.17

Units of angular measure

There are two systems of angular measure that are widely used: degrees-minutes-seconds and radians. Scientific calculators frequently include a third unit of angle measure, the *grad*.* Since this unit is rarely encountered, it will not be used in this text.

Degrees-minutes-seconds

If the initial side of an angle is rotated counterclockwise one complete revolution, the measure of the corresponding angle is defined to be 360 degrees, denoted by 360°. Thus an angle of 1° is one in which the initial side is rotated counterclockwise 1/360 of a revolution. For more refined measurements the units of minutes and seconds are used, which are defined by:

60 minutes equals one degree, denoted by $60' = 1°$,

60 seconds equals one minute, denoted by $60'' = 1'$.

When a calculator is used, minutes and seconds must be entered as a decimal part of a degree.

For example, $30°15' = 30.25°$, and $42°12'45'' = 42.2125°$.

Figure 1.18 illustrates degree measure of several angles. For brevity we write $A = 90°$ to denote that the measure of angle A is 90° and similarly for other angles.

EXAMPLE 1   Express $36°16'23''$ in decimal form rounded off to four decimal places.

---

*A grad is 1/100 of a right angle; that is, 400 grads is equivalent to a complete revolution.

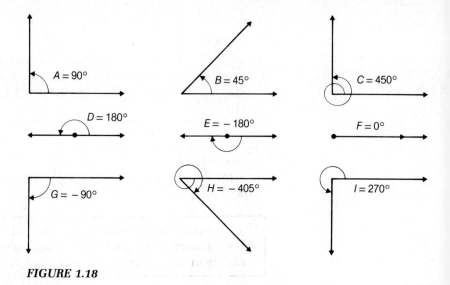

FIGURE 1.18

Solution   Since $60' = 1°$, then $16' = 16/60$ degrees. Also since $3600'' = 1°$, then $23'' = 23/3600$ degrees. Therefore

$$36°16'23'' = \left(36 + \frac{16}{60} + \frac{23}{3600}\right)° \approx 36.2731°.$$

The computation here is easily done by using a calculator.*   ∎

EXAMPLE 2   Express $64.276°$ in degrees, minutes, and seconds (to the nearest second).

Solution

$$64.276° = 64° + (0.276)(60') = 64° + 16.56'$$
$$= 64° + 16' + (0.56)(60'') \approx 64°16'34''.$$

Note: In order to get maximum accuracy we suggest the following steps: Record $64°$, enter 0.276 into the calculator and multiply by 60 to get minutes, then record the whole number part of the result (16); then subtract 16 from the display, multiply the result by 60, and this gives the number of seconds.   ∎

Radians

Although the measure of angles in degrees is useful in some fields of application (such as surveying and navigation), it is more convenient to use another unit of measure for theoretical work in

---

* Throughout the text it is assumed that a calculator is used to do most of the arithmetic computations. Appendix A includes instructions for readers not familiar with calculator use.

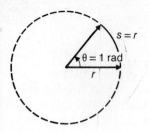

FIGURE 1.19

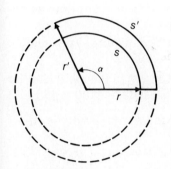

FIGURE 1.20

mathematics as well as applied areas. This unit is the *radian* and is defined as follows:

> An angle (with its vertex at the center of a circle) subtending an arc whose length is equal to the radius of the circle, has a measure of *one radian.*

An angle of measure 1 radian is shown in Fig. 1.19. In this case we write $\theta = 1$ rad.

In general, the radian measure of any angle is defined as follows:

> If α is an angle (with vertex at the center of a circle of radius r) that subtends an arc of length s (where r and s are measured in the same units), then the *radian measure of α* is defined as $\alpha = s/r$ radians.*

Note that this definition is independent of the size of the circle. In Fig. 1.20 s/r and s'/r' are equal (a fact from geometry).

If $r = 4$ cm and $s = 3$ cm, then $\alpha = 3$ cm$/4$ cm $= \frac{3}{4}$. Since the centimeters units cancel, the result is a real number, and it is not necessary to write "radians" after $\frac{3}{4}$. In this text we shall write $\alpha = \frac{3}{4}$ or $\alpha = \frac{3}{4}$ rad (for emphasis) to mean that α is an angle having radian measure $\frac{3}{4}$.

> When the measure of an angle is given as a real number (with no unit designation), it will be understood that the unit of measure is the radian.

For example, $\theta = 15$ means that θ is an angle whose measure is 15 radians.

We mentioned that the degree measure of an angle corresponding to one complete revolution is 360°. Since the arc length is the circumference of the circle, $s = 2\pi r$, the radian measure corresponding to 360° is given by $s/r = 2\pi r/r = 2\pi$ radians. We conclude that 360° and 2π radians are measures in different systems of the same angle. Thus 180° and π radians are equivalent measures (just as one yard and three feet are equivalent measures of length). The basic degree–radian relationship is given by

$$180° = \pi \text{ radians.} \qquad \text{[1.12]}$$

---

* In trigonometry, angles are frequently indicated by Greek letters. See footnote in Section 1.1 for a list of Greek letters.

From Eq. (1.12) we conclude the following:

$$1° = \pi/180 \approx 0.017453 \text{ rad},$$
$$1 \text{ rad} = 180°/\pi \approx 57.296° \approx 57°17'45''.$$

We can apply these results to convert from one unit to the other, but the decimal numbers are difficult to remember. We suggest that the equality given in Eq. (1.12) be used as a starting point for conversions. This suggests the following proportion:

$$\frac{\alpha}{\pi} = \frac{\beta}{180},$$

[1.13]

where α is the *number of radians* and β is the corresponding *number of degrees* for a given angle.

EXAMPLE 3 Draw a sketch to illustrate each of the following angles, whose measures are given in radians.

a] $\alpha = \pi/2$　　　　**b]** $\beta = 2\pi/3$　　　　**c]** $\theta = -5\pi/4$.

Solution First let us express each of the angles in the more familiar measure of degrees. (From Eq. (1.13) we get $\theta \text{ deg} = (180/\pi)(\theta \text{ rad})$.)

a] $\alpha = \left(\dfrac{180}{\pi}\right)\left(\dfrac{\pi}{2}\right) \text{ deg} = 90°$. Thus $\alpha = 90°$.

b] $\beta = \left(\dfrac{180}{\pi}\right)\left(\dfrac{2\pi}{3}\right) \text{ deg} = 120°$.

c] $\theta = \left(\dfrac{180}{\pi}\right)\left(-\dfrac{5\pi}{4}\right) \text{ deg} = -225°$.

We can now sketch the given angles as shown in Fig. 1.21.　■

EXAMPLE 4 Convert 130° to radian measure. Give your answer as a decimal approximation rounded off to two decimal places.

FIGURE 1.21

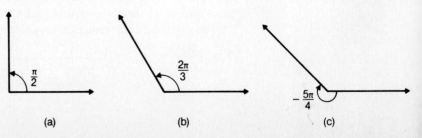

(a)　　　　　　(b)　　　　　　(c)

Solution   From Eq. (1.13) we have θ rad $= (\pi/180)(\theta$ deg). Let $\theta = 130°$; then $\theta = (\pi/180)(130)$ rad $= (13\pi/18)$ rad ≈ 2.27 rad.   ∎

EXAMPLE 5   Express 2.5 radians in terms of degrees. Give your answer as a decimal approximation rounded off to two decimal places.

Solution   Since 1 radian $= (180/\pi)$ deg, we have

$$2.5 \text{ rad} = 2.5\,(180/\pi) \text{ deg} \approx 143.24°.$$

Hence 2.5 rad $\approx 143.24°$.   ∎

EXAMPLE 6   Convert $13\pi/4$ to degree measure.

Solution   This is similar to Example 5 except that we can get the result in exact form:

$$\frac{13\pi}{4} = \left(\frac{13\pi}{4}\right)\left(\frac{180°}{\pi}\right) = 585°.$$

Thus $13\pi/4$ radians $= 585°$.   ∎

The number pi

The number π occurs frequently in mathematics. Although the student may have some familiarity with this number, it is worthwhile recalling some facts about it. More than 2000 years ago the Greeks were aware of an interesting property of circles. That is, in *any* two given circles (one with diameter d_1 and circumference c_1 and the other with diameter d_2 and circumference c_2), the ratios c_1/d_1 and c_2/d_2 are equal (Fig. 1.22). The common ratio is denoted by π (the Greek letter pi).

Scientific calculators have a key labeled ⬭π. When this key is pressed, the display shows some portion of 3.141592654, depending upon the number of digits displayed. Actually, this is an approximation to the value of π rounded off to nine decimal places. The number 22/7 is frequently used as a value of π. It is important to

FIGURE 1.22

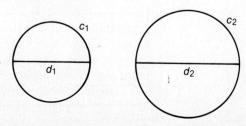

realize that this is also an approximation. In decimal form, 22/7 is given by the repeating decimal $3.\overline{142857}$, which approximates π correctly to two decimal places.

Another ancient approximation to π is $355/113 \approx 3.14159292$. We see that this agrees with π (as given in Eq. (1.14) below) in the first six decimal digits.

Even though approximations to π by fractions or decimals are useful in practical applications, it is important to understand that it is impossible to represent π exactly as a rational number, that is, a quotient of two integers. The number π is an irrational number, and so its decimal representation is nonterminating and nonrepeating. Calculating π correctly to several decimal places requires a representation in terms of an infinite process (such as infinite series), and discussion of this must be delayed until the study of calculus. The decimal approximation correct to 24 decimal places is

$$\pi \approx 3.1415\ 92653\ 58979\ 32384\ 62643. \qquad \textbf{[1.14]}$$

Number representation: Exact form and approximate decimal form _____

Numbers can be represented in various ways. For instance, 1/2 and 0.5 are names for the same number, and we write $1/2 = 0.5$. Similarly, $2/\sqrt{2}$ and $\sqrt{2}$ represent the same number, and we write $2/\sqrt{2} = \sqrt{2}$, but it is impossible to express this number in finite decimal form. However, we can find finite decimal expressions that *approximate* $\sqrt{2}$ to any degree of accuracy we wish, and it is these approximations that are usually used in applications. For example, 1.41 and 1.414 are approximations to $\sqrt{2}$, *rounded off* to three and four significant digits, respectively.* We write $\sqrt{2} \approx 1.414$ to indicate that $\sqrt{2}$ is *approximately equal* to 1.414 rounded off to four significant digits (or three decimal places).

Note: In this book, when we ask for a numerical result in *exact form*, we expect the answer to be given in terms of appropriate symbols, which may involve $\sqrt{}$, $\sqrt[3]{}$, π, and so on. For instance, $2\pi/3$ is an exact form for the radian measure of an angle of 120°, while 2.09440 is the approximate decimal form rounded off to six significant digits (or to five decimal places). We write $2\pi/3 \approx 2.09440$.

> In general, exact form answers are not given by a calculator. When we say, "Give the answer in exact form," we are implying "do not use your calculator." When we ask for an answer to a given number of significant digits or decimal places, you should expect that this calls for use of a calculator.

---

* See Appendix B for a discussion of computing with approximate numbers.

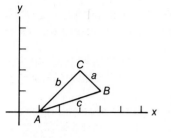

FIGURE 1.23

EXAMPLE 7   There are situations in which we need results in exact form. As an illustration, consider the triangle with vertices $A(1,0)$, $B(4,1)$, $C(3,2)$ as shown in Fig. 1.23. Show that $\triangle ABC$ is a right triangle.

Solution   The lengths of the sides are given by $a = \sqrt{2}$, $b = 2\sqrt{2}$, and $c = \sqrt{10}$ (exact form). Since $a^2 + b^2 = (\sqrt{2})^2 + (2\sqrt{2})^2 = 2 + 8 = 10$ and $c^2 = (\sqrt{10})^2 = 10$, we can conclude that $\triangle ABC$ is a right triangle.

However, if we use approximate values, say $a \approx 1.414$, $b \approx 2.828$, and $c \approx 3.162$, then for these values $a^2 + b^2 \approx 9.996980$ and $c^2 \approx 9.998244$. From these computations it is not clear whether $\triangle ABC$ is a right triangle or not. We need a, b, and c in exact form (as above) to conclude from the Pythagorean Theorem that $\triangle ABC$ is actually a right triangle.   ■

EXERCISE 1.2

In Problems 1 and 2, illustrate each of the given angles by a sketch. A protractor is useful if one is available; otherwise a reasonable approximate drawing is sufficient.

1. a] $\alpha = 135°$ **b]** $\beta = -60°$ **c]** $\gamma = 540°$ **d]** $\theta = 67°30'$

2. a] $\alpha = 450°$ **b]** $\beta = -225°$ **c]** $\gamma = -180°$ **d]** $\theta = 337°30'$

In Problems 3 and 4, angles are given in radian measure. Illustrate each by drawing a sketch.

3. a] $\alpha = \dfrac{\pi}{3}$ **b]** $\beta = -\dfrac{3\pi}{4}$ **c]** $\gamma = \dfrac{9\pi}{4}$

4. a] $\alpha = \dfrac{2\pi}{3}$ **b]** $\beta = -\pi$ **c]** $\gamma = -\dfrac{5\pi}{2}$

In Problems 5–8, two angles of a triangle are given. Determine the third angle and tell whether the triangle is acute or obtuse. A triangle is said to be *acute* if all three angles are less than 90°. It is *obtuse* if one of its angles is greater than 90°.

5. $\alpha = 37°$, $\beta = 43°$ **6.** $\beta = 64°27'$, $\gamma = 42°16'$

7. $\beta = 28°15'33''$, $\gamma = 67°24'37''$ **8.** $\alpha = \dfrac{\pi}{4}$, $\gamma = \dfrac{\pi}{3}$

In Problems 9 and 10, angles α, β, and θ satisfy the given inequalities. In each case, sketch *any* such angle. In your sketch, illustrate the range of positions for the terminal side with broken line rays.

9. a] $90° < \alpha < 135°$ **b]** $\dfrac{5\pi}{4} < \beta < \dfrac{3\pi}{2}$ **c]** $-\pi < \theta < -\dfrac{\pi}{2}$

10. a] $270° < \alpha < 315°$ **b]** $\dfrac{3\pi}{4} < \beta < \dfrac{5\pi}{4}$ **c]** $-\dfrac{3\pi}{2} < \theta < -\dfrac{5\pi}{4}$

In Problems 11 and 12, express each of the given angles as a decimal number of degrees. Give answers rounded off to two decimal places.

11. a] $156°37'$ **b]** $247°23'$

12. a] $215°43'$ **b]** $65°37'$

In Problems 13 and 14, express each of the given angles in terms of degrees-minutes-seconds. Round off answers to the nearest ten seconds.

13. a] $24.364°$ **b]** $149.375°$

14. a] $234.178°$ **b]** $32.143°$

15. Complete the following table. Enter appropriate values in exact form, using π where necessary.

Deg	30°		90°	120°			210°	270°	300°			x
Rad		$\pi/4$			$11\pi/12$	π				$11\pi/6$	2π	y

In Problems 16 and 17, express each of the given angles in radian measure. Give answers in two forms: exact (involving the number π) and approximate decimal rounded off to three decimal places.

16. a] $60°$ **b]** $-135°$ **c]** $22°30'$

17. a] $210°$ **b]** $-67.5°$ **c]** $420°$

In Problems 18 and 19, the given numbers represent measures of angles in radians. Convert to degrees and express answers in two forms: as a decimal number of degrees (to two decimal places) and in degrees and minutes (to the nearest minute).

18. a] 1.15 **b]** 0.542

19. a] 1.48 **b]** -3.45

In Problems 20 and 21, complete the tables by entering the corresponding missing values. Express entries to one decimal place for degrees and two decimal places for radians.

20.

Deg	23.4°		140.3°			532.4°
Rad		1.56		4.36	5.13	

21.

Deg	64.7°	241.4°			453.4°	
Rad			2.45	3.58		6.47

Angles can be ordered in size by first expressing them in the same units of measure and then comparing. In Problems 22–25, order the three given angles from smallest to largest. Give answers using the "less than" symbol, for instance, $\alpha < \beta < \theta$.

22. $\alpha = 47°36'$, $\beta = 47.54°$, $\theta = 0.829$

23. $\alpha = 154°17'$, $\beta = 154.26°$, $\theta = 2.692$

24. $\alpha = \pi$, $\beta = \dfrac{22}{7}$, $\theta = \dfrac{355}{113}$

25. $\alpha = 179°56'$, $\beta = 179.90°$, $\theta = \dfrac{179}{57}$

In Problems 26–28, measures of two of the three angles (α, β, and γ) of a triangle are given. Find the third in exact form.

26. $\alpha = 42°$, $\beta = 63°$

27. $\alpha = 27°47'$, $\gamma = 82°35'$

28. $\alpha = \dfrac{\pi}{5}$, $\beta = \dfrac{4\pi}{15}$

29. Use your calculator to express the fraction $\frac{333}{106}$ as a decimal. Obtain a sufficient number of decimal digits to determine how closely $\frac{333}{106}$ approximates π. See the value of π given in Eq. (1.14).

30. Use your calculator to get a decimal approximation to the rational number 208341/66317. To how many decimal digits does this number agree with that of π? It will be necessary to get more decimal digits than the full display will show. Determine a way to do this.

1.3 Applications of Radian Measure: Arc Length and Area

In Section 1.2 we used the fact that for a fixed central angle, the ratio of arc length to the radius of the circle (s/r) is independent of the size of the circle (see Fig. 1.20). The early Greeks were aware of this and other properties related to central angles. For instance, if α and β are any two central angles of a given circle and s and s_1 are the corresponding arc lengths, then the ratio of arc lengths is equal to that of the corresponding angles. In Fig. 1.24(a) we have $s_1/s = \beta/\alpha$. If $\beta = 360°$, then $s_1 = c$ (circumference of the circle), and so $c/s = 360/\alpha$. That is, $c = 360s/\alpha$.

As early as 200 B.C., Eratosthenes applied this fact to get an approximation to the circumference of the earth. This was done as follows. He observed that in Syene (point P in Fig. 1.24(b)) on June 21 the sun is directly overhead at noon. At exactly the same time in

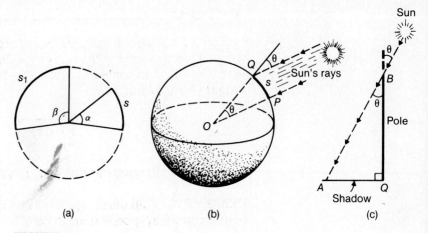

(a)   (b)   (c)

FIGURE 1.24

Alexandria (point Q), directly north of Syene, the sun makes an angle of 7.2° (1/50 of a complete revolution) with the vertical direction (angle θ at Q). If it is assumed that rays from the sun to P and Q are parallel, then the central angle at the center of the earth must also be 7.2°. To determine the circumference, all that was needed was the arc length from P to Q.

Eratosthenes was told that a camel caravan takes 50 days to travel from Syene to Alexandria; estimating that a camel travels 100 stadia in a day, he concluded that the arc length is 5000 stadia. Thus the circumference of the earth is given by

$$C = \frac{360(5000)}{7.2} = 250{,}000 \text{ stadia.}$$

In modern units this is $C = 46{,}250$ km $\approx 28{,}700$ miles.

How did Eratosthenes determine the central angle (7.2°)? At Point Q a vertical pole $\overline{QB}$ casts a shadow $\overline{QA}$, determining angle θ as in Fig. 1.24(c). The lengths of $\overline{QB}$ and $\overline{QA}$ are measured, and a similar triangle is drawn from which θ can be measured.

Based on modern measuring tools the earth's circumference is about 23,200 miles. Although the result obtained by Eratosthenes is about 20% off, it is still a remarkable achievement to come that close, considering that distance was measured by camels and Alexandria is not directly north of Syene (about 3° off), as well as other crude measurements. The important fact is that his theoretical solution (the geometry) is completely correct (assuming that the earth is spherical).

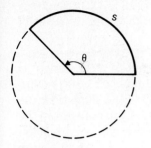

FIGURE 1.25

Arc length

In Section 1.2, radian measure of an angle θ is defined as $\theta = s/r$, where θ has its vertex at the center of a circle of radius r and s is the length of arc intercepted as shown in Fig. 1.25. Thus we have a formula for arc length,

$$s = r\theta, \qquad\qquad [1.15]$$

where we must remember that θ is in radians.

EXAMPLE 1   Find the arc length of a circle with radius 64.87 m that is intercepted by a central angle 23°37′.

Solution   We first express the given angle in radians,

$$\theta = 23°37' = \left(23 + \frac{37}{60}\right) \cdot \left(\frac{\pi}{180}\right) \text{ rad;}$$

substituting into Eq. (1.15), we get

$$s = 64.87 \left(23 + \frac{37}{60}\right) \cdot \left(\frac{\pi}{180}\right) \approx 26.74 \text{ m.}$$

The final computations are done by calculator and then rounded to four significant digits.   ∎

EXAMPLE 2   The distance from the earth to the moon is approximately 384,000 km. If the angle subtended by the moon from a point on the earth is measured as 30′50″, then we can approximate the diameter of the moon by assuming it to be the arc of a circle, as shown in Fig. 1.26.

Solution

$$s = r\theta = 384{,}000 \left[\left(\frac{30}{60} + \frac{50}{3600}\right) \cdot \frac{\pi}{180}\right] \text{ km} \approx 3444 \text{ km.}$$

Thus the diameter of the moon is approximately 3400 km.   ∎

FIGURE 1.26

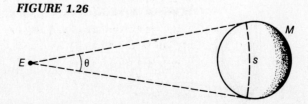

FIGURE 1.27

EXAMPLE 3   Find the angle between the minute and hour hands of a clock at time 1:25.

Solution   Suppose the desired angle is denoted by θ as shown in Fig. 1.27. The minute hand rotates through an angle of $5(30°) = 150°$ in going from the 12 position to the 5 position. At the same time the hour hand rotates $1/12$ as much (or $150°/12 = 12.5°$) in going from the 1 position to its location at 1:25. Therefore as we can see from the diagram, $30° + 12.5° + \theta = 150°$. Solving for θ gives $\theta = 107.5°$.   ∎

Statute miles—nautical miles

The intersection of a sphere and a plane is a circle. If the plane passes through the center of the sphere, then the intersection is called a *great circle*. For example, the equator is a great circle on the surface of the earth; so is any circle of longitude.

A *nautical mile* (nm) is the length of arc on a great circle intercepted by a central angle of one minute. A *statute mile* (mi) is the familiar mile length equal to 5280 feet. If we assume that the radius of the earth is 3960 statute miles, then one nautical mile is given by

$$s = r\theta = 3960 \left(\frac{1}{60} \cdot \frac{\pi}{180} \right) \approx 1.1519.$$

Hence 1 nautical mile ≈ 1.15 statute miles.*

EXAMPLE 4   Suppose P and Q are two points on the surface of the earth having the same longitude and P has latitude 23°N while the latitude of Q is 56°N. Assuming that the earth is spherical with a radius of 3960 miles, find the great circle distance between P and Q. Give answer in

a] miles   **b]** nautical miles.

Solution   Suppose C is the center of the earth. $\theta =$ angle $PCQ = 56° - 23° = 33°$, and $r = 3960$ miles (see Fig. 1.28).
   a] Hence $s = 3960 [33(\pi/180)] \approx 2280$ miles.
   b] Since $33° = 33 \times 60' = 1980'$ and each minute is one nautical mile, $s = 1980$ nautical miles.   ∎

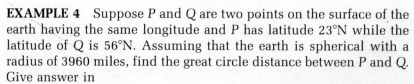
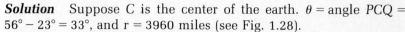

FIGURE 1.28

Area of a sector of a circle

A *sector of a circle* is defined as a region bounded by two radial lines and the intercepted arc of the circle. Figure 1.29 shows two regions bounded by the same radial lines. In order to distinguish between these two, we always indicate the central angle of the sector. In Fig.

* When we say "miles," we mean statute miles.

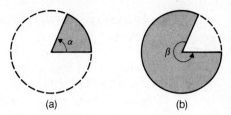

(a)   (b)

FIGURE 1.29

1.29(a) the sector has central angle α, while in Fig. 1.29(b) the central angle is β.

From the study of geometry we know that in any given circle the areas of two sectors are proportional to the corresponding central angles. That is, in the diagrams shown in Fig. 1.30,

$$\frac{\text{Area of sector } AOB}{\theta} = \frac{\text{Area of sector } COD}{\alpha}.$$

In particular, if $\alpha = 2\pi$, then the sector COD is the entire circle whose area is πr^2. In this case the preceding equality becomes

$$\frac{\text{Area of sector } AOB}{\theta} = \frac{\pi r^2}{2\pi} = \frac{r^2}{2}.$$

Hence we have

$$\text{Area of sector } AOB = \frac{\theta r^2}{2}.$$

Therefore the area of the sector of a circle of radius r and central angle θ in *radians* is given by

$$\boxed{\text{Area} = \theta r^2/2.} \qquad \text{[1.16]}$$

We emphasize that θ must be in *radians*.

FIGURE 1.30

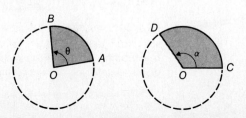

EXAMPLE 5   Find the area of the sector of a circle of radius 2.54 cm and central angle 73°24′. Give answer rounded off to three significant digits.

Solution   We first convert 73°24′ to radians and then substitute into Eq. (1.16):

$$73°24′ = \left(73 + \frac{24}{60}\right) \cdot \frac{\pi}{180} \text{ rad.}$$

Therefore

$$\text{Area} = \frac{1}{2} \cdot \left(73 + \frac{24}{60}\right) \cdot \frac{\pi}{180} \cdot 2.54^2 \approx 4.13 \text{ cm}^2. \quad ■$$

EXAMPLE 6   The area of a circular sector is 24.6 cm², and the length of its arc is 8.32 cm. Find the radius and central angle (in radians) of the sector. Give answers rounded off to three significant digits.

Solution   We are given s = 8.32 cm and area = 24.6 cm². Substitute these quantities into Eqs. (1.15) and (1.16) to get

$$\begin{cases} r\theta = 8.32, \\ \frac{1}{2}\theta r^2 = 24.6. \end{cases} \qquad \textbf{[1.17]}$$

We can solve this system of equations by first eliminating θ. From the first equation of Eq. (1.17) we get $\theta = 8.32/r$, and substituting this into the second equation gives $\frac{1}{2}r^2 (8.32/r) = 24.6$, or $r = 49.2/8.32 \approx 5.91$. Hence $r \approx 5.91$ cm. Replacing r by 49.2/8.32 in $\theta = 8.32/r$ gives $\theta = (8.32)^2/49.2 \approx 1.41$. Thus $\theta \approx 1.41$ radians. ■

EXERCISE 1.3

Answers given in approximate decimal form should be rounded off to the number of significant digits consistent with the given data. See Appendix B.
   In Problems 1–4, the radius and central angle of a circular sector are given. Determine the corresponding length of arc.

1. **a]** r = 37 cm, θ = 45°   **b]** r = 3.45 m, θ = 73°20′
2. **a]** r = 4.5 cm, θ = 135°   **b]** r = 21.53 cm, θ = 125°24′
3. **a]** r = 12.4 cm, θ = 1.56   **b]** r = 1.57 m, θ = 5π/23
4. **a]** r = 7.53 cm, θ = 3.58   **b]** r = 2.64 m, θ = 1.7π

In Problems 5 and 6, the radius and the arc length of a circular sector are given. Determine the central angle in radians.

5. **a]** r = 3.15 cm, s = 2.41 cm   **b]** r = 5.42 cm, s = 0.0842 m
6. **a]** r = 24.56 m, s = 63.51 m   **b]** r = 24.5 cm, s = 0.834 m

In each of Problems 7 and 8, the table includes the radius r, measure of the central angle θ, and the corresponding arc length of a circular sector s. Complete the tables.

7.

r	4.76 cm	8.63 cm	____ m	5.62 cm	____ cm
θ	2.45 rad	____ rad	$2\pi/5$ rad	____ rad	24.6°
s	____ cm	6.45 cm	124 cm	6.47 cm	5.64 cm

8.

r	2.16 cm	64.7 cm	____ m	7.21 cm	____ cm
θ	1.42 rad	____ rad	$17\pi/12$ rad	____ deg	125°20′
s	____ cm	0.0836 m	6.21 m	8.53 cm	6.83 cm

In Problems 9 and 10, a point P moves on a circle with center at O and radius 3.57 m.

9. Find the total distance traveled by P if the radial line segment $\overline{OP}$ sweeps out the given angle.

    a] 257°30′   **b]** $\dfrac{9\pi}{2}$ **c]** 8.53 revolutions

10. Point P travels a distance of 47.5 m. Through what angle does $\overline{OP}$ sweep? Give your answer in radians and in number of revolutions.

11. Find the area of a sector of a circle of radius 17.3 cm with the given central angle.

    a] 24°10′   **b]** 37°50′   **c]** $\dfrac{2\pi}{5}$ radians

12. If the radius of a circle is 1.26 m and the area of a circular sector is 0.876 m², find the central angle in
    a] radians   **b]** degrees.

13. What is the radius of a circular sector having central angle 67°30′ and area 8.64 cm²?

14. Determine the area of a circular sector having a radius of 3.4 m and an arc length of 5.3 m.

15. Find the area of a circular sector having a central angle of 1.32 radians and an arc length of 52.3 cm.

16. In Fig. 1.31, three concentric circles are shown with radii 2.3, 5.6, and 7.1, and θ is a central angle for each of them. If the length of arc intercepted by θ for the circle of radius 2.3 is 3.6,
    a] determine the lengths of arc intercepted by θ for the other two circles,
    b] find the areas of each of the three circular sectors.

FIGURE 1.31

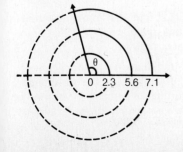

17. If we assume that the earth's orbit about the sun is a circle, what angle will a radial line from the sun to the earth sweep out in 73 days? Give your answer in radians
    a] in exact form (involving π),
    b] as an approximate decimal rounded off to two significant digits.
    Assume 365 days in a year.

18. Through how many radians does a radial line of a pulley of 8-cm diameter turn when a 6 meter section of rope has been pulled through without slipping?

19. A pulley of diameter 31.6 cm is driven by a belt. If 3.24 meters of belt passes around the pulley (without slipping), through what angle does a radial line on the pulley turn? Give answer in radians.

20. Assuming that the earth E is 93 million miles from the sun S and that it travels in a circular path about the sun, making one complete orbit in 52 weeks, through what angle in radians does the radial line $\overline{SE}$ sweep in one week? Give answer in exact form (involving π). What is the length of arc in miles (to two significant digits) traveled by the earth in 8 weeks? Actually, the earth's orbit about the sun is elliptic, but circular is a good first approximation.

21. Assume that the moon travels about the earth in a circular path of radius 240,000 miles and that it makes one complete orbit every 28 days. What is the distance traveled by the moon in 5 days in its rotation about the earth?

22. Through how many radians does a spoke of a bicycle wheel turn when the wheel makes
    a] 3 revolutions?   **b]** 4.8 revolutions?   **c]** x revolutions?

23. The following table gives measurements related to circular sectors, where r = radius, θ = central angle, s = arc length, and A = area. Two of the four quantities are given. Complete the table by entering the missing quantities in *exact form*, involving π in some cases.

r	6 ft	___ cm	1.5 m	2.5 cm	___ cm	___ cm
θ	30°	2.4 rad	___ rad	___ rad	___ rad	x rad
s	___ ft	3.6 cm	4.5 m	___ cm	4 cm	y cm
A	___ ft²	___ cm²	___ m²	8.6 cm²	10 cm²	___ cm²

24. What is the measure in degrees of the smaller angle between the hour and minute hands of a clock at
    a] 1:30?   **b]** 1:45?

25. If the length of the hour hand of a clock from the pivot point to the tip is 5.40 cm, how far will its tip travel in the given time? Give answers in exact form (involving π) and also rounded off to three significant digits.
    a] 2 hr   **b]** 3 hr 40 min

26. The time is between one and two o'clock, and the angle measured clockwise from the hour hand to the minute hand is 64°15′. What time is it? Give the answer correct to the nearest minute.

27. The front wheel of a tricycle is 51.4 cm in diameter, and the diameter of each of the rear wheels is 23.5 cm. If the tricycle travels along a straight path for a distance of 48.5 m, through how many revolutions will each wheel turn? Also express each answer in number of radians the wheel will turn.

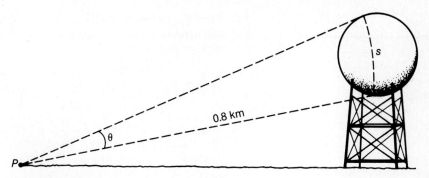

FIGURE 1.32

28. A spherical water tank is located 800 m from point P, and the angle it subtends at P is measured to be 17.5 minutes. (See Fig. 1.32.) Using this information, obtain a reasonable approximation to the volume of the tank in cubic meters.
    (Hint: The diagram shows a vertical plane through the center of the tank and P. Assume that P is the center of a circle of radius 800 m and that θ is a central angle of measure 17.5'. Calculate the arc length s and use this as an approximation to the diameter D of the tank. The formula for calculating the volume of a sphere is $V = (\pi/6)D^3$.)

In Problems 29–33, assume that the earth is spherical with radius 3960 miles. In each case, "distance between two locations" means the shortest distance along a great circle path. Give answers involving distances rounded off to three significant digits.

29. Santa Barbara, California, and Reno, Nevada, both have a longitude of 119.6°W. The latitude of Santa Barbara is 34.5°N, and that of Reno is 39.4°N.
    a] Find the distance (in miles) between the two cities.
    b] What is the distance in nautical miles (nm)?

30. What is the distance from New York (latitude 40.7°N) to the equator? Express the answer in
    a] miles   **b]** nautical miles (nm).

31. What is the distance from Chicago (latitude 41.8°N) to the North Pole? Give answer in
    a] miles   **b]** nautical miles (nm).

32. The distance between Salt Lake City (latitude 40.3°N) and a point P directly north is 634 miles. What is the latitude (in degrees) of point P?

33. The latitude of Rio de Janeiro is 23.1°S and P is a point 2150 miles directly north. What is the latitude (in degrees) of P?

34. A sector with central angle of 90° is cut out of a circular piece of tin of radius 24 cm, and the edges of the remaining piece are joined together to form a cone as shown in Fig. 1.33. Determine the volume of the cone

FIGURE 1.33

by applying the formula $V = \frac{1}{3}\pi r^2 h$, where r is the radius of the base of the cone and h is its height.

a] First find r. (Hint: The circumference of the base is the length of arc of the piece of tin folded.)

b] Now apply the Pythagorean Theorem to get h (leave it in exact form).

c] Using the above results, find V in exact form first, and then evaluate to get an answer rounded off to two significant digits.

35. The distance between the earth and the sun is approximately 93 million miles. When viewed from the earth, the opposite edges of the sun subtend an angle of approximately 0°30′. The diameter of the sun is approximately equal to the length of arc, where $\theta = 0°30′$ and $r = 93$ million miles. Find the approximate diameter in miles and give the answer rounded off to two significant digits.

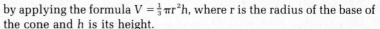

1.4 Applications of Radian Measure: Linear and Angular Velocity

In this section we discuss two types of velocities (or speeds*) *due to rotation.* The basic ideas are introduced by considering the following example.

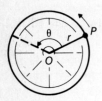

FIGURE 1.34

Suppose we have a circular wheel of radius r = 10 cm rotating about its center O and P is a point on the circumference (Fig. 1.34). Suppose also that point P travels a distance of s = 20 cm each second. We say the *linear velocity* of P is 20 cm per second and write v = 20 cm/sec. During each second the radial line $\overline{OP}$ rotates through an angle $\theta = s/r = 20$ cm/10 cm = 2 rad. We say that the *angular velocity* of rotation is 2 radians per second, and we denote this by ω = 2 rad/sec (ω is the Greek letter omega).

This example illustrates the problem of a point P moving in a circular path. We distinguish two types of velocity: Linear velocity v tells us how fast P is moving, while angular velocity ω tells us how fast the central angle θ is changing (that is, how fast the radial line $\overline{OP}$ is rotating). Both v and ω are measures of how fast P is moving at any given instant. In general, v and ω are functions of time. In the special case when P is moving at a *constant speed* we call the motion uni-*form circular motion.* We shall limit our discussion to this case and leave the general case (when v varies with time) for calculus.

---

* Velocity implies a vector quantity and so has a direction associated with "how fast" an object is moving. In this section we are merely interested in "how fast," which is referred to as *speed.* However, most textbooks use the word "velocity" here, and we shall follow tradition.

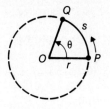

FIGURE 1.35

We wish to determine an equation that gives the relationship between v and ω. Suppose that point P moves to point Q, covering distance s in time t (see Fig. 1.35). Then $v = s/t$. During the same time, the radial line $\overline{OP}$ rotates through a central angle θ, and so $\omega = \theta/t$. Since $s = r\theta$, we get

$$v = \frac{s}{t} = \frac{r\theta}{t} = r \cdot \frac{\theta}{t} = r\omega.$$

Thus we have

$$\boxed{v = r\omega,}$$

[1.18]

where ω is in *radians per unit of time.*
Solving Eq. (1.18) for ω gives

$$\boxed{\omega = \frac{v}{r}.}$$

[1.19]

In applying the formulas given in Eq. (1.18) or Eq. (1.19), keep in mind that ω *must be in radians per unit of time.*

EXAMPLE 1   A wheel of diameter 24 cm rotates at a constant angular velocity of 5 rad/sec.
  a] What is the linear velocity of a point P on the rim of the wheel?
  b] How far does point P travel in one minute? Give your answer in meters.

Solution

  a] Substituting the given data $(r = \frac{1}{2}(24) = 12$ cm and $\omega = 5$ rad/sec) into Eq. (1.18) gives

$$v = (12)(5) \text{ cm/sec} = 60 \text{ cm/sec}.$$

  b] Since point P travels at a constant velocity of 60 cm per second, in one minute it will travel a distance d given by

$$d = (60)(60) \text{ cm} = \frac{3600}{100} \text{ m} = 36 \text{ m}.$$

Thus P will travel 36 meters in one minute.   ∎

EXAMPLE 2   A wheel of radius 5.0 cm is rotating at a constant angular velocity. If the linear velocity of a point on the rim is 8.5 cm/sec, what is the angular velocity in
  a] rad/sec?   **b]** rev/sec?

Solution

a] Here we are given $r = 5.0$, $v = 8.5$ cm/sec. Substituting into Eq. (1.19), we get

$$\omega = \frac{8.5}{5} \text{ rad/sec} = 1.7 \text{ rad/sec.}$$

b] Since 1 revolution = 2π radians, we divide 1.7 rad/sec by 2π to get

$$\omega = \frac{1.7}{2\pi} \text{ rev/sec} \approx 0.27 \text{ rev/sec.} \qquad \blacksquare$$

EXAMPLE 3   A shaft of radius 4.8 cm is rotating at 3600 rpm (revolutions per minute). How fast is a point P on the shaft traveling in its rotation? Give answer in
a] cm/min   **b]** m/sec.

Solution

a] We are given $r = 4.8$ cm and $\omega = 3600$ rev/min. In order to apply Eq. (1.18) we must first express ω in rad/min. Thus $\omega = 3600(2\pi)$ rad/min, and so

$$v = r\omega = 4.8[3600(2\pi)] \text{ cm/min}$$
$$= 34560\pi \text{ cm/min} \approx 108600 \text{ cm/min.}$$

b] To convert v to m/sec, we must divide the result in (a) by 100 (cm to m) and by 60 (min to sec). Thus

$$v = \frac{34560\pi}{100(60)} = 5.76\pi \approx 18.1.$$

Thus point P is traveling at approximately 18 meters per second. $\blacksquare$

EXAMPLE 4   A wheel of diameter 4.8 m is rotating at a constant speed. If a point P on its rim travels a distance of 60 meters in 15 seconds, find
a] the linear velocity of P in m/sec,
b] the angular velocity of the wheel in rev/sec.

Solution

a] Since velocity = distance/time, we have

$$v = \frac{60}{15} \text{ m/sec} = 4 \text{ m/sec.}$$

b] Substituting 4 for v and 2.4 for r in Eq. (1.19) gives

$$\omega = \frac{4}{2.4} \text{ rad/sec} = \frac{5}{3} \text{ rad/sec.}$$

Since we want ω in rev/sec, we divide by 2π to get

$$\omega = \frac{(5/3)}{2\pi} = \frac{5}{(6\pi)} \approx 0.27.$$

Hence $\omega \approx 0.27$ rev/sec. ∎

EXAMPLE 5 The moon travels around the earth in an orbit that is approximately circular with radius 240,000 miles. If we assume that it takes 28 days to make one complete orbit:

a] What is its angular velocity? Give your answer in rad/hr.
b] What is the linear velocity of the moon in its rotation about the earth? Give your answer in mi/hr.

Solution

a] The moon makes one orbit (or 2π radians) in 28 days (or 28×24 hours). Thus

$$\omega = \frac{2\pi \text{ rad}}{28(24) \text{ hr}} \approx 0.0093 \text{ rad/hr}.$$

b] Substituting the exact form result in (a) for ω and 240,000 for r in Eq. (1.18) gives

$$v = 240{,}000 \left(\frac{2\pi}{28(24)} \right) \text{ mi/hr} \approx 2244 \text{ mi/hr}.$$

Thus the moon travels at a speed of approximately 2200 miles per hour. ∎

EXAMPLE 6 The diameter of each wheel of a bicycle is 80 cm. Suppose a person riding the bicycle travels at a constant speed and is timed at 4 min over a distance of two city blocks, where the length of a block is 200 m. Find the angular velocity of a spoke of a wheel.

Solution Each time the wheel (or a spoke) makes one revolution, the bicycle moves forward a distance equal to the circumference of the wheel, that is, 80π cm. Therefore when the bicycle travels two blocks (400 m or 40,000 cm), the number of revolutions of a wheel is $40000/(80\pi)$. It takes 4 min to make this number of revolutions, and so

$$\omega = \frac{40000}{80\pi} \div 4 = \frac{125}{\pi} \frac{\text{rev}}{\text{min}}.$$

Expressing ω in radians per second, we have

$$\omega = \frac{125}{\pi} \frac{\text{rev}}{\text{min}} \cdot \frac{2\pi \text{ rad}}{1 \text{ rev}} \cdot \frac{1 \text{ min}}{60 \text{ sec}} = \frac{25}{6} \frac{\text{rad}}{\text{sec}} \approx 4.167 \frac{\text{rad}}{\text{sec}}.$$

Thus the angular velocity of a spoke is approximately 4 rad/sec. ∎

EXERCISE 1.4

Answers that are given in approximate decimal form should be rounded off to the number of significant digits consistent with the given data. (See Appendix B.)

1. Find the velocity v of a point on the rim of a wheel of radius 24 cm if it is rotating at the given angular velocity:
   a] $\omega = 5.4$ rad/sec   **b]** $\omega = 6.4$ rev/sec   **c]** $\omega = 120$ deg/sec

2. A wheel of diameter 124 cm is rotating at a constant rate. Find the angular velocity if a point on the rim is moving at the given speed. Give answers in rad/sec and rev/sec.
   a] $v = 348$ cm/sec   **b]** $v = 2.75$ m/sec

3. Find the angular velocity of the minute hand of a clock in each of the following units:
   a] rev/hr   **b]** deg/min   **c]** rad/min

4. Find the angular velocity of the second hand of a watch in
   a] rev/min   **b]** deg/hr   **c]** rad/sec

5. If the length of the minute hand of a clock from the pivot point to the tip is 6.5 cm, find the linear velocity of its tip in each of the following units
   a] cm/hr   **b]** cm/min

6. Find the linear velocity of the tip of a propeller blade that is 2.48 m from the pivot point and is rotating at 640 rev/min.

7. The length of the minute hand of a clock is 8.4 cm, and the length of the hour hand is 6.2 cm.
   a] How far in meters will the tip of the minute hand travel in a year? Assume 365 days in a year.
   b] How far in meters will the tip of the hour hand travel in a year?

8. Assume that the earth is spherical with radius 6400 km and that its period of rotation about an axis passing through the north and south poles is 24 hours. How fast is a point on the equator moving in km/hr in its rotation?

9. A satellite travels around the earth and makes one revolution every 4.5 hours. Assuming that the orbit is a circle of radius 7200 km, find how fast it is traveling in its rotation.

10. Assume that the earth travels about the sun in a circular orbit and the distance between the earth and sun is 149 million kilometers. A radial line is drawn from the sun through the earth.
    a] What is the angle (in radians) swept out by the radial line in a day? Assume that it takes 365.25 days to travel once around the sun.
    b] What is the angular velocity of the radial line in radians per hour?
    c] What is the linear velocity of the earth in its rotation in kilometers per hour?

11. The diameter of each of the tires of a car is 30 inches. When the car is traveling at 50 mph, what is the angular velocity of the wheels in
   a] rad/sec?   **b]** rev/sec?
   Give answers rounded off to two significant digits.

12. The following table includes values of r, v, and ω for an object that is moving in a circular path at a constant speed. Two of the three quantities are given. Complete the table by entering the appropriate missing value rounded off to two significant digits.

r	3.6 cm	6.3 m	___ cm	15 m
ω	1.5 rad/sec	___ rad/min	4.2 rev/sec	72 deg/sec
v	___ cm/sec	12 m/min	12 cm/sec	___ m/sec

13. The radius of a phonograph record is 15.0 cm. It is rotating at 33 ⅓ rpm.
   a] Give its angular velocity in rad/min.
   b] How fast is a point on the outside edge moving?

14. The diameter of a phonograph record is 16 cm and is rotating at 45 rpm. A speck of dust is halfway between the spindle and the outside edge of the record. How fast is it moving in cm/sec (to two significant digits)?

15. A pulley of diameter 8.6 cm is driven by a belt. Suppose 4.0 meters of the belt passes around the pulley in 15 seconds.
   a] How fast in cm/sec is a point on the outside edge of the pulley moving? (Give your answer in exact form.)
   b] What is the angular velocity (to two significant digits) of the pulley in rad/sec?

16. The planet Venus travels around the sun in a path that is approximately circular of radius 67.2 million miles. It makes one complete orbit in 225 days. Give answers to the following, rounded off to three significant digits.
   a] What is the angular velocity of Venus in rad/hr?
   b] What is its linear velocity in miles per hour?

17. A pickup truck comes factory equipped with standard-size tires having a diameter of 29 inches. The speedometer is calibrated with this size tire.
   a] If the truck travels for one hour at a constant speed with the speedometer reading 55 mph, how many revolutions will each wheel make?
   b] The owner of the truck prefers larger tires and replaces the originals with tires of diameter 30.75 inches. Now he travels for one hour at a constant speed with the speedometer reading 55 mph (thus each wheel makes the same number of revolutions as in (a)). How fast is he actually going during that hour? Give answers rounded off to two significant digits.

18. The diagram in Fig. 1.36 shows the larger gear rotating counterclock-

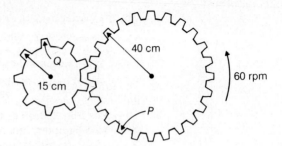

FIGURE 1.36

wise at the rate of 60 rpm and driving the smaller gear. The radii dimensions are as shown.

a] In what direction is the smaller gear rotating?

b] What is the angular velocity in rpm of the smaller gear?

c] P and Q are points on the two gears as shown. How fast is P moving in cm/sec? How fast is Q moving in cm/sec?

d] What is the distance (in cm) traveled by P in 20 seconds? What is the distance traveled by Q in 20 seconds?

19. In Fig. 1.37, wheel A is driving wheel B by a belt without slipping. Suppose wheel A is rotating at 4 rev/sec.

a] What is the linear velocity of a point P on the rim of wheel A, a point Q on the belt, a point T on the rim of wheel B?

b] What is the angular velocity of wheel B?

20. A treadle sewing machine is driven by two wheels with a belt passing around them, as shown in Fig. 1.38. The sewing machine used by

FIGURE 1.37

FIGURE 1.38

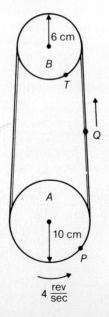

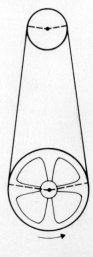

Motel the tailor has the following measurements: the diameter of the larger wheel is 31 cm, while that of the smaller wheel is 7.0 cm. If Motel treadles his machine at a fixed rate, so that in 45 seconds the larger wheel turns through 63 revolutions, find the angular velocity of each wheel (assume that the belt does not slip). Express each answer in

a] rev/sec   **b]** rad/sec

21. Using the information of Problem 20, find the linear velocity of point P on the belt, in cm/sec. Also determine how far point P travels when the sewing machine is operated at the given rate for 10 seconds.

Summary

Angle measures

Degrees: 1 rev $= 360°$, $1° = 60'$, $1' = 60''$.
Radians: $180° = \pi$ radians.
Convert: From degrees to radians, multiply by $\pi/180$.
From radians to degrees, multiply by $180/\pi$.

Arc length and area of circular sector

$s = r\theta$ and area $= \frac{1}{2}r^2\theta$ where θ is in *radians*.

Linear and angular velocity

For an object moving in a circular path of radius r at a constant speed,

$$v = r\omega \qquad \text{or} \qquad \omega = \frac{v}{r},$$

where ω is in *radians* per unit of time.

Computer Problems (Optional)

Section 1.1

1. A Pythagorean triple consists of three *positive integers* a, b, and c that satisfy the equation $a^2 + b^2 = c^2$. Write a program that will list all Pythagorean triples for which c is less than 200.

2. Write a program that will determine whether or not a triangle is determined when you enter three positive numbers that are to be the lengths of the sides. (Hint: The sum of the lengths of any two sides must be greater than the length of the third side.)

Section 1.2

3. Write a program that will round off any given positive number to k decimal places where k is 0, 1, 2, 3, or 4.

4. Write a program that will round off any given positive number between 0.001 and 1000 to two significant digits.

5. Write a program that will convert angular measure in degrees to radians. Have your program list two columns (degrees, radians) where the degree values are 0–90 in five-degree increments and the radian values are rounded off to three decimal places.

Section 1.3

6. Suppose θ is the central angle of a circle of radius r. The arc length s is given by $s = r\theta$, and the area A of the circular sector is given by $A = (1/2)r^2\theta$ where θ is in radians. Write a program that gives the values of s and A for any θ and r. Have your program accept θ in any of the forms.
   a] radians   **b]** decimal degrees   **c]** degrees-minutes-seconds

Section 1.4

7. The angular velocity of the minute hand of a clock is 6 deg/min, and that of the hour hand is $(1/12)(6) = 0.5$ deg/min. Let t be the *number of minutes after 12 o'clock noon*. The angle through which the minute hand rotates in t minutes is 6t degrees, while the hour hand rotates 0.5t degrees. Let θ be the angle measured clockwise from the hour hand to the minute hand as shown in Fig. 1.39, and so $\theta = 6t - 0.5t = 5.5t$ deg.
   The two hands will coincide at noon $(t = 0)$. The next time will be at 1:5:27.
   a] Write a program that will give the times on the clock (hour, minutes, and seconds) when the two hands will coincide.
   b] Write a program that will give the times on the clock when the two hands are perpendicular to each other.

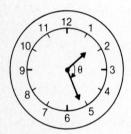

FIGURE 1.39

Review Exercises

1. Express the following angles in decimal number of degrees rounded off to three significant digits.
   a] 37°42′   **b]** 1.43 rad   **c]** $15\pi/23$ rad

2. Give the following angles in radian measure rounded off to two decimal places:
   a] 175°   **b]** 23°16′   **c]** 137°16′37″

3. Make a sketch illustrating the given angles (a reasonable approximation is sufficient):
   a] 150°   **b]** −250°   **c]** $2\pi/3$ **d]** $7\pi/5$

4. The central angle of a circular sector is 64°20′. If the radius of the circle is 24.6 cm, find the length of arc of the sector.

5. In Problem 4, find the area of the circular sector.

6. The measures of three angles α, β, and γ are: $\alpha = 0.935$, $\beta = 5\pi/17$, $\gamma = 3\pi/10$. Determine which is the largest angle and which is the smallest angle.

7. The measures of four angles α, β, γ, and θ are:

$$\alpha = 126°27', \qquad \beta = 126.43°, \qquad \gamma = 2.21, \qquad \theta = 7\pi/10.$$

Order these according to size from the smallest to the largest.

8. An arc of a circle of radius 37.63 m has length equal to 12.37 m. Find the measure of the central angle subtended by this arc in degree measure to the nearest minute.

9. Find the area of the circular sector described in Problem 8. Give answer in square meters rounded off to four significant digits.

10. Determine the smaller angle between the hour and minute hands of a clock when the time is 3:45. Express your answer in degree measure.

11. The area of a circular sector is 35.61 cm², and its central angle is 34.63°. Find the length of arc of the sector.

12. A particle travels in a circular path of radius 3.45 cm at a constant speed. It takes 1 min 36 sec to make 84 revolutions.
a] Find its angular velocity in radians per second.
b] If it travels at the given rate for 3 min 20 sec, what is the total distance traveled?

13. If both the radius and central angle of a circular sector are doubled, by what factor is the area increased?

14. Figure 1.40 illustrates part of a machine in which the larger wheel drives the smaller wheel by a belt around the two wheels. The diameter of the larger wheel is 63.4 cm, while that of the smaller wheel is 25.8 cm; the distance between their centers is 124.3 cm.
a] If the larger wheel rotates at a constant rate of 250 rev/min, find the rate at which the smaller wheel rotates (in rev/min).
b] If P is a point on the circumference of the larger wheel, what is the linear velocity of P in m/min?
c] If Q is a point on the circumference of the smaller wheel, find the linear velocity of Q in m/min.
d] If T is a point on the belt, how far will T travel in 1.5 minutes? Give answer in meters.

15. A circular pizza is cut into four pieces by making two straight cuts across through the center. Two of the pieces are smaller, each having a central angle 10° narrower than that of each larger piece. Find the ratio of the area of a larger piece to that of a smaller piece.

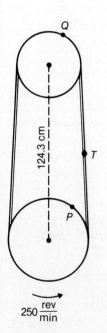

124.3 cm

Q

T

P

$250 \dfrac{\text{rev}}{\text{min}}$

FIGURE 1.40

Trigonometric Functions
2

2.1 Review of Functions from Algebra (Optional)

The reader has already dealt with numerous functions in algebra courses. Each of the functions encountered in the context of algebra and trigonometry is defined on a set of real numbers called the *domain D* of the function, and we have a *rule of correspondence* according to which each number in D is associated with a unique real number. This rule of correspondence is most often given by a formula such as $y = 4x$ or $y = x^3 + x^2 - 1$, but it may as easily be given by a table listing the ordered pairs, by a graph, or by a verbal statement. The set of y-values corresponding to x-values from the domain is called the *range* of the function. The important fact is that *for each element in the domain* there is *precisely one corresponding element in the range.* When it is necessary to distinguish between domains and ranges of several functions, we may use $D(f)$ and $R(f)$, to denote the domain and range, respectively, of the function f.

This idea of correspondence yields a *set of ordered pairs*:

$$\{(x, y) | x \in D, \text{ and } y \text{ is the number corresponding to } x\}.$$

It is often convenient to denote functions by letters such as f, g, etc., and we write $y = f(x)$ to mean that y *is the number that the function* f *associates with the number x.* The number $f(x)$ is often called the "value" of the function f "at x," and in this case we speak of x as the *independent variable* and y as the *dependent variable.* While the symbols x and y are often used, we will use other symbols as well. Trigonometry traditionally uses a wider variety of symbols (including Greek letters), but the *independent variable* (from the domain) *is always the first entry* in an ordered pair, and the *dependent variable is the second* (the corresponding) entry, coming from the range.

Considering a function as a collection of ordered pairs provides a convenient definition for the *graph of the function*. We define the graph as a picture of the ordered pairs belonging to the function; that is, the point (x, y) belongs to the graph of the function f if and only if (x, y) is one of the pairs in f, which is the same as saying that $y = f(x)$.

> Graphs are very valuable tools for every student of mathematics because they provide a visual and conceptual framework for understanding basic ideas that are not readily apparent from an algebraic description. *It is worth considerable effort to learn as much as possible about the graph of any function that is used regularly.*

Among other things, the graph shows the domain and range of a function immediately; the portion of the x-axis for which the graph exists is the domain, and the corresponding portion of the y-axis is the range.

EXAMPLE 1   Let f be the function $\{(x, y)|y = x^2 - 2x, x \geq 0\}$. Sketch the graph and find the domain and range of f.

FIGURE 2.1

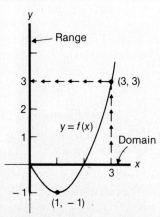

Solution   From algebra we recognize that the graph of any function that can be written in the form $y = ax^2 + bx + c$ is a *parabola* with vertical axis $x = -b/2a$ and vertex at the point with y-value corresponding to $x = -b/2a$. Since the domain of our function is restricted to nonnegative numbers, however, the graph will be *only a portion of* the parabola; in this case the graph is the part of the parabola to the right of the y-axis. Using this information, we find that the parabola has its vertex at the point $(1, -1)$, and plotting a few other points such as $(0, 0)$ (since $f(0) = 0$), $(2, 0)$, and $(3, 3)$ allows us to sketch a fairly accurate graph. The domain is specified in the definition, so $D(f) = \{x|x \geq 0\}$. From the graph we may readily see that the low point of the graph is the vertex, so $R(f) = \{y|y \geq -1\}$. (See Fig. 2.1.)   ∎

Not every equation defines a function. Graphs can also be helpful in distinguishing which equations do define functions. Since a function associates a *unique* number $f(x)$ to each x in the domain, *no vertical line can intersect the graph of a function in more than one point.*

EXAMPLE 2   Explain why the equation $x^2 + y^2 = 1$ does not define y as a function of x.

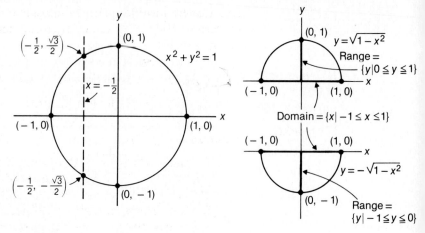

FIGURE 2.2

Solution   The set of points whose coordinates satisfy the equation $x^2 + y^2 = 1$ is called the *unit circle*, the circle with center at the origin and radius 1. (See Fig. 2.2.) Each vertical line between $x = -1$ and $x = 1$ intersects the graph at two points as shown. The fact that two pairs have the same x-value violates the definition of function.

If we solve the equation $x^2 + y^2 = 1$ for y, we get $y = \pm\sqrt{1 - x^2}$, which also shows that there is not a unique number y paired with a given x. Each of the equations $y = \sqrt{1 - x^2}$ and $y = -\sqrt{1 - x^2}$ does define a function. The graphs of these two functions are the upper and lower semicircles, respectively, in Fig. 2.2. The domain of each function is the interval $[-1, 1]$. The ranges are $[0, 1]$ and $[-1, 0]$, respectively.* ■

A look at domains

When we are not familiar with the graph of a function, it may be difficult to determine the domain and range. Evaluating several function values, by calculator for instance, may be of some help, but point plotting without some guidance is generally not enlightening. One rule of thumb can be very helpful in determining the domain, at least. Since the domain is the set of all real numbers for which the function is *defined*, we may say intuitively that "whatever is not forbidden is allowed." Thus when the function is given by a formula or equation, we may be able to ask simply, "What numbers do not work?" Num-

---

* Interval notation: For $a < b$, the notation $[a, b]$ means the set $\{x \mid a \leq x \leq b\}$. Similarly, (a, b) means $\{x \mid a < x < b\}$.

bers that are forbidden, in the sense that they would require division by zero or taking the square root of a negative number, are not part of the domain.

EXAMPLE 3   Find the domain of each of the functions:

a] $f(x) = \sqrt{1 - x^2}$ **b]** $F = \{(x, y)|2x - 3y = 5\}$ **c]** $g(x) = \sqrt{1 + 1/x}$

Solution

a] Setting $y = \sqrt{1 - x^2}$ and squaring both sides of the equation, we obtain $x^2 + y^2 = 1$, the equation of the unit circle. In the previous example we saw that the unit circle is not the graph of a function. Here, however, y cannot be negative (square roots are never negative), so the equation defines the *upper semicircle pictured in Fig. 2.2*, for which we have already found the domain and range.

Even if we do not recognize the equation as defining a familiar graph, we may still find the domain by observing that numbers that make the quantity $(1 - x^2)$ negative are "not allowed." Solving the inequality $1 - x^2 \geq 0$ gives us the same domain, $D = \{x|-1 \leq x \leq 1\}$.

b] The graph of any equation of the form $ax + by = c$ is a straight line, so $2x - 3y = 5$ defines a line. Since there is nothing in the definition that explicitly limits the x-values we may use, the graph of $y = F(x) = (2x - 5)/3$ is the line shown in Fig. 2.3, and the domain is the set of all real numbers. The range is also the set of real numbers.

c] There are two things not allowed in the formula $g(x) = \sqrt{1 + 1/x}$. First, we must rule out $x = 0$, since we may not divide by zero. Second, the argument of the radical must be nonnegative: $1 + 1/x \geq 0$. This inequality is satisfied for all positive x and for all $x \leq -1$, so $D(g) = \{x|x > 0\} \cup \{x|x \leq -1\}$.   ∎

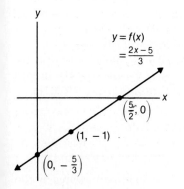

FIGURE 2.3

The calculator as function machine

Our calculators may be considered as remarkable collections of functions. We enter a number x and press one or more function keys, and the calculator displays the number $f(x)$ corresponding to the number entered. If a number is entered that is not in the domain of the function, the calculator will display some sort of "ERROR" message.

In addition to performing difficult functional operations with speed and accuracy, calculators can also help remind us of some of the properties of functions. To illustrate, some students want to use a symbol such as $\sqrt{2}$ to stand for two numbers, probably because they know that there are two numbers satisfying the equation $x^2 = 2$. The square root key on the calculator, however, will give us a *positive number* whenever we enter 2 and press $(\sqrt{x})$, namely, 1.414213562. The square root function *is a function* and has only one value corresponding to any number x in its domain. The calculator is programmed to display the *nonnegative number whose square is x.**

Rationalizing

In calculus there are many situations in which it is necessary to rationalize the numerator or the denominator of algebraic expressions. For instance, we would rationalize the numerator of $(\sqrt{x + 3} - \sqrt{3})/x$ by multiplying both numerator and denominator by $\sqrt{x + 3} + \sqrt{3}$ and then simplifying to obtain $1/(\sqrt{x + 3} + \sqrt{3})$. This skill should be practiced regularly.

While this kind of rationalization is needed constantly, the custom of requiring answers with rational denominators is much less important. This was often done for computational reasons before the advent of calculators. In evaluating $1/\sqrt{2} = \sqrt{2}/2$, it is easier to divide 2 into 1.4142136 by long division than it is to divide 1.4142136 into 1. With a calculator there is no essential difference.

In this book we may leave the exact form of answers involving radicals in any of several convenient equivalent forms. If your answer appears to differ from the one given in the answer section, you should first check to see if possibly yours is another, equivalent name for the same number.

Combining functions

If f and g are functions, we may combine them in obvious ways by the normal algebraic operations of addition, subtraction, multiplication, and division (being careful to avoid division by zero). Another operation that is used throughout mathematics is *composition of functions*, taking a *function of a function.* All students have used composition of functions, whether or not the idea has been formally studied.

In evaluating the function $F(x) = \sqrt{1 - x^2}$ on a calculator, for instance, we first find $1 - x^2$ and then use the square root key. If f is the square root function, $f(x) = \sqrt{x}$, and $g(x) = 1 - x^2$, then we want "f of g(x)," which we write as $f(g(x)) = f(1 - x^2) = \sqrt{1 - x^2}$. Function

---

* The calculator $(\sqrt{x})$ function is technically different from the mathematical function given by $f(x) = \sqrt{x}$. For instance, when $x = 2$, the calculator gives a finite decimal, while the value of $f(2)$ is $\sqrt{2}$, a nonterminating (and nonrepeating) decimal.

F is the *composition*, f *of* g: $F(x) = f(g(x))$. It is important to recognize that the *order of composition* is critical; $g(f(x))$ is usually not the same function as $f(g(x))$.

EXAMPLE 4   Let $f(x) = \sqrt{x}$ and $g(x) = 1 - x^2$.

a] Evaluate $f(g(\tfrac{1}{2}))$ and $g(f(\tfrac{1}{2}))$.
b] Write formulas for $f(g(x))$ and $g(f(x))$.
c] Find the domains of $f(g(x))$ and $g(f(x))$.

Solution

a] $f(g(\tfrac{1}{2})) = f(1 - (\tfrac{1}{2})^2) = f(3/4) = \sqrt{3/4} \approx 0.866$.
$g(f(\tfrac{1}{2})) = g(\sqrt{\tfrac{1}{2}}) = 1 - (\sqrt{\tfrac{1}{2}})^2 = 1 - \tfrac{1}{2} = 0.5$.

b, c] $f(g(x)) = f(1 - x^2) = \sqrt{1 - x^2}$. The domain of this function was shown in Example 2 to be $\{x \mid -1 \le x \le 1\}$.

$$g(f(x)) = g(\sqrt{x}) = 1 - (\sqrt{x})^2 = 1 - x$$
as long as $\sqrt{x}$ *is defined.*

The composition is defined only if $f(x)$ is defined. In this case we cannot evaluate $f(x)$ unless $x \ge 0$. The domain of $g(f(x))$ is $\{x \mid x \ge 0\}$ even though $1 - x$ is defined for all x.   ■

Warning: The domain of the composition $g(f(x))$ is the set of numbers such that

1. $f(x)$ is defined, *and*
2. $g(f(x))$ is also defined.

EXERCISE 2.1

In Problems 1–14, determine the domain of the given functions. It may be helpful to sketch a graph.

1. $f(x) = 3 - 2x$ **2.** $g(x) = \dfrac{x - 3}{2}$

3. $F = \{(x, y) \mid 2x + y = 3\}$ **4.** $h = \{(x, y) \mid y = 1 - x^2\}$

5. $f = \{(x, y) \mid y = 1 - x^2, -1 \le x \le 1\}$

6. $f(x) = x^2 + 2x, \; x < 0$

7. $h(x) = |x - 1|$ **8.** $f(x) = \dfrac{x}{|x|}, \; x > 0$

9. $F(x) = \dfrac{x}{|x|}$ **10.** $f(x) = |x| - x$

11. $f(x) = \sqrt{x} - 1$ **12.** $H(x) = \sqrt{x - 1}$

13. $y = \dfrac{\sqrt{x + 1}}{x}$ **14.** $y = \sqrt{4 - x^2}$

In each of Problems 15–20, a function and two values of x are given. Use a calculator to evaluate the function at each of the two values of x (to three decimal place accuracy). If your calculator indicates an ERROR, explain why.

15. $F(x) = \dfrac{x}{|x|}$; $x = -3.21, \dfrac{1 - \sqrt{2}}{2}$

16. $G(x) = |x| - x$; $x = -2.792, \dfrac{\pi + 1}{2}$

17. $f(x) = \sqrt{x} - 1$; $x = \dfrac{3}{7}, -\dfrac{\pi}{10}$

18. $H(x) = \sqrt{x - 1}$; $x = 4.311, 0.998$

19. $y = \dfrac{\sqrt{x - 1}}{x}$; $x = 1.004, \dfrac{\sqrt{2} - \pi}{4}$

20. $y = \sqrt{4 - x^2}$; $x = -1.252, \dfrac{\pi - 8}{2}$

In Problems 21–26, find a formula for each composition of functions and find the domain of each. Use the functions

$$f(x) = 1 + \dfrac{1}{x}, \qquad g(x) = \sqrt{x}, \qquad h(x) = 1 + 2x.$$

21. $f(g(x))$ **22.** $g(f(x))$ **23.** $f(h(x))$

24. $h(f(x))$ **25.** $g(h(x))$ **26.** $g(f(h(x)))$

In Problems 27–30, a verbal description of function f is given. Write a formula for $f(x)$ and find the domain.

27. f sends each *natural number* x to the number that is 1 less than twice x. The set of natural numbers is $\{1, 2, 3, \ldots\}$.

28. f adds 1 to the reciprocal of x and then takes the square root of the result.

29. f divides each number by its absolute value.

30. f squares the number, subtracts the square from 4, and then takes the square root of the result.

In Problems 31–34, rationalize the portion of the fraction that now involves radicals. (All of these expressions are taken from calculus.)

31. $\dfrac{\sqrt{1 + h^2} - h}{h}$ **32.** $\dfrac{\sqrt{x + t} - \sqrt{x}}{t}$

33. $\dfrac{1}{\sqrt{x+1}+\sqrt{x}}$

34. $\dfrac{x-4}{\sqrt{x}+2}$

35. $\dfrac{1}{t}\left(\dfrac{1}{\sqrt{x}}-\dfrac{1}{\sqrt{x+t}}\right)$. Add fractions and rationalize the numerator.

2.2 Trigonometric Functions (Acute Angles)

There are six functions, collectively called the *trigonometric functions*, which are the central concern of this course. The six functions are the sine, cosine, tangent, cotangent, secant, and cosecant. We usually use the abbreviations *sin*, *cos*, *tan*, *cot*, *sec*, and *csc*, respectively.

The trigonometric functions may be correctly defined in several different ways, some of which have little to do with triangles. Our choice is to follow the historical development in terms of right triangles. Later we will generalize our definitions.

We label a right triangle using the convention introduced in Section 1.1, as shown in Fig. 2.4. It is also convenient to identify the legs of the triangle as being *adjacent to* and *opposite* the acute angles of the triangle. The definitions displayed in Eq. (2.1) will be used so frequently that, when working with triangles, the automatic response to "sine" will be "opposite over hypotenuse," and so on.

$$
\begin{aligned}
\sin \alpha &= \frac{\text{opp } \alpha}{\text{hyp}} = \frac{a}{c}, & \csc \alpha &= \frac{\text{hyp}}{\text{opp } \alpha} = \frac{c}{a}, \\[2mm]
\cos \alpha &= \frac{\text{adj } \alpha}{\text{hyp}} = \frac{b}{c}, & \sec \alpha &= \frac{\text{hyp}}{\text{adj } \alpha} = \frac{c}{b}, \\[2mm]
\tan \alpha &= \frac{\text{opp } \alpha}{\text{adj } \alpha} = \frac{a}{b}, & \cot \alpha &= \frac{\text{adj } \alpha}{\text{opp } \alpha} = \frac{b}{a}.
\end{aligned}
$$

[2.1]

Reciprocal identities

The functions defined in (2.1) are displayed to emphasize the *reciprocal relations*:

$$
\csc \alpha = \frac{1}{\sin \alpha}, \qquad \sec \alpha = \frac{1}{\cos \alpha}, \qquad \cot \alpha = \frac{1}{\tan \alpha}.
$$

[2.2]

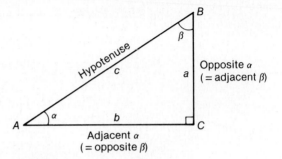

FIGURE 2.4

Figure 2.4 also shows that the side adjacent to angle α is opposite to angle β, and vice versa. The same definitions, in terms of "opposite" and "adjacent" apply to the trigonometric functions of the angle β:

$$\sin \beta = \frac{\text{opp } \beta}{\text{hyp}} = \frac{b}{c}, \qquad \csc \beta = \frac{\text{hyp}}{\text{opp } \beta} = \frac{c}{b},$$

$$\cos \beta = \frac{\text{adj } \beta}{\text{hyp}} = \frac{a}{c}, \qquad \sec \beta = \frac{\text{hyp}}{\text{adj } \beta} = \frac{c}{a}, \qquad [2.3]$$

$$\tan \beta = \frac{\text{opp } \beta}{\text{adj } \beta} = \frac{b}{a}, \qquad \cot \beta = \frac{\text{adj } \beta}{\text{opp } \beta} = \frac{a}{b}.$$

Complementary angle identities

Two acute angles whose sum is 90° are said to be complements of each other, or *complementary*. This means that the two acute angles of any right triangle are complementary. In symbols, the complement of angle α is $(90° - \alpha)$. This leads to another pairing among the trigonometric functions, which is also reflected in their names. Each function has a co- (or complementary) function: sine and cosine, tangent and cotangent, secant and cosecant. For each of these pairs the following relation holds: *a function of the complementary angle equals the cofunction of the angle.* These relations are summarized in the *complementary angle identities*:

$$\begin{array}{ll}
\sin (90° - \alpha) = \cos \alpha, & \cos (90° - \alpha) = \sin \alpha, \\
\tan (90° - \alpha) = \cot \alpha, & \cot (90° - \alpha) = \tan \alpha, \\
\sec (90° - \alpha) = \csc \alpha, & \csc (90° - \alpha) = \sec \alpha.
\end{array} \qquad [2.4]$$

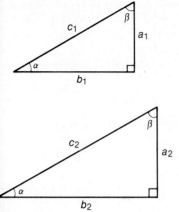

FIGURE 2.5

Observations

The following observations can be made from the above definitions:

1. There are many right triangles that contain a given angle, such as α in Fig. 2.4, and so it may appear that the above definitions depend upon the particular right triangle used. However, this is not the case, since we recall from geometry that any two such triangles are similar and the ratios of corresponding sides are always equal. For example, in Fig. 2.5 we have two similar right triangles, and so $a_1/c_1 = a_2/c_2$. Thus $\sin\alpha$ is equal to a_1/c_1 or a_2/c_2.

2. It should be clear that the definitions do actually define *functions*. For example, to each acute angle the sine function assigns the unique real number given by the ratio a/c. The situation is similar for each of the other five functions. The domain of all the functions as we have defined them is the set of *acute angles*. In Section 2.6 we shall extend our definitions to angles of any size, and then finally to the set of real numbers in Section 2.7.

Trigonometric functions of 30°, 45°, and 60° angles

FIGURE 2.6

The ratios of sides of two special right triangles (45°–45° and 30°–60°) were described in Section 1.1. These two triangles allow us to evaluate trigonometric functions in exact form for a whole family of angles as we shall see in Section 2.6. We may label the lengths of the sides as shown in Fig. 2.6, from which we may read the values of all six functions. The results are displayed in *exact form* in Table 2.1. Decimal approximations may be obtained from the exact form by the use of a calculator or by evaluating the functions directly as discussed in Section 2.3.

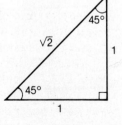

relationship

memorize

Table 2.1 Trigonometric Functions of Special Angles

x	30°	45°	60°
sin x	$1/2$	$1/\sqrt{2}$	$\sqrt{3}/2$
cos x	$\sqrt{3}/2$	$1/\sqrt{2}$	$1/2$
tan x	$1/\sqrt{3}$	1	$\sqrt{3}$
csc x	2	$\sqrt{2}$	$2/\sqrt{3}$
sec x	$2/\sqrt{3}$	$\sqrt{2}$	2
cot x	$\sqrt{3}$	1	$1/\sqrt{3}$

> We *strongly* urge each student to learn to quickly sketch the triangles in Figure 2.6. These figures contain *all* the information about the trigonometric functions of these special angles.

EXAMPLE 1 If θ is an angle for which $\sin \theta = \frac{1}{3}$, find $\tan \theta$ and $\sin (90° - \theta)$:

 a] in exact form,   **b]** in decimal form (to four places).

Solution Consider a right triangle with θ as an acute angle. Since $\sin \theta = \frac{1}{3}$, we can use the side opposite θ as one unit and the hypotenuse as three units, as shown in Fig. 2.7. The length of the third side will be $\sqrt{3^2 - 1^2} = \sqrt{8} = 2\sqrt{2}$. Therefore

a] $\tan \theta = \dfrac{\text{opp }(\theta)}{\text{adj }(\theta)} = \dfrac{1}{2\sqrt{2}} = \dfrac{\sqrt{2}}{4}$ (exact)

$\sin (90° - \theta) = \dfrac{\text{opp }(90° - \theta)}{\text{hyp}} = \dfrac{2\sqrt{2}}{3}$ (exact)

b] $\tan \theta \approx 0.3536$, $\sin (90° - \theta) \approx 0.9428$
(decimal approximation). ∎

EXAMPLE 2 In a right triangle, $a = 5.24$ cm and $c = 16.36$ cm (Fig. 2.8). Find:

 a] the length of side b (to three significant digits),
 b] $\tan \alpha$ (remember, α is the angle opposite side a).

Solution

 a] From the Pythagorean Theorem,

$$b = \sqrt{16.36^2 - 5.24^2} \approx 15.5 \text{ cm.}$$

 b] $\tan \alpha = \dfrac{a}{b} = \dfrac{5.24}{\sqrt{16.36^2 - 5.24^2}} \approx 0.338.$

Note that we did not use the rounded off value of b in determining $\tan \alpha$. ∎

FIGURE 2.7

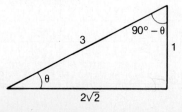

FIGURE 2.8

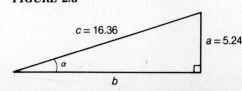

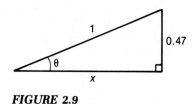

FIGURE 2.9

EXAMPLE 3   If $\sin \theta = 0.47$, find the remaining five trigonometric functions of θ (correct to two decimal places).

Solution   Since $\sin \theta = 0.47/1$, we can use a right triangle with hypotenuse 1 and side opposite θ as 0.47 (Fig. 2.9). Let x represent the length of the adjacent side; then $x = \sqrt{1^2 - (0.47)^2} \approx 0.8827$. Thus

$$\cos \theta \approx 0.88, \qquad \tan \theta \approx 0.53, \qquad \cot \theta \approx 1.88,$$
$$\sec \theta \approx 1.13, \qquad \csc \theta \approx 2.13. \quad \blacksquare$$

EXAMPLE 4   In a right triangle we are given that $c = 15.72$ and $\sin \beta = 3/5$ (Fig. 2.10). Find

a] the length of side a,   **b]** $\tan \alpha$.

Solution

a] Since $\cos \beta = a/15.72$, we have $a = 15.72 \cos \beta$. Thus we need to determine $\cos \beta$. Since $\sin \beta = 3/5$, we draw a second triangle as shown in Fig. 2.10 in which we first determine $x = \sqrt{5^2 - 3^2} = 4$. From this triangle we have $\cos \beta = 4/5$. Therefore

$$a = 15.72 \cos \beta = (15.72)\left(\frac{4}{5}\right) \approx 12.58.$$

b] From the second triangle we have $\tan \alpha = 4/3 \approx 1.333$.

FIGURE 2.10

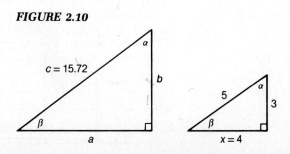

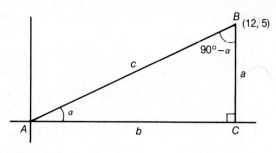

FIGURE 2.11

Alternative solution

a] Without drawing a second triangle we are given that $\sin \beta = 3/5 = b/c = b/15.72$, from which $b = (15.72)(3/5) = 9.432$. From the Pythagorean Theorem, $a = \sqrt{c^2 - b^2} \approx 12.58$.

b] Using the exact values of b and a we get $\tan \alpha = a/b \approx 1.333$.

∎

EXAMPLE 5 A right triangle is placed on the coordinate plane with vertex A at the origin and C on the positive x-axis. If the coordinates of B are (12, 5), find:

a] the length of the hypotenuse c, **b]** $\sin \alpha$, **c]** $\cos(90° - \alpha)$. (See Fig. 2.11.)

Solution

a] In the first quadrant the coordinates of B give the lengths of sides b and a, respectively. Thus $b = 12$, $a = 5$, and $c = \sqrt{12^2 + 5^2} = 13$.

b] $\sin \alpha = a/c = 5/13$.

c] $\cos(90° - \alpha) = \sin \alpha = 5/13$. ∎

EXERCISE 2.2

1. We gave the trigonometric functions for the special angles 30°, 45°, and 60° in degree measure. The corresponding radian measures are $\pi/6$, $\pi/4$, and $\pi/3$. Draw a right triangle having an acute angle of $\pi/4$ and one having an acute angle of $\pi/6$. From your triangles, complete the following table (using *exact form*):

	sin	cos	tan	cot	sec	csc
$\pi/6$						
$\pi/4$						
$\pi/3$						

2. If $\cos \theta = 3/5$, find in exact form

    a] $\tan \theta$ **b]** $\cot \theta$ **c]** $\csc \theta$

3. If $\tan \alpha = 4/3$, find in exact form

    a] $\sin \alpha$ **b]** $\cos \alpha$ **c]** $\sec \alpha$

4. If $\sin \theta = 3/4$, find the answers correct to two decimal places:

    a] $\cos \theta$ **b]** $\tan \theta$

5. If $\sin \alpha = 2/7$, determine each of the following in exact form:

    a] $\cos \alpha$ **b]** $\sin (90° - \alpha)$

    c] $\tan \alpha$ **d]** $\sec (90° - \alpha)$

6. If $\cos \theta = 8/17$, find in exact form:

    a] $\tan \theta$ **b]** $\tan (90° - \theta)$

    c] $\sec (90° - \theta)$ **d]** $\csc \theta$

7. If $\sec \theta = 1.5$, find in exact form:

    a] $\sin \theta$ **b]** $\tan \theta$ **c]** $\cos (90° - \theta)$

8. If $\cos \theta = 0.63$, find the remaining five trigonometric functions of θ. Give results correct to two decimal places.

9. In a right triangle, $\alpha = \pi/4$ and $a = 1.32$. Find

    a] b **b]** the hypotenuse c **c]** $\sin (\pi/2 - \alpha)$

In Problems 10–13, we assume that a right triangle is placed on the coordinate plane as in Example 5.

10. If B is the point $(3, 8)$, find (in exact form)

    a] $\tan \alpha$ **b]** $\sec \beta$

11. If B is the point $(3, 3)$, find

    a] the measure of α (in radians)   **b]** the hypotenuse c (exact)

12. If C is the point $(5, 0)$ and $\alpha = 30°$, find (in exact form)

    a] the length of a **b]** the coordinates of B

13. If B is the point $(2\sqrt{3}, 2)$, find

    a] $\tan \alpha$ **b]** the hypotenuse c **c]** the measure of α (in degrees)

In Problems 14–17, let ℓ be the line through the origin with slope 2/3. Then points $P(3, 2)$ and $Q(6, 4)$ are on the line ℓ. Drop perpendiculars $\overline{PR}$ and $\overline{QS}$ to the x-axis as shown in Fig. 2.12. Then right triangles OPR and OQS are similar.

14. Find the coordinates of R and S.

15. Find the lengths of $\overline{OP}$ and $\overline{OQ}$.

16. Use $\triangle OPR$ to find the entries in the table:

	sin	cos	tan
α			
β			

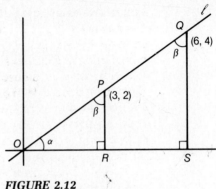

FIGURE 2.12

17. Use ΔOQS to find the entries in the table:

	sin	cos	tan
α			
β			

18. A cat stranded on a telephone pole has found secure footing at a point where the guy wire meets the pole. If the distance from the foot of the pole to the foot of the guy wire is 3 m and the wire makes an angle of 60° with the ground, how high above the ground is the cat?

19. In a right triangle $a = 2.36$, $b = 5.63$. Find
a] the length of c **b]** $\sin \alpha$ **c]** $\cot \beta$

20. In a right triangle we are given that $c = 6.47$ and $\sin \alpha = 5/17$. Find correct to two decimal places:
a] the length of a **b]** the length of b **c]** $\tan \beta$

21. Find the height of the Washington Monument if it casts a shadow of 290 m when the sun is 30° above the horizon.

22. Lighthouse *BC* is located on the edge of a cliff, as shown in Fig. 2.13. From point *A* (which is 67 m from the base of the cliff *D*), angles α and β are measured and found to be 60° and 45°, respectively. Find the height *h* of the lighthouse.

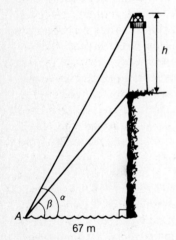

FIGURE 2.13

67 m

2.3 Using a Calculator to Find Values of Trigonometric Functions*

While the *definitions* of the trigonometric functions are perfectly straightforward, *evaluation* is another matter. In the preceding section we considered some examples of *special angles* for which geom-

---

* See Appendix A for basic calculator instruction.

etry provides enough information to evaluate the trigonometric functions *exactly*. Such angles are relatively scarce. Ratios of lengths of triangle sides is simply not an efficient means of getting values for trigonometric functions. If we needed to determine sin 37°, as an example, we would need a right triangle having an acute angle of 37°. Even if we were able to construct such a triangle with some accuracy, there is no simple relationship (as in the case of 45°–45° or 30°–60° right triangles) among the lengths a, b, c. We could measure the lengths and evaluate sin 37° as the ratio a/c but could hope for no more than a reasonable approximation at best.

The development of techniques for the evaluation of trigonometric functions is a remarkable chapter in the history of ideas. As refinements were made in earlier approximations, the information was collected in tables. Trigonometric tables have been traced to the time of Hipparchus (150 B.C.). Three hundred years later, Ptolemy put together tables that served astronomers for a thousand years. In the early 1600s, Henry Briggs compiled tables containing values of sines, tangents, and their logarithms for every one hundredth of a degree from 0° to 90°. By the latter part of that century, calculus techniques had been invented that gave infinite series methods for function evaluation. Whatever methods were used to find values, however, the results were available to users only through tables. Slide rules, an indispensable tool for the scientist and engineer of a few years ago, were really just mechanical devices for making the information from tables more convenient. With the advent of hand-held calculators, there has finally been a fundamental change in function evaluation. Every time you push the buttons for sin 30°, the calculator goes through all of its programmed steps to display 0.5, and the calculations giving sin 30.01° ≈ 0.500151142 and sin 30.001° ≈ 0.500015115 are no more difficult than the one for sin 30°.

Until very recently, every trigonometry course required the mastery of tables, and every textbook included at least a short table to teach principles of using and interpolating from tables.* Since tables are no longer necessary, there is no need for their study as part of trigonometry. Faster, easier, and more accurate information is literally in our hands.

Calculator design dictates multiple function usage for calculator keys. This is compounded for the trigonometric function keys. Scientific calculators have keys labeled (sin), (cos), and (tan) for the evaluation of these functions, and in many cases the inverse trigonometric functions are given by the same key, usually preceded by another key labeled (f), (2nd), or (INV). In addition, there are keys to put the calculator into *degree mode* or *radian mode* (and calculation in grads, which we will not use in this book).

---

* See Appendix C for a discussion of tables and interpolation.

The reason for different modes is the use of different methods of measuring angles. In a right triangle such as that shown in Fig. 2.4, the trigonometric functions are defined for the *angle* α, independently of how it might be measured. But since an angle has one measure in degrees and another measure in radians, there are effectively two sets of trigonometric functions. Each angle in an equilateral triangle is the same, one sixth of a revolution, measured as 60° or $\pi/3$ radians. In degree mode a calculator will give 1.732 (to three places) when we press 60 $\boxed{\tan}$. In radian mode we also get 1.732 when we enter $\pi/3$ into the display and then press $\boxed{\tan}$.

Sometimes, to emphasize that the function value does depend on the mode, we may distinguish the modes by using $\sin^R t$, to indicate that the angle t is measured in radians, or $\sin^\circ t$ when t is measured in degrees. More often, though, the appropriate mode will be clear from the context. When working with triangles, we will most often use degree measure. When we consider the trigonometric ratios *as functions* (as they are used in calculus), there are important reasons always to use radian measure.

EXAMPLE 1   Evaluate sin 37°.

Solution   First be certain that your calculator is in degree mode. Then merely press the following keys: 3, 7, $\boxed{\sin}$. The display will read 0.601815023 if your calculator shows ten digits. Thus sin 37° $\approx$ 0.60182 to five decimal places.   ■

EXAMPLE 2   Evaluate cot 64°.

Solution   The calculator does not have a key labeled $\boxed{\cot}$. However, as we observed in Section 2.1, the cotangent function is the reciprocal of the tangent, and so we have cot 64° = 1/tan 64°. Therefore with the calculator in degree mode, press the following keys: 6, 4, $\boxed{\tan}$, $\boxed{1/x}$. The display will give cot 64° $\approx$ 0.487732589. The reader should note at this point that 1/tan 64° and tan (1/64)° are not equal. That is, the $\boxed{1/x}$ key should be pressed after the $\boxed{\tan}$ key.

Alternative solution   As was pointed out in Section 2.2, cot θ = tan $(90° - \theta)$, and so we have cot 64° = tan (90° − 64°) = tan 26°. Thus pressing the keys 2, 6, $\boxed{\tan}$, gives cot 64° $\approx$ 0.487732589.   ■

EXAMPLE 3   Evaluate cos 24°31′43″ correct to five decimal places.

Solution   We first convert 24°31′43″ into a decimal number of degrees as follows:

$$24°31'43'' = \left(24 + \frac{31}{60} + \frac{43}{3600}\right)^\circ.$$

Be sure your calculator is in degree mode and carry out the following sequence of steps: first evaluate $24 + 31/60 + 43/3600$; then press $\boxed{\cos}$, and the answer will appear in the display. That is, $\cos 24°31'43'' \approx 0.90975$. ∎

EXAMPLE 4 Compare $\sin°x$ and $\sin^R x$ when $x = 1.2$. (Give answers to five decimal places.)

Solution Note that for $\sin°x$ we normally write $\sin 1.2°$ and evaluate by placing the calculator in degree mode and pressing 1.2 $\boxed{\sin}$; the display gives $\sin 1.2° \approx 0.02094$. For $\sin^R x$ we write $\sin 1.2$. With the calculator in radian mode the same keys (1.2, $\boxed{\sin}$) give us $\sin 1.2 \approx 0.93204$. ∎

Notation: In Example 4 we wrote $\sin 1.2°$ to indicate that degree measure is being used for angle measure. We also see $\sin 1.2$, which indicates that radian measure is being used.

Whenever there are no units shown, we are to understand that radian measure is intended.

EXAMPLE 5 Evaluate $\tan (3\pi/11)$ correct to eight decimal places.

Solution Place the calculator in radian mode; calculate $3\pi/11$ (use the $\boxed{\pi}$ key on the calculator), and then press $\boxed{\tan}$: $\tan (3\pi/11) \approx 1.15406152$. ∎

EXERCISE 2.3

In Problems 1–15, use a calculator to evaluate the given function and express your answer to four decimal places.

1. $\sin 28°$
2. $\tan 49°$
3. $\cos 72°$
4. $\cot 78°$
5. $\sec 35°$
6. $\csc 17°$
7. $\sin 43°21'$
8. $\sec 57°16'$
9. $\cos 12°37'41''$
10. $\sin 0.4$
11. $\cos 1.25$
12. $\tan (\pi/3)$
13. $\cot (3\pi/8)$
14. $\sec (\pi/4)$
15. $\tan (\pi/4)$

For Problems 16–18, express your answers to five significant digits.

16. Evaluate $\sin^R x$ and $\sin°x$ if $x = 0.234$.
17. Evaluate $\tan^R x$ and $\tan°x$ if $x = 1.57$.
18. Evaluate $\cos^R x$ and $\cos°x$ if $x = 3/4$.

19. Complete the table. Give entries to three significant digits. Observe that $\pi/2 \approx 1.570796$.

x	1.5	1.55	1.56	1.57	1.5707
$\tan^R x$					

20. Complete the table with entries rounded off to five decimal places. Observe that $(\sin x)/x$ can have meaning only if x is a number (hence radian mode is required).

x	0.1	0.05	0.01	0.005	0.001
$\sin^R x$	.				
$(\sin x)/x$					

For Problems 21–25, let $\alpha = 30°$, $\beta = 20°$, $\gamma = \pi/6$, and $\delta = \pi/4$ and evaluate the given expressions to four decimal places. Compare the two answers in each problem.

21. $(1/2) \sin \alpha$; $\sin (\alpha/2)$ **22.** $2 \cos \beta$; $\cos 2\beta$

23. $\tan (\alpha + \gamma)$; $\tan \alpha + \tan \gamma$ (Note that you must convert α and γ to the same measure before you can evaluate $\tan (\alpha + \gamma)$.)

24. $\sec (\delta - \gamma)$; $\sec \delta - \sec \gamma$ **25.** $\csc \alpha$; $\csc \gamma$

In Problems 26–43. use a calculator to evaluate the given expression. Give answers correct to two decimal places. (If necessary, see Appendix A for a review.)

26. $(2.48) \sin 73°16'$

27. $\dfrac{3.56 \sin 24°17'}{\sin 47°21'}$

28. $\dfrac{2 \tan 35°12'}{1 - (\tan 35°12')^2}$

29. $65.48 \csc 43°18'$

30. $\tan \dfrac{3\pi}{13}$

31. $\sec 1.47$

32. $\cos \dfrac{7\pi}{17}$

33. $\dfrac{8.54 \sin (5\pi/11)}{\sin (3\pi/7)}$

34. $(\sin 23°48')^2 + (\cos 23°48')^2$

35. $\sec 31°12'36''$

36. $\cot 72°15'41''$

37. $\dfrac{1}{\csc (3\pi/8)} + \dfrac{1}{\sec (3\pi/8)}$

38. $\sin \left(\dfrac{1 + \sqrt{2}}{5} \right)$

39. $\sin 37° \cos 56° - \sin 56° \cos 37°$

40. $\left(\dfrac{1 + \sqrt{5}}{3} \right) \sin \left(\dfrac{5\pi}{12} \right)$

41. $-\cos \left(\dfrac{\pi}{2} - 0.43 \right)$

42. $\sqrt{3^2 + 5^2 - 2(3)(5) \cos 37°}$

43. $\sqrt{3.56^2 + 4.73^2 - 2(3.56)(4.73) \cos 38.4°}$

44. How tall is a flagpole that casts a shadow of 23 m when the sun is 37° above the horizon?

45. The distance from the base to the top of the Leaning Tower of Pisa is 54.6 m, and it makes an angle of 84°45′ with the horizontal. How far does the top overhang the base?

2.4 Solving Right Triangles

As we noted earlier, the word *trigonometry* implies the study of measurements related to triangles. Historically, the development of the subject was indeed motivated by the practical needs of surveying, navigation, and architecture (among other things), and these involved problems of determining certain unknown parts of a triangle from known information about it.

We first describe a problem that involves triangles for its solution. Suppose we wish to determine the height of a mountain peak and there is no convenient way to measure it directly. One approach is to locate two points A and B on the ground, as shown in Fig. 2.14, and measure the distance between them. Also we can measure the angles α and β. With this much information we can determine the height h by using trigonometric properties of triangles that will be developed in this chapter. We postpone further discussion of this example until such properties are at our disposal (see Problem 35 of this section).

A triangle has six parts—three angles and three sides. When we say "angle of a triangle," we mean the angle formed by the two rays that contain two sides of the triangle and have the vertex as their common endpoint. To "solve a triangle" means that measurements of some of these parts are given (usually sufficient to determine a unique triangle) and we determine the remaining parts from the given information. We study the solution of right triangles in this section and delay the solution of general triangles until Chapter 7.

FIGURE 2.14

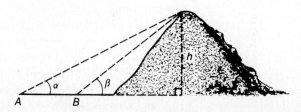

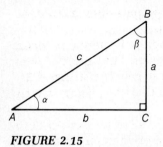

FIGURE 2.15

Figure 2.15 illustrates again the standard notation (introduced in Section 1.1) for a right triangle. As we have done previously, we shall use a letter interchangeably to denote a side (segment) of the triangle or to represent the length of the side. Similarly, the letter α may represent either the angle α or the measure of the angle.

If, in addition to the right angle, the measures of two of the remaining five parts are known and at least one of these is a, b, or c, then a unique triangle is determined, and we can find the remaining parts. This will involve only the use of the definitions of trigonometric functions (as given in Section 2.2), the Pythagorean Theorem, and the calculator. We illustrate by considering some examples. Solution of the first example is discussed in some detail. The others involve similar considerations, not all of which are recorded. In each case a calculator is used for numerical computations. Before proceeding with examples we should say something about accuracy of measurement and computations.

Accuracy of measurements and computations _____

The world of mathematics is an ideal world of exact numbers and complete precision. In the physical universe we deal with questions such as "How far?" and "How fast?" and "How much?," questions that seldom have precise or exact answers. Even though the real world is almost never exact, mathematics does help us understand real-world relationships, and it models and predicts physical phenomena remarkably well. Sometimes it is surprising how well an idealized model does predict; people have even written about the "unreasonable" applicability of mathematics.

Whole number counting and arithmetic are usually precise, but rational numbers already take us into the realm of approximations. It is questionable whether there has ever been such a thing as "half of a pie," notwithstanding the pictures that appear in every elementary arithmetic book. "Half" as a mental concept is precise; "half" of a physical quantity must be measured and can never be more than approximate.

The same thing is certainly true throughout trigonometry. The Pythagoreans knew that the diagonal of a square is not "commensurable with its side," a fact that we describe today by saying that $\sqrt{2}$ is an *irrational number*, one whose decimal expansion is nonterminating and nonrepeating. Any physical shape resembling a square has a diagonal measurement that is only approximated, however well or poorly, by the number 1.41421.... We have some sort of mental assumption that if we were to measure more and more precisely, the ratio of diagonal to the side would get closer and closer to $\sqrt{2}$.

A major source of approximate numbers is applications involving measurements. The accuracy of the results is necessarily limited by the degree of accuracy of the measuring instruments. These approximations are then used in formulas to compute other quantities, with the possibility of compounding errors.

Even exact numbers can give rise to approximations, as when we use decimal names for fractions, for instance. The number 4/3 is exact; its decimal name is 1.333 . . . with a nonending string of 3s. Because we cannot easily deal with infinite decimals, we are forced to terminate and "round off."

In this text we are concerned about manipulation with both exact and approximate numbers. The use of exact numbers emphasizes the principles and relationships involved. When answers are requested in *exact form*, there will be symbols involved, fractions (as quotients of integers) or radicals or π. It must be understood that in the abstract world of mathematics, answers are exact numbers. We may require *exact form* (meaning "keep the symbols and *do not use a calculator*") or we may specify any degree of calculator accuracy without being restricted to rules from the applied world.

Important as it is to know how to handle exact numbers, it is just as important to be able to know when to use a calculator and to understand the significance of the results displayed by the machine. We must also recognize that not all measurements are equally reliable, either in terms of the accuracy of measurement or the consistency of significant digits given. Consider an extreme example.

Suppose we are asked for the length of the diagonal of a square field that is 1 mile on a side; the Pythagorean Theorem says that the diagonal has length $\sqrt{2}$ miles. If we were to strictly follow the guideline that we are entitled to no more significant digits than the given data, we would be forced to round off our answer to 1 mile and conclude that it is no further between diagonally opposite corners of the square than it is along the edges.

A more natural assumption, perhaps, might be that the side of the field is measured more accurately than the one significant digit would indicate. We might feel entirely justified in saying that the diagonal of the field is approximately 1.4 miles.

To assert that the diagonal is 1.414214 miles just because the calculator displays that approximation is at least as ridiculous as giving a rounded value of 1 mile, nor are we justified in expressing the answer as 7467 feet, even though the exact relationship of 5280 feet per mile would give us that figure when multiplied by $\sqrt{2}$. There is just no reasonable basis for assuming that the stated 1-mile side of the field has been measured to the nearest foot.

Appendix B includes a more detailed discussion of computing with approximate numbers and rounding off results. The guidelines may be summarized in oversimplified form as follows:*

---

* This rule (as all others) must be used with discretion; exceptions do occur. For example, if we calculate $\sqrt{215}$ (understanding that the given 215 lies between 214.5 and 215.5), then we should expect the square root to lie between $\sqrt{214.5}$ and $\sqrt{215.5}$. However, when $\sqrt{215}$ is rounded to three digits, the result is 14.7, which is *larger* than $\sqrt{215.5} \approx 14.68$.

> In working problems involving *applications* we are not justified in recording final answers with more significant digits than the least precise numbers given.

We want to emphasize that such rules are meant to provide *direction and assistance* to make computation less burdensome for the student.

EXAMPLE 1   In a right triangle, $a = 32.4$ cm, $\alpha = 40.3°$. Find b, c, and β.

Solution   We draw a right triangle and denote the given parts (a and α), as shown in Fig. 2.16. To determine side b, the first step is to look for an equation that involves b and the given parts. We could use either $\tan \alpha = a/b$, which gives $b = a / \tan \alpha$, or $\cot \alpha = b/a$ to get $b = a \cot \alpha$. Since the calculator does not have a $\boxed{\text{cot}}$ key, we shall choose the first equation:

$$b = \frac{a}{\tan \alpha} = \frac{32.4}{\tan 40.3°} \approx 38.2 \text{ cm.}$$

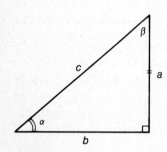

FIGURE 2.16

To determine the hypotenuse c, we could use any of the three equations: $\sin \alpha = a/c$; $\csc \alpha = c/a$; $c = \sqrt{a^2 + b^2}$. In general, it is a good practice to use a relationship that involves only the given parts, if possible. That is, the third option has a slight disadvantage in case we make an error in solving for b. The second has the disadvantage of involving cosecant, and our calculator does not have a $\boxed{\text{csc}}$ key. Therefore we decide upon the first expression:

$$c = \frac{a}{\sin \alpha} = \frac{32.4}{\sin 40.3°} \approx 50.1 \text{ cm.}$$

We know from geometry that the sum of the three angles of a triangle is 180°: $\alpha + \beta + 90° = 180°$. Therefore we have

$$\beta = 180° - 90° - \alpha = 90° - 40.3° = 49.7°. \quad \blacksquare$$

EXAMPLE 2   Given $\alpha = 15°20'$ and $c = 3.59$ m, find a, b, and β (Fig. 2.17).

Solution   Since $\sin \alpha = a/c$, then $a = c \sin \alpha$, and

$$a = 3.59 \sin 15°20' \approx 0.95 \text{ m.}$$

FIGURE 2.17

For b we use $\cos \alpha = b/c$, and so $b = c \cos \alpha$:

$$b = 3.59 \cos 15°20' \approx 3.46 \text{ m.}$$

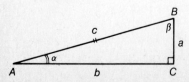

To find β, we use $\beta = 90° - \alpha$:

$$\beta = 89°60' - 15°20' = 74°40'. \quad \blacksquare$$

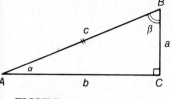

FIGURE 2.18

EXAMPLE 3 Given $c = 16.25$ cm and $\beta = 68°24'$, find the area of the triangle (Fig. 2.18).

Solution The area is equal to $ab/2$, so we first need to find sides a and b. From $\sin \beta = b/c$ we get $b = c \sin \beta$, and from $\cos \beta = a/c$ we get $a = c \cos \beta$. Therefore

$$\text{Area} = \frac{1}{2} (c \cos \beta)(c \sin \beta) = \frac{c^2 \sin \beta \cos \beta}{2}$$
$$= \frac{(16.25)^2 (\sin 68°24' \cos 68°24')}{2} \approx 45.19 \text{ cm}^2. \quad \blacksquare$$

Inverse functions by calculator

In the next example we see a need for evaluation of inverse trigonometric functions by calculator. A formal study of inverse functions is presented in Chapter 3. For solving right triangles the following example illustrates all that is needed.

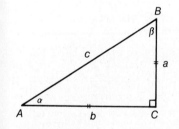

FIGURE 2.19

EXAMPLE 4 Given $a = 37.4$ cm, $b = 63.3$ cm, find c, α, and β (Fig. 2.19).

Solution $c = \sqrt{a^2 + b^2} = \sqrt{(37.4)^2 + (63.3)^2} \approx 73.5$ cm. For angle α we use $\tan \alpha = a/b = 37.4/63.3 \approx 0.59084$.

We are now confronted with the problem of finding α when we know $\tan \alpha$. This is the inverse of the problem of finding $\tan \alpha$ when α is given. The subject of inverse trigonometric functions will be discussed formally in Chapter 3; here we merely point out that scientific calculators can be used to find an angle corresponding to a given value of a trigonometric function. Calculator keys for inverse functions are usually labeled as (sin⁻¹), (cos⁻¹), (tan⁻¹); or there is an (INV) key that is to be followed by the appropriate (sin), (cos), (tan) key. We illustrate by completing the above problem where we have $\tan \alpha = 0.59084$ and we wish to determine α.

If the calculator has an (INV) key, then enter the number 0.59084 into the display and, with calculator in degree mode, press the (INV), (tan) keys in that order. The display will read 30.57° (to four significant digits).

If the calculator has a (tan⁻¹) key, then, with 0.59084 in the display and the calculator in degree mode, press (tan⁻¹). The display will read 30.57°. Expressing this result in degrees-minutes and rounding off to the nearest ten minutes, we get $\alpha \approx 30°30'$. To find β, we apply $\beta = 90° - \alpha$ to get $\beta \approx 59°30'$. $\blacksquare$

EXAMPLE 5

a] If $\sin \alpha = 0.4835$, find α in degrees correct to two decimal places.

b] If $\cos \alpha = 0.6897$, find α in radians correct to three decimal places.

Solution

a] Place the calculator in degree mode, enter the number 0.4835 into the display, and then press $\boxed{\sin^{-1}}$ or $\boxed{\text{INV}}$, $\boxed{\sin}$. The display will show 28.914243°. Thus $\alpha \approx 28.91°$.

b] Place the calculator in radian mode, enter the number 0.6897 into the display, and then press $\boxed{\cos^{-1}}$ or $\boxed{\text{INV}}$, $\boxed{\cos}$. The display will show 0.8097217. That is, $\alpha \approx 0.810$ rad. ■

EXAMPLE 6 If $a = 8.31$ cm and $\beta = 21.6°$, find the area of the right triangle (Fig. 2.20).

Solution The area is equal to $ab/2$, and since $b = a \tan \beta$, we have

$$\text{Area} = \frac{1}{2} \cdot a^2 \tan \beta = \frac{1}{2} (8.31)^2 \tan 21.6° \approx 13.7 \text{ cm}^2. \quad ■$$

Angle of elevation–angle of depression

In certain applications it is necessary to measure angles from a horizontal line of sight. An angle formed by a horizontal ray and the observer's line of sight to an object above the horizontal is called the *angle of elevation*. If the object is below the horizontal, the angle between the horizontal and the line of sight is called the *angle of depression* (Fig. 2.21).

EXAMPLE 7 From a window 25 meters above the ground the angle of elevation to the top of a nearby building is 24°, and the angle of depression to the bottom of the building is 15° (Fig. 2.22). Find the height of the building.

FIGURE 2.20

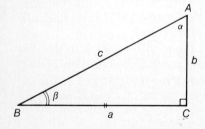

FIGURE 2.21

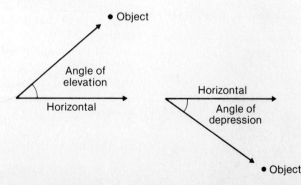

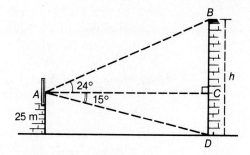

FIGURE 2.22

Solution In the diagram we wish to find $h = |\overline{BC}| + |\overline{CD}|$. We know that $|\overline{CD}| = 25$ m, so $h = |\overline{BC}| + 25$ m. By using triangle ACD we have

$$|\overline{AC}| = |\overline{CD}| \cot 15° = 25 \cot 15°.$$

Therefore from triangle ABC, we get

$$|\overline{BC}| = |\overline{AC}| \tan 24° = (25 \cot 15°) \tan 24°.$$

Thus

$$h = 25 + 25 \cot 15° \tan 24°$$
$$= 25 + \frac{25 \tan 24°}{\tan 15°} \approx 66.54 \text{ m.} \quad \blacksquare$$

Hence we conclude that $h \approx 67$ m.

EXAMPLE 8 Given Fig. 2.23, where $DECB$ is a square of side a, each of the segments may be expressed in terms of a and trigonometric functions of angle α. Find $|\overline{DF}|$, $|\overline{FE}|$, and $|\overline{AF}|$.

Solution First note that $\angle DBF = \alpha$. Using triangle $\triangle BFD$, $\tan \alpha = |\overline{DF}|/a$, so $|\overline{DF}| = a \tan \alpha$. Since $\overline{DE}$ is a side of the square, its length is a, and $|\overline{FE}| = a - |\overline{DF}| = a - a \tan \alpha$. To express $|\overline{AF}|$ in terms of a

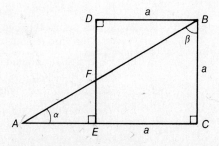

FIGURE 2.23

and α, we observe that $\overline{AF}$ is the hypotenuse of ΔAFE; so in that triangle, $\sin \alpha = |\overline{FE}|/|\overline{AF}|$, or $\csc \alpha = |\overline{AF}|/|\overline{FE}|$. Thus $|\overline{AF}| = |\overline{FE}|$ $\csc \alpha = (a - a \tan \alpha) \csc \alpha$. Thus we conclude that

$$|\overline{DF}| = a \tan \alpha, \qquad |\overline{FE}| = a - a \tan \alpha,$$
$$\text{and} \qquad |\overline{AF}| = (a - a \tan \alpha) \csc \alpha. \qquad \blacksquare$$

EXERCISE 2.4

Answers given in approximate decimal form should be rounded off to the number of significant digits consistent with the given data.

In each of Problems 1–10, two parts of a right triangle are given, and you are asked to *find the remaining parts*. The letters denote the sides and angles as in Fig. 2.15 at the beginning of this section.

1. $a = 4.3$, $\alpha = 35°$
2. $a = 7.4$, $\beta = 48°$
3. $b = 87$, $\alpha = 23°$
4. $c = 143$, $\alpha = 24°20'$
5. $c = 24.5$, $\beta = 63°40'$
6. $c = 24.32$, $\alpha = 64°48'$
7. $a = 53$, $b = 47$
8. $a = 0.743$, $b = 1.24$
9. $a = 25.4$, $c = 43.5$
10. $b = 1648$, $c = 2143$

11–14. For each of the triangles in Problems 1–4, find the area of the triangle.

15. The area of a right triangle is 6.8 cm^2, and one of its angles is 36°. Find the length of the hypotenuse.

16. The perimeter of a right triangle is 8.56 m, and one of its angles is 23°30'. Find the lengths of the two sides.

17. One angle of a right triangle is 48°35', and its perimeter is 15.48 cm. Determine the area of the triangle.

18. A right triangle is inscribed in a circle of radius 5.6 cm. One angle of the triangle is 64°. Find the lengths of the two sides.

19. Complete Example 8 by expressing the lengths of segments $\overline{AE}$, $\overline{AC}$, $\overline{AB}$ and $\overline{FB}$ in terms of a and α. (See Fig. 2.23.)

20. If, in Fig. 2.23, $\overline{EF}$ is half as long as $\overline{ED}$ (that is, $|\overline{EF}| = \frac{1}{2}a$), find the radian measure of angle α.

21. A line passes through two points (5, 2) and (8, 15). Find the acute angle that the line makes with the x-axis.

22. From a tower 27 meters tall the angle of depression of a boat on a lake is 56°. How far is the boat from the base of the tower? Assume that the base of the tower is in the same horizontal plane as the lake.

23. You wish to fence a piece of land that is in the shape of a right triangle with dimensions $a = 230$ m and $\alpha = 70°$. Find the total amount of fencing you must purchase.

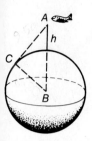

FIGURE 2.24

24. Assuming that the earth is a sphere with a radius of 6400 km, find the minimum height of an airplane above the surface, at which the pilot will be able to see an object on the ground 250 km away. In Fig. 2.24, point B is the center of the earth, A is the position of the plane, and object C is on the horizon ($|\overline{AC}| = 250$ km).

25. Find the area of an equilateral triangle with a side of length 12.56 cm.

26. Find the area of an isosceles triangle with equal sides 2.47 m long and an angle 41°30′ opposite one of them.

27. The sides of a parallelogram are 38.4 cm and 64.8 cm, and an interior angle is 115.6°. Find the area of the parallelogram.

28. A regular polygon is inscribed in a circle of radius 57 cm. Find the area of the polygon if it has
  a] four sides (a square)   **b]** six sides (a hexagon)
  c] eight sides (an octagon)   **d]** n sides

29. In Problem 28, find the areas of the polygons when $n = 100$ and when $n = 500$. Compare your answers with the area of the circle of radius 57 cm.

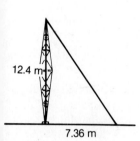

12.4 m

7.36 m

FIGURE 2.25

30. You wish to mount an antenna and have purchased a tower 12.4 meters tall. The tower is to be anchored from the top by three guy wires at a distance of 7.36 meters from the base (Fig. 2.25). How much guy wire do you need?

31. In Fig. 2.26, line segment $\overline{AB}$ is a diameter of the circle with radius 24.4 cm, C is a point on the circle, and arc AC is 27.3 cm long. Find the length of chord $\overline{AC}$. (Hint: Let θ be the central angle shown in the diagram; use the definition of radian measure to find θ. Recall facts from geometry about measures of central and inscribed angles in a circle.) (See Section 1.1.)

32. If the elevation of the sun is 17.5° at 5 P.M. on December 21, how far east of a retaining wall 5.48 meters tall should one locate plants requiring year-round full sun?

33. A segment of a circle of radius 4.56 cm is shown as a shaded region between chord $\overline{AB}$ and arc AB (Fig. 2.27). If the central angle θ is 1.15 radians, find the area of the segment.

FIGURE 2.26

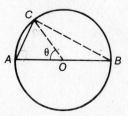

FIGURE 2.27

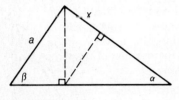

FIGURE 2.28

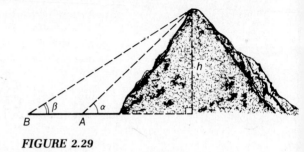

FIGURE 2.29

34. In Fig. 2.28, side a and angles α and β are given. Show that $x = a \sin \alpha \sin \beta$.

35. A surveyor wishes to determine the height of a mountaintop above the horizontal ground. He observes the angles of elevation from two points A and B on the ground and in line with the mountaintop. He measures the distance from A to B. These measurements are $\alpha = 43°30'$, $\beta = 32°20'$, $AB = 256$ m. Find the height of the mountaintop above the horizontal ground level (Fig. 2.29).

36. In Fig. 2.30, line segments $\overline{AD}$ and $\overline{BC}$ are parallel, the length of $\overline{AD}$ is 8.47 cm, and $\theta = 41°30'$. Find the lengths of $\overline{BC}$ and $\overline{CD}$.

37. A triangular piece of land is bounded by two farm roads intersecting at right angles and a highway intersecting one of the roads at an angle of 24.5°. You wish to purchase the property and know that the previous owner required 843 meters of fencing to enclose it. Land sells at $2.50 per square meter in this region. Find the cost of the property to the nearest hundred dollars.

38. From point A that is 8.15 meters above the horizontal level of the ground, the angle of elevation of the top of a tower (point B) is $\alpha = 32°30'$ and the angle of depression of its base (point C) is $\beta = 16°40'$ (Fig. 2.31). Find the height of the tower.

FIGURE 2.30

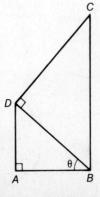

FIGURE 2.31

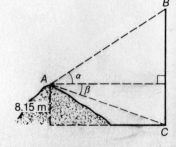

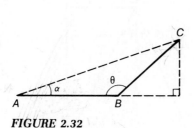

FIGURE 2.32

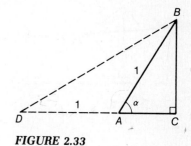

FIGURE 2.33

39. A surveyor starts at point A and measures $|\overline{AB}| = 41.32$ m, $|\overline{BC}| = 37.53$ m, $\theta = 137.44°$ (Fig. 2.32). Find the distance from A to C and angle α.

40. A sector with central angle $72°$ is cut out of a circular piece of tin of radius 16 cm. The edges of the remaining piece are joined together to form a cone. Find the volume of the cone (see inside front cover for volume formula).

41. Suppose A, B, C are vertices of a right triangle and α is the acute angle at A, as shown in Fig. 2.33. Also suppose the length of $|\overline{AB}|$ is 1. Extend side $\overline{CA}$ to point D so that the length of $\overline{AD}$ is also 1.
   a] Show that the angle CDB is equal to $\alpha/2$.
   b] Use right triangle BCD to find $\tan \alpha/2$. Specifically, show that it can be expressed in the form $\tan \alpha/2 = (\sin \alpha)/(1 + \cos \alpha)$. This useful identity will be seen again in Chapter 4.

42. In Problem 20 of Exercise 1.4, Motel's treadle sewing machine was described (Fig. 2.34). The radii of the two wheels are $r_1 = 3.50$ cm and $r_2 = 15.5$ cm. The distance between the centers is $|\overline{EF}| = 56.0$ cm. Find the length of the belt that goes around the two wheels. In the diagram, E and F are centers of the wheels, points A, B, C, and D are points at which the belt is tangent to the respective wheels, and we construct line BG through B parallel to EF.

FIGURE 2.34

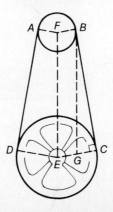

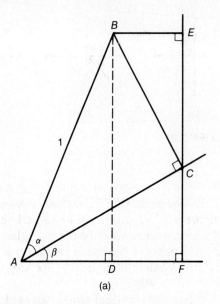
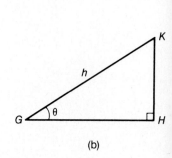

(a)   (b)

FIGURE 2.35

43. Suppose that α, β, and also $\alpha + \beta$ are acute angles. We wish to derive formulas for sin $(\alpha + \beta)$ and cos $(\alpha + \beta)$. First construct the diagram shown in Fig. 2.35(a) where $|\overline{AB}| = 1$ and angles at E, C, D, and F are right angles as indicated.

a] In Fig. 2.35(b), show that for given θ and h, $|\overline{HK}| = h \sin \theta$ and $|\overline{GH}| = h \cos \theta$. Use these results where appropriate in the following.

b] Show that $\angle BCE = \beta$. Now show the following:
In right triangle ABC, $|\overline{BC}| = \sin \alpha$ and $|\overline{AC}| = \cos \alpha$.
In right triangle BCE, $|\overline{BE}| = \sin \alpha \sin \beta$ and $|\overline{CE}| = \sin \alpha \cos \beta$.
In right triangle ACF, $|\overline{CF}| = \cos \alpha \sin \beta$ and $|\overline{AF}| = \cos \alpha \cos \beta$.

c] Finally, from right triangle ABD, show that
$|\overline{BD}| = \sin (\alpha + \beta) = \sin \alpha \cos \beta + \cos \alpha \sin \beta$.
$|\overline{AD}| = \cos (\alpha + \beta) = \cos \alpha \cos \beta - \sin \alpha \sin \beta$.

2.5 Angles in Standard Position

In order to define trigonometric functions for angles of any size, it is convenient to consider angles in a standard position. We shall say that *an angle is in standard position* when the vertex of the angle coincides with the origin of a rectangular coordinate system and the initial side coincides with the positive x-axis.

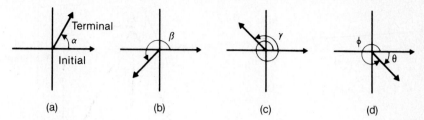

FIGURE 2.36

   Figure 2.36 illustrates five angles in standard position. The small arrows indicate the amount and direction of "turning" from the initial to the terminal side of the angles. Angles α, β, γ, and ϕ are all positive, and θ is negative. Angle γ is greater than one complete rotation (greater than 360°). Its degree measure appears to be $360° + 135° = 495°$.

   When the terminal side of an angle in standard position lies in a given quadrant, we say that *the angle is in that quadrant*. For the angles in Fig. 2.36, α is in quadrant I, β is in quadrant III (or is a *third-quadrant angle*), γ is in quadrant II, and both θ and ϕ are fourth-quadrant angles. If the terminal side of an angle δ coincides with one of the coordinate axes, then δ is said to be a *quadrantal angle* and is not said to be in any quadrant.

Coterminal angles

When two angles in standard position (in the same coordinate system) have terminal sides that coincide, we say that the two *angles are coterminal*. In Fig. 2.36(d), ϕ and θ are coterminal. Note that without an indication of the amount and direction of turning, coterminal angles are indistinguishable. We will see in the next section that the trigonometric functions also coincide for coterminal angles so that, for example, $\csc \phi = \csc \theta$. Examples of coterminal angles include 45° and 405°, since $405° = 360° + 45°$. Similarly, 210° and $-150°$ are coterminal, since $210° = 360° + (-150°)$. The angles that are coterminal with a given angle θ are all those which differ from θ by any integer number of rotations. In degree measure they are all angles of the form $\theta + k \cdot 360°$; in radian measure, $\theta + k \cdot 2\pi$, where k is any integer.

EXAMPLE 1   Sketch (by a freehand approximation or using a protractor) each of the following angles in standard position and determine the quadrant in which each angle is located:

   a] 64°   **b]** $-155°$ **c]** 248°   **d]** 450°   **e]** $-180°$

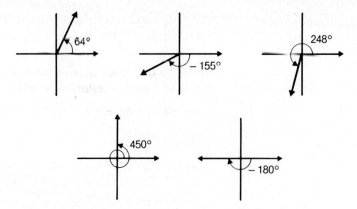

FIGURE 2.37

Solution   The given angles are shown in Fig. 2.37. We see that 64°
is in quadrant I, and −155° and 248° are in quadrant III, while 450°
and −180° are quadrantal angles.   ■

EXAMPLE 2   Sketch the given angles in standard position and determine the quadrant in which each lies.

a] $\pi/2$ b] $-\dfrac{3\pi}{4}$ c] 3.05   d] −5   e] 7.5

Solution   The radian measure of quadrantal angles may be expressed in terms of multiples of π or as decimal approximations...,
$-\pi \approx -3.14$, $-\pi/2 \approx -1.57$, 0, $\pi/2 \approx 1.57$, $\pi \approx 3.14$, $3\pi/2 \approx 4.71$,
$2\pi \approx 6.28,\ldots$. The given angles are shown in Fig. 2.38. We see that

FIGURE 2.38

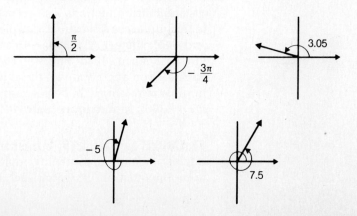

$\pi/2$ is a quadrantal angle, $-3\pi/4$ is in quadrant III, 3.05 is in quadrant II, and -5 and 7.5 are in quadrant I. ∎

EXAMPLE 3 Sketch each angle in standard position; then draw the smallest *positive* angle that is coterminal and determine its measure.

    a] $-48°$ **b]** -2.48 **c]** $-4\pi/3$ **d]** $-450°$

Solution

    a] $\alpha = 360° + (-48°) = 312°$ is coterminal with $-48°$
    b] $\beta = 2\pi + (-2.48) \approx 6.28 - 2.48 \approx 3.80$ (to two decimal places)
    c] $\gamma = 2\pi + (-4\pi/3) = 2\pi/3$
    d] $\delta = 720° - 450° = 270°$

These angles are shown in Fig. 2.39. ∎

EXAMPLE 4 Determine the set of all angles coterminal with
    a] $120°$ **b]** $\pi/4$

Solution   Since any number of complete revolutions added or subtracted gives a coterminal angle, the set of angles coterminal with $120°$ is given by

    a] $\{\ldots, -600°, -240°, 120°, 480°, 840°, \ldots\}$
                           $= \{120° + k \cdot 360° \,|\, k \text{ is any integer}\}$
    b] Adding or subtracting multiples of 2π to $\pi/4$ gives the set of angles coterminal with $\pi/4$,

$$\left\{\ldots, -\frac{7\pi}{4}, \frac{\pi}{4}, \frac{9\pi}{4}, \frac{17\pi}{4}, \ldots\right\} = \left\{\frac{\pi}{4} + k \cdot 2\pi \,\big|\, k \text{ is any integer}\right\}.$$ ∎

FIGURE 2.39

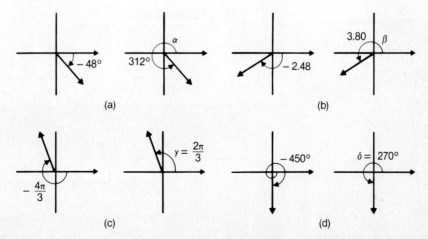

        (a)   (b)

        (c)   (d)

EXERCISE 2.5

For Problems 1–4, sketch the given angles in standard position and determine the quadrant in which each angle is located.

1. a] $40°$ b] $5\pi/4$ c] $-220°$ d] 3.41

2. a] $-\dfrac{5\pi}{6}$ b] -2.1 c] $210°$ d] $-270°$

3. a] $375°$ b] $-600°$ c] 8.47 d] 3π

4. a] $-\dfrac{9\pi}{2}$ b] $570°$ c] $840°$ d] 4.32

For Problems 5–10, first determine and then sketch the smallest positive angle that is coterminal with the given angle.

5. $-100°$ 6. $-\dfrac{9\pi}{2}$ 7. 8.47

8. -2.1 9. $570°$ 10. $-5\pi/6$

For Problems 11–16, determine whether or not the two angles are coterminal.

11. $60°, 240°$ 12. $-45°, 315°$ 13. $-3\pi/4, 5\pi/4$

14. $-\pi, \pi$ 15. $4.14, -2.24$ 16. $3\pi/2, -3\pi/2$

17. Find three angles coterminal with $\theta = 90°$.

18. Find three angles coterminal with $\theta = -\pi/6$.

For Problems 19–24, determine the set of all angles that are coterminal with the given angle. For Problems 21–24, each angle is in standard position.

19. $-2\pi/3$ 20. $30°$

21. The terminal side of θ contains the point $(1, 1)$. Give answers in radian measure.

22. The terminal side of θ contains the point $(\sqrt{2}, -\sqrt{2})$. Give answers in degree measure.

23. The terminal side of θ contains the point $(-5, 0)$ (radian measure).

24. The terminal side of θ contains the point $(2\sqrt{3}, 2)$ (degree measure).

2.6 Trigonometric Functions of Angles of Any Size

Given any acute angle, there is essentially one right triangle whose acute angles are the given angle and its complement. That is, all such right triangles are *similar*, and for similar triangles, ratios of corresponding sides are equal. That fact allowed us to define the trigonometric functions for acute angles in Section 2.2.

Definitions of trigonometric functions of any angle

To define the trigonometric functions of more general angles, we make use of standard position. Let θ be an angle in standard position. Choose any point $P(x, y)$ (other than the origin) on the terminal side of θ and let r be the distance from P to the origin (that is, $r = \sqrt{x^2 + y^2}$). See Fig. 2.40, in which we show two different choices on the terminal side of θ, $P(x, y)$ and $P_1(x_1, y_1)$. The six trigonometric functions are defined by ratios of x, y, and r:

$$
\begin{array}{lll}
\sin \theta = y/r, & \cos \theta = x/r, & \tan \theta = y/x, \\
\csc \theta = r/y, & \sec \theta = r/x, & \cot \theta = x/y.
\end{array}
$$

[2.5]

These definitions should be compared with the definitions (2.1) in Section 2.2 and should be memorized as thoroughly as those right triangle definitions, with the emphasis on coordinates rather than symbols (that is, "tan θ is the y-coordinate of P divided by the x-coordinate").

There are two important questions we must consider in defining the trigonometric functions of θ in terms of the coordinates of P.

1. Are the definitions independent of the choice of the point P on the terminal side of θ?

2. How are these definitions related to the right triangle definitions?

For both of these questions it is helpful to introduce the ideas of *reference triangle* and *reference angle*.

FIGURE 2.40

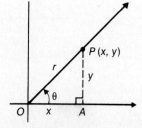 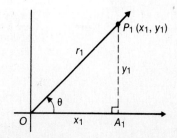

Reference triangle and reference angle _____

If θ is not a quadrantal angle, then from any point $P(x, y)$ on the terminal side, we may draw a *perpendicular to the x-axis* at the point $A(x, 0)$. Then the right triangle ΔPOA is called a *reference triangle* for θ, and *the acute angle* at the origin O (angle $\angle POA$) is the *reference angle* for θ. See Fig. 2.41, in which we show three angles in standard position and the corresponding reference triangles and reference angles.

The critical observations are that different points P on the terminal side of θ determine *similar* reference triangles and that the *reference angle is always the same for any choice of P*.

We may answer the second question first, by noting that for an acute angle θ, the reference triangle is *a right triangle containing θ*, so the definitions (2.5) are *identical* with the right triangle definitions. (See Fig. 2.40.)

The situation is similar for angles in other quadrants, but the differences should be examined carefully. The angle θ is never an angle of the reference triangle unless θ is acute. Furthermore, in all quadrants except the first, one or both of the coordinates of P are negative, while r *is always positive*. It follows that in these three quadrants, some of the trigonometric functions are *negative*. Even more important, however, is the following fact:

> The trigonometric functions of θ are the same as the functions of the reference angle (determined from the reference triangle), with *appropriate choice of signs.*

FIGURE 2.41

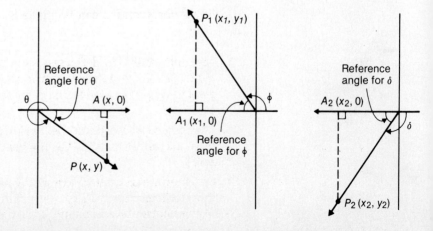

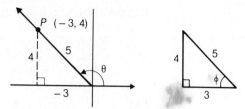

FIGURE 2.42

Because different points on the terminal side of θ determine the *same* reference angle and *similar* reference triangles, the values of the trigonometric functions of θ do not depend on the choice of the point P on the terminal side. This may be illustrated by the following example.

EXAMPLE 1   Suppose θ is an angle in standard position and point $(-3, 4)$ is on the terminal side of θ. Find the values of six trigonometric functions of θ.

Solution   The diagram in Fig. 2.42 shows a reference triangle for θ, in which point P is taken as $(-3, 4)$, and so $r = \sqrt{(-3)^2 + 4^2} = 5$. Therefore

$$\sin \theta = 4/5, \qquad \cos \theta = -3/5, \qquad \tan \theta = 4/-3,$$
$$\csc \theta = 5/4, \qquad \sec \theta = 5/-3, \qquad \cot \theta = -3/4.$$

Note that the values of $\sin \theta$ and $\csc \theta$ are the same as for the reference angle ϕ but that the other four functions have opposite signs.   ∎

Some properties of the trigonometric functions

Additional observations can be made from the definitions given in (2.5).

1. The *definitions do define six functions*; that is, each function associates a unique real number (the indicated ratio) with each given angle θ (whenever the ratio does not involve division by zero).

2. The *reciprocal relations* observed in Section 2.2 for acute angles (Eqs. 2.2) are still valid (again whenever there is no division by zero).

3. For *quadrantal angles*, one of the coordinates of any point on the terminal side will be zero. This means that some of the trigonometric functions are undefined for each quadrantal an-

gle. When the terminal side of θ lies on the x-axis, the y-coordinate is zero, and hence $\csc \theta = r/0$ and $\cot \theta = x/0$ are undefined. Similarly, for all angles that are coterminal with $\pm 90°$, $\sec \theta = r/0$ and $\tan \theta = y/0$ are undefined.

4. Pythagorean identity: A very important fact is an immediate consequence of definitions (2.5). Since $r^2 = x^2 + y^2$, we may see that

$$(\sin \theta)^2 + (\cos \theta)^2 = (y/r)^2 + (x/r)^2 = \frac{y^2 + x^2}{r^2} = 1.$$

Thus for any angle θ, we have

$$(\sin \theta)^2 + (\cos \theta)^2 = 1.$$

Calculator evaluations

The information in Section 2.3 on calculator evaluation of trigonometric functions of acute angles is still valid for angles of any size. This includes the use of radian and degree modes and the distinction between $\sin° \theta$ and $\sin^R \theta$. The calculator is programmed to automatically take care of the signs in various quadrants. For example, if $\theta = 318°$ is a fourth-quadrant angle, the y-coordinate of any point on the terminal side will be negative, so $\tan 318°$ should be negative, and the calculator displays -0.90040 (to five places) for $\tan 318°$. Because the trigonometric functions of some angles are not defined, the calculator must accommodate that as well. For the quadrantal angle $\pi/2$, the tangent function is undefined, and if we put the calculator in radian mode, enter $\pi/2$, and press $\boxed{\tan}$, the calculator should indicate an ERROR. It is instructive to note that the tangent is defined for angles extremely close to $\pi/2$. See Problem 19 in Exercise 2.3 and decide what happens to the values of $\tan \theta$ as θ gets closer and closer (from below) to $\pi/2$. We urge you to complete a similar table in which the angles are slightly greater than $\pi/2$.

Since the location of the terminal side of an angle θ determines the trigonometric functions of θ, *coterminal angles have the same corresponding trigonometric function values*. This observation leads to another interesting fact about calculators. Because of the nature of the computations required, there are limitations to the size of numbers for which the calculator will evaluate trigonometric functions. Most calculators will not give a value for $\sin 18\pi$, for example. But since $18\pi = 9(2\pi)$, 18π is coterminal with 2π (and with 0), and the calculator will give us values for 0 or 2π. Since every angle is coterminal with some positive angle no larger than $360°$, *by finding an appropriate coterminal angle we may get calculator evaluation for any angle.*

Evaluating trigonometric functions in exact form

If the angle is given in radians or degrees, the calculator may be used to get an approximate value. When coordinates of a point on the terminal side can be found, definitions (2.5) give *exact values*. The procedure for evaluating trigonometric functions exactly is as follows:

1. Sketch the angle in standard position (this includes determining the quadrant in which it is located).
2. Take a convenient point on the terminal side (from which the numbers x, y, and r can be found, including the signs of x and y).
3. Use the appropriate definitions from Eqs. (2.5).

Sign patterns are often helpful. In Fig. 2.43 we are shown which of the three functions sine, cosine, and tangent are positive in each quadrant, as may be checked by considering where the x- and y-coordinates are positive or negative. Some students like to use a mnemonic device to assist recall for positive values, such as "ACTS," from the initial letters moving clockwise from the first quadrant, or "CAST" (counterclockwise from quadrant IV).

The signs for cosecant, secant, and cotangent agree with the signs of sine, cosine, and tangent functions, respectively.

EXAMPLE 2   Evaluate the six trigonometric functions for 315°. Express each answer in exact form and in decimal form (correct to four places).

Solution   In the diagram of Fig. 2.44 we see that the reference triangle for 315° is a 45° right triangle. It is therefore convenient to take $(1, -1)$ as point P, and so $r = \sqrt{1^2 + (-1)^2} = \sqrt{2}$. Thus

$$\sin 315° = -\frac{1}{\sqrt{2}} = -\frac{\sqrt{2}}{2} \quad \text{(exact form)}.$$

FIGURE 2.43

```
                    y
Sine                        All
x < 0                       x > 0
y > 0     (−, +)  |  (+, +)  y > 0
                  |
─────────────────────────────── x
                  |
Tangent                     Cosine
x < 0                       x > 0
y < 0     (−, −)  |  (+, −)  y < 0
```

FIGURE 2.44

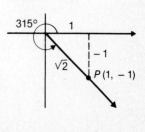

Using the calculator to evaluate $-\sqrt{2}/2$, we get $\sin 315° \approx -0.7071$ (to four decimal places). Similarly,

$$\cos 315° = \frac{1}{\sqrt{2}} = \frac{\sqrt{2}}{2}, \qquad \cos 315° \approx 0.7071$$

$$\tan 315° = \frac{-1}{1} = -1, \qquad \tan 315° = -1.0000,$$

$$\cot 315° = \frac{1}{-1} = -1, \qquad \cot 315° = -1.0000,$$

$$\sec 315° = \frac{\sqrt{2}}{1} = \sqrt{2}, \qquad \sec 315° \approx 1.4142.$$

$$\csc 315° = \frac{\sqrt{2}}{-1} = -\sqrt{2}, \qquad \csc 315° \approx -1.4142. \quad ■$$

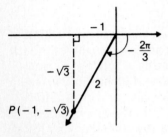

FIGURE 2.45

EXAMPLE 3 Evaluate $\sin (-2\pi/3)$ and $\tan (-2\pi/3)$. Express answers in exact form.

Solution Sketch $\theta = -2\pi/3$. The reference triangle for $\theta = -2\pi/3$ is a 30°–60° right triangle, so we can take P as $(-1, -\sqrt{3})$ (see Fig. 2.45). Thus

$$\sin \left(-\frac{2\pi}{3}\right) = -\frac{\sqrt{3}}{2} \quad \text{and} \quad \tan \left(-\frac{2\pi}{3}\right) = \frac{-\sqrt{3}}{-1} = \sqrt{3}. \quad ■$$

EXAMPLE 4 Evaluate the following functions of quadrantal angles (see Fig. 2.46).
a] $\sin 180°$ **b]** $\cos 180°$ **c]** $\tan 90°$ **d]** $\sec (-540°)$

Solution

a] Take point P as $(-1, 0)$, so $r = 1$. Then $\sin 180° = y/r = 0/1 = 0$.
b] Take P as in (a), then $\cos 180° = x/r = -1/1 = -1$.
c] Let P be $(0, 1)$, so $r = 1$. Then $\tan 90° = y/x = 1/0$. Since division by zero is not defined, we say that $\tan 90°$ is undefined.

FIGURE 2.46

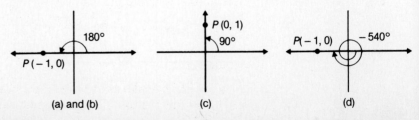

(a) and (b) (c) (d)

d] In the diagram of $-540°$ in standard position we see that the terminal side coincides with the negative x-axis. Therefore we can take point P as $(-1, 0)$ and so $r = 1$. Thus

$$\sec(-540°) = \frac{r}{x} = \frac{1}{-1} = -1. \quad \blacksquare$$

EXAMPLE 5   If angle θ is in the second quadrant and $\cos \theta = -7/10$, find the other five trigonometric functions of θ. Express each result **a]** in exact form,   **b]** in decimal form correct to three places.

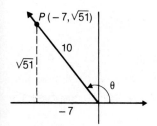

$P(-7, \sqrt{51})$

10

$\sqrt{51}$

-7

θ

FIGURE 2.47

Solution   Since θ is in the second quadrant and $\cos \theta = -7/10$, we get a reference triangle as shown in Fig. 2.47, by taking $x = -7$, $r = 10$. Then $y = \sqrt{10^2 - (-7)^2} = \sqrt{51}$. Using definitions given in Eqs. (2.5), we have

a] $\sin \theta = \dfrac{y}{r} = \dfrac{\sqrt{51}}{10}$, $\tan \theta = \dfrac{y}{x} = \dfrac{\sqrt{51}}{-7}$,

$\cot \theta = \dfrac{x}{y} = \dfrac{-7}{\sqrt{51}}$, $\sec \theta = \dfrac{r}{x} = \dfrac{10}{-7}$,

$\csc \theta = \dfrac{r}{y} = \dfrac{10}{\sqrt{51}}$.

b] Using the calculator to evaluate the expressions in (a), we get

$\sin \theta \approx 0.714$, $\tan \theta \approx -1.020$, $\cot \theta \approx -0.980$,
$\sec \theta \approx -1.429$, $\csc \theta \approx 1.400$. $\blacksquare$

EXAMPLE 6   If θ is an angle in the third quadrant and $\tan \theta = 3/4$, find the remaining five trigonometric functions of θ (see Fig. 2.48).

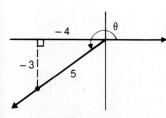

-4

θ

-3

5

FIGURE 2.48

Solution   Since $\tan \theta = 3/4 = -3/-4$ and θ is in quadrant III, we can take $(-4, -3)$ as the point to determine a reference triangle as shown in the diagram, and so $r = \sqrt{(-3)^2 + (-4)^2} = \sqrt{25} = 5$. Therefore

$$\sin \theta = \frac{y}{r} = -\frac{3}{5}, \quad \cos \theta = \frac{x}{r} = -\frac{4}{5}, \quad \cot \theta = \frac{x}{y} = \frac{-4}{-3} = \frac{4}{3},$$

$$\sec \theta = \frac{r}{x} = -\frac{5}{4}, \quad \csc \theta = \frac{r}{y} = -\frac{5}{3}. \quad \blacksquare$$

EXAMPLE 7   Evaluate the given expressions. Give answers rounded off to four decimal places.

a] $\sin(-237°)$ **b]** $\cos\left(\dfrac{\pi}{4} + 3.6\right)$ **c]** $\sec(24\pi + 3.57)$

Solution

a] With the calculator in degree mode, enter -237 in the display and then press (sin). This gives $\sin(-237°) \approx 0.8387$.

b] Place the calculator in radian mode. Enter the number $(\pi/4 + 3.6)$ into the display and press $\boxed{\text{COS}}$. This gives $\cos(\pi/4 + 3.6) \approx -0.3212$.

c] Since calculators will not evaluate trigonometric functions for angles as large as $24\pi + 3.57$, we first note that this angle is coterminal with 3.57. Therefore $\sec(24\pi + 3.57) = \sec 3.57 = 1/\cos 3.57$. With the calculator in radian mode, enter 3.57 into the display and then press $\boxed{\text{COS}}$, $\boxed{1/x}$. This gives $\sec(24\pi + 3.57) \approx -1.0993$. ■

EXERCISE 2.6

In each of Problems 1–6, P is a point on the terminal side of angle θ. Find the six trigonometric functions of θ in *exact form*.

1. $P(4, -3)$ **2.** $P(2, 3)$ **3.** $P(-2, -4)$

4. $P(\sqrt{3}, -1)$ **5.** $P(\sqrt{2}, \sqrt{3})$ **6.** $P(-\sqrt{5}, 2)$

In Problems 7–18, evaluate each of the given expressions in *exact form*.

7. a] $\sin 60°$ **8. a]** $\tan 30°$ **9. a]** $\cot(-45°)$
  b] $\cos 210°$ **b]** $\sec 300°$ **b]** $\csc 405°$

10. a] $\sin(-270°)$ **11. a]** $\tan 90°$ **12. a]** $\cos 720°$
   b] $\sec 180°$ **b]** $\cos 540°$ **b]** $\tan 225°$

13. a] $\sin\dfrac{2\pi}{3}$ **14. a]** $\tan\left(-\dfrac{3\pi}{4}\right)$ **15. a]** $\sin\left(\pi + \dfrac{5\pi}{6}\right)$

   b] $\cos\dfrac{5\pi}{3}$ **b]** $\sec\left(-\dfrac{5\pi}{4}\right)$ **b]** $\cos\left(\pi - \dfrac{\pi}{6}\right)$

16. a] $\cos\left(-\dfrac{9\pi}{4}\right)$ **17. a]** $\tan\left(-\dfrac{2\pi}{3}\right)$ **18. a]** $\sin\left(\dfrac{\pi}{2} + \dfrac{3\pi}{4}\right)$

   b] $\sec\left(-\dfrac{19\pi}{4}\right)$ **b]** $\csc\left(-\dfrac{5\pi}{6}\right)$ **b]** $\cos\left(2\pi - \dfrac{3\pi}{4}\right)$

19. In the accompanying table write a "+" sign or a "−" sign indicating the sign of the corresponding entry:

	sin	cos	tan	cot	sec	csc
124°						
−320°						
3.04						
−1.16						

In Problems 20–28, evaluate each of the given expressions and give answers rounded off to *three decimal places*.

20. **a]** $\sin 65°$ **21.** **a]** $\tan 124°$ **22.** **a]** $\tan (-243°)$
     b] $\cos 173°$ **b]** $\sec 347°$ **b]** $\cot 748°$

23. **a]** $\sin 48°16'$ **24.** **a]** $\tan 216°15'$ **25.** **a]** $\sin 2.43$
     b] $\cos 235°23'$ **b]** $\sec 437°20'$ **b]** $\cos 5.48$

26. **a]** $\tan (-2.5)$ **27.** **a]** $\cos \dfrac{3\pi}{7}$ **28.** **a]** $\tan (16\pi + 2.7)$

     b] $\cot (-4.3)$ **b]** $\tan \dfrac{9\pi}{5}$ **b]** $\cos (2\pi - 3.6)$

29. Find the value of

$$\frac{\cos \dfrac{2\pi}{3} - \sin \dfrac{4\pi}{3} + \tan \dfrac{5\pi}{4}}{\sin \dfrac{\pi}{2} - \tan \dfrac{5\pi}{3} + \sec \dfrac{2\pi}{3}}$$

in exact form and also in decimal form correct to three decimal places.

In Problems 30–33, verify that the given equations are satisfied for the specified angles. In each case, evaluate in *exact form* the left-hand side and the right-hand side and show that the resulting numbers are equal.

30. Verify that $\sin (\alpha - \beta) = \sin \alpha \cos \beta - \cos \alpha \sin \beta$ for each of the following pairs of values of α and β:

    a] $\alpha = \dfrac{2\pi}{3}, \beta = \dfrac{\pi}{6}$ **b]** $\alpha = \dfrac{\pi}{2}, \beta = \pi$

    c] $\alpha = \dfrac{3\pi}{2}, \beta = \dfrac{\pi}{2}$ **d]** $\alpha = \dfrac{5\pi}{4}, \beta = 3\pi$

31. Verify that $(\sin \theta)^2 + (\cos \theta)^2 = 1$ for each of the given values of θ:
    a] $\theta = 60°$ **b]** $\theta = 150°$ **c]** $\theta = \pi$

32. Verify that $\sin (2\theta) = 2 (\sin \theta)(\cos \theta)$ for the given values of θ:

    a] $\theta = 90°$ **b]** $\theta = 30°$ **c]** $\theta = \dfrac{2\pi}{3}$

33. Verify that $(\sec \theta)^2 - (\tan \theta)^2 = 1$ for the given values of θ:

    a] $\theta = \dfrac{3\pi}{4}$ **b]** $\theta = 225°$ **c]** $\theta = 495°$

34. For which of the given angles α and β is $\cos (\alpha + \beta) = \cos \alpha + \cos \beta$?

    a] $\alpha = \pi, \beta = 0$ **b]** $\alpha = 0, \beta = \dfrac{\pi}{2}$

    c] $\alpha = 45°, \beta = 45°$ **d]** $\alpha = 120°, \beta = 30°$

In Problems 35–39, give each answer in exact form and in decimal form correct to three decimal places.

35. If θ is an angle in the second quadrant and $\cos \theta = -3/5$, find the other five trigonometric functions of θ.

36. If $\sin \alpha = -3/4$ and the terminal side of α is in the fourth quadrant, find the remaining five trigonometric functions of α.

37. If $\cot \beta = 3/4$ and β is in the third quadrant, find the other five trigonometric functions of β.

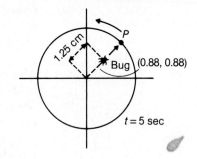

FIGURE 2.49

38. If $\tan \gamma = -6/5$ and the terminal side of γ is in the second quadrant, find the remaining five trigonometric functions of γ.

39. If $\sin \theta = -1/4$ and $\tan \theta$ is negative, find the remaining five functions of θ.

40. A bug starts at the center of a turntable of radius 20 cm and always heads towards a point P on the outside edge of the turntable. The turntable rotates in the direction shown in Fig. 2.49 at the rate of 1.5 rpm (revolutions per minute), and the bug travels at a constant speed of 15 cm/min. For instance, at the end of 5 seconds the bug is $15(5/60) = 1.25$ cm from the center, and at the same time the turntable has rotated through an angle of $1.5(5/60)$ rev $= 1/8$ rev $= 45°$. The location of the bug at the end of 5 seconds is given by

$$x = 1.25 \cos 45° \approx 0.88 \text{ cm}, \qquad y = 1.25 \sin 45° \approx 0.88 \text{ cm}.$$

a] Find the location (the coordinates) of the bug at the end of 10 sec; 25 sec; 1 min; t seconds.

b] What are the coordinates of point P at each of the times in (a)?

c] How far is the bug from point P at each of the times in (a)?

d] How long will it take for the bug to reach point P? How many revolutions does the turntable make during that time?

e] Make a sketch of the path of the bug as it travels from the origin to P. Indicate in your drawing its positions at the end of 10 sec, 25 sec, 1 min.

2.7 Trigonometric Functions for Real Numbers (Circular Functions)

We now have trigonometric functions defined for all *angles*, but we have already mentioned that many of the most important applications of trigonometry depend on functional properties that have no direct reference to angles at all. The trigonometric functions are also called *circular functions*, in part to emphasize their periodic behavior. This aspect is emphasized in Chapter 3, in which we examine the graphs of these functions.

Circles of any size could be used to develop properties of the trigonometric functions, but things are greatly simplified by the use of the *unit circle*, the circle with radius 1 centered at the origin. Each point $P(x, y)$ on the unit circle may also be considered as a point on the terminal side of an angle t in standard position. See Fig. 2.50 and compare with Fig. 2.40. Since P is at a distance 1 from the origin, the definitions of sine and cosine (Eqs. 2.5) give

$$\cos t = \frac{x}{1} = x, \quad \text{and} \quad \sin t = \frac{y}{1} = y. \qquad \textbf{[2.6]}$$

That is, *the coordinates of every point on the unit circle have the form (cos t, sin t), where t is the number of radians in the central angle.*

Recall from Section 1.2 that the arc length intercepted by a central angle is equal to the *length of the radius* times the measure of the central angle in *radians*; that is, $s = rt$, where r is the radius and t is the number of radians. This is even simpler in the unit circle where $r = 1$: The arc length from $A(1, 0)$ to $P(\cos t, \sin t)$ is numerically equal to the number of radians of the central angle (See Fig. 2.50).

What this means for circular functions is that for points on the unit circle the coordinates are given by (cos t, sin t), where t is the *real number that measures the arc length from the positive x-axis to the point.*

General arc length

We can generalize the concept of central angle and its corresponding arc length to allow t to be *any real number.* If we move from point A to P in the counterclockwise direction (even more than one complete

FIGURE 2.50

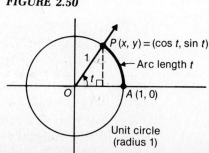

$P(x, y) = (\cos t, \sin t)$

Arc length t

$A(1, 0)$

Unit circle
(radius 1)

revolution), then t is considered to be positive, while a clockwise direction gives t as a negative number. Hence $(\cos t, \sin t)$ is a point on the unit circle corresponding to *any real number t.*

Equations (2.6) may actually be used as *definitions* for the functions sine and cosine for *any real number t.* It is in this context that sine and cosine are called circular functions, and if that were to be the starting point for defining circular functions, then we would define four other functions in terms of ratios and reciprocals of these two:

$$
\begin{array}{ll}
\tan t = \dfrac{\sin t}{\cos t}, & \cot t = \dfrac{\cos t}{\sin t}, \\[2ex]
\sec t = \dfrac{1}{\cos t}, & \csc t = \dfrac{1}{\sin t}.
\end{array}
\qquad \textbf{[2.7]}
$$

Because these functions look and behave exactly like our previously defined trigonometric functions, we shall make no effort to consider them as conceptually different.

What we do want to emphasize is twofold:

1. The trigonometric functions are actually real-valued *functions of a real number.* That is, the domains of all six functions are sets of real numbers, and there need be neither "angles" nor "arc length" connected with the functions unless it is convenient to do so.

2. The coordinates of each point on the unit circle have the form $(\cos t, \sin t)$. This means that we may observe what happens to the coordinates of a point moving along the circle and thereby learn about the behavior of the sine and cosine (and hence about all of the trigonometric functions). We will use this point of view to graph sine and cosine functions in the next chapter.

EXAMPLE 1 For each real number t, sketch a unit circle showing the point P at an arc length t from point $(1,0)$, find the coordinates of P, and evaluate all six circular functions at t, using exact form if possible and three decimal place accuracy otherwise.

a] $t = 2.5$ b] $t = -\dfrac{3\pi}{2}$ c] $t = 8$

Solution

a] $t = 2.5$, so t is a number between $\pi/2$ and π, and P is a point in the second quadrant. Using the radian mode for the calculator, we can find the coordinates of P by evaluating $\cos t \approx -0.8011$ and $\sin t \approx 0.59847$. See Fig. 2.51, in which we show all three unit circles.

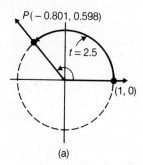
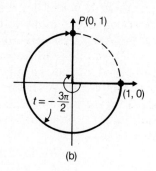
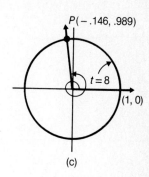

FIGURE 2.51

Using Eqs. (2.7), we have, for $t = 2.5$,

$$\sin t \approx 0.598, \qquad \cos t \approx -0.801,$$

$$\tan t = \frac{\sin t}{\cos t} \approx -0.747, \qquad \cot t = \frac{\cos t}{\sin t} \approx -1.339,$$

$$\csc t = \frac{1}{\sin t} \approx 1.671, \qquad \sec t = \frac{1}{\cos t} \approx -1.248.$$

b] The arc of length $3\pi/2$ directed negatively from $A(1,0)$ terminates on the positive y-axis at the point $P(0,1)$, from which we have exactly

$$\sin(-3\pi/2) = 1, \qquad \cos(-3\pi/2) = 0,$$
$$\tan(-3\pi/2) \text{ is undefined,}$$
$$\csc(-3\pi/2) = 1, \qquad \sec(-3\pi/2) \text{ is undefined,}$$
$$\cot(-3\pi/2) = 0.$$

c] The arc of length 8 wraps more than one time around the unit circle. It may be helpful in visualizing P to subtract $2\pi \cong 6.2832$ from 8 and convert to degrees: $(8 - 2\pi)(180/\pi) \approx 98.366°$ (see Fig. 2.51(c)). But the calculator will give us values directly (no need to subtract 2π) in radian mode:

$$\sin 8 \approx 0.989, \qquad \cos 8 \approx -0.146, \qquad \tan 8 \approx -6.800,$$
$$\csc 8 \approx 1.011, \qquad \sec 8 \approx -6.873, \qquad \cot 8 \approx -0.147. \quad \blacksquare$$

EXAMPLE 2 Describe the points $P(\cos t, \sin t)$ near $Q(0,1)$ (that is, for t near $\pi/2$) where $\tan t$ is greater than 5.

Solution Let $P(\cos t, \sin t)$ be the point where $\tan t = (\sin t)/(\cos t)$ equals 5. Since P is on the unit circle, its coordinates satisfy the equation for the unit circle: $x^2 + y^2 = 1$. Using $u = \cos t$ and $v = \sin t$, we have $v/u = 5$ and $u^2 + v^2 = 1$. Substituting $v = 5u$ into the equation for the circle, we have $u^2 + (5u)^2 = 26u^2 = 1$, from which $u = \pm 1/\sqrt{26}$. For t near $\pi/2$, the y-coordinate of P must be near 1 (hence positive), so we must use $u = 1/\sqrt{26} \approx 0.1961$ and $v = 5/\sqrt{26} \approx 0.9806$.

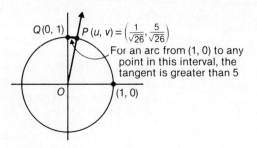

FIGURE 2.52

From Fig. 2.52, it should be clear that for any point on the unit circle between $P(1/\sqrt{26},\ 5/\sqrt{26})$ and $Q(0,1)$, we have $\tan t > 5$. The distance between P and Q is much exaggerated in Fig. 2.52. We should also note that there are no points near Q in the second quadrant where $\tan t = (\sin t)/(\cos t) > 5$ because in the second quadrant $\cos t$ is negative. Compare $\tan 8 = -6.800$ in the preceding example. ∎

EXAMPLE 3 Find the coordinates of the points on the unit circle $R(\cos(\pi - t),\ \sin(\pi - t))$ and $S(\cos(-t),\ \sin(-t))$ in terms of $\cos t$ and $\sin t$. See Fig. 2.53.

Solution We may consider R and S as points on the terminal side of the angles $\pi - t$ and $-t$, respectively. Both reference triangles are congruent to triangle OPA in Fig. 2.53. It is easy to see that R has the same y-coordinate as P and that the x-coordinate of R is the negative of that of P. Thus the coordinates of R are given by $(\cos(\pi - t),\ \sin(\pi - t))$ or by $(-\cos t, \sin t)$.

Similarly, S and P have the same x-coordinate and the y-coordinate of S is the negative of that of P. Therefore the coordinates of S are given by $(\cos(-t), \sin(-t))$ or by $(\cos t, -\sin t)$. ∎

FIGURE 2.53

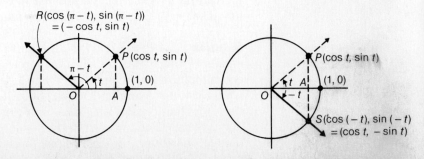

EXERCISE 2.7

In Problems 1–12, t is a real number. In each case, sketch a unit circle and show the point P associated with the arc length t. Use (2.6) to find the coordinates of P, and then apply (2.6) and (2.7) to evaluate the six circular functions at t, giving results in exact form if possible and three decimal place accuracy otherwise. (See Example 1.)

1. $t = 1$ **2.** $t = 14.5$ **3.** $t = \pi/6$

4. $t = \sqrt{\pi}$ **5.** $t = \pi + 1$ **6.** $t = 0.662$

7. $t = -0.523$ **8.** $t = -8\pi/3$ **9.** $t = 9\pi/4$

10. $t = 8.53$ **11.** $t = 0$ **12.** $t = -\sqrt{2}$

For Problems 13–18, refer to Fig. 2.54, where we have unit circles and an arc of length t, where $0 < t < \pi/2$. We know that the coordinates of P are $(\cos t, \sin t)$. Line AM is a vertical line in Fig. 2.54(a), and line ST is tangent to the circle at P in Fig. 2.54(b).

Find the lengths of the given line segments in terms of trigonometric functions of t.

13. $\overline{OQ}$ and $\overline{PQ}$. Use right triangle OQP.

14. $\overline{AM}$. Use right triangle OAM.

15. $\overline{OM}$. Use right triangle OAM.

16. Find the measures of angles $\angle SOP$, $\angle PSO$, and $\angle PTO$, in terms of t.

17. $\overline{OS}$ and $\overline{OT}$.   **18.** $\overline{PS}$ and $\overline{PT}$.

19. Describe the points $P(\cos t, \sin t)$ near $R(0, -1)$ (that is, near $t = 3\pi/2$ or $t = -\pi/2$) where $\tan t$ is greater than 10. (See Example 2.)

20. Describe the points $P(\cos t, \sin t)$ near $S(-1, 0)$ (that is, near $t = \pi$) where $\csc t$ is greater than 5. (See Example 2.)

21. Find the coordinates of the point on the unit circle $R(\cos(\pi + t), \sin(\pi + t))$ in terms of $\cos t$ and $\sin t$. (See Example 3.)

FIGURE 2.54

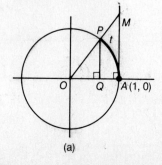

(a)

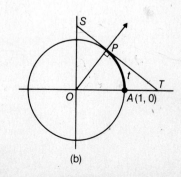

(b)

22. Find the coordinates of the point on the unit circle $R(\cos(\pi/2 + t), \sin(\pi/2 + t))$ in terms of $\cos t$ and $\sin t$. (See Example 3.)

23. For what values of t between 0 and 2π is $\sec t$ negative?

24. For what values of t between π and 3π is $\csc t$ positive?

25. For what values of t between 0 and 2π is $\tan t$ not defined?

26. For what values of t between 0 and 2π is $\cot t$ not defined?

27. For what values of t between 0 and 2π is $\sec t$ not defined?

28. For what values of t between 0 and 2π is $\csc t$ not defined?

Summary

We discussed three different ways of defining trigonometric functions:

1. As ratios of sides of right triangles (acute angles only).
2. As ratios of coordinates of points on the terminal side of an angle in standard position (all angles).
3. As ratios of coordinates of points on the unit circle (all real numbers).

These concepts are closely related (and coincide for acute angles) as may be seen from Fig. 2.55.

The six equations in Fig. 2.55(a) and the fact that $\beta = 90° - \alpha$ permit us to use any two of a, b, c, α, β (where at least one is $a, b,$ or

FIGURE 2.55

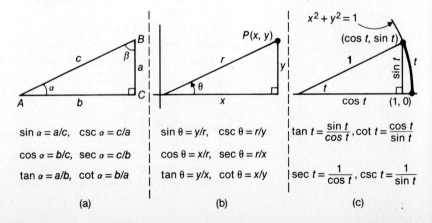

$$\sin \alpha = a/c, \quad \csc \alpha = c/a$$
$$\cos \alpha = b/c, \quad \sec \alpha = c/b$$
$$\tan \alpha = a/b, \quad \cot \alpha = b/a$$

(a)

$$\sin \theta = y/r, \quad \csc \theta = r/y$$
$$\cos \theta = x/r, \quad \sec \theta = r/x$$
$$\tan \theta = y/x, \quad \cot \theta = x/y$$

(b)

$$\tan t = \frac{\sin t}{\cos t}, \cot t = \frac{\cos t}{\sin t}$$
$$\sec t = \frac{1}{\cos t}, \csc t = \frac{1}{\sin t}$$

(c)

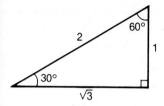

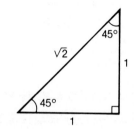

FIGURE 2.56

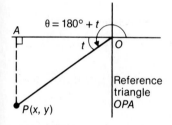

FIGURE 2.57

c) to solve for the other three quantities. This process is called *solving the triangle.*

The trigonometric functions for 30°, 45°, and 60° angles may be read in exact form from the two special triangles (Fig. 2.56).

The use of the reference angle and reference triangle make it possible to relate the trigonometric functions of angles in any quadrant to the corresponding functions of an acute angle (the reference angle), and the signs of coordinates in each quadrant give the signs of the functions. If *t* denotes the reference angle shown in Fig. 2.57, then $\theta = 180° + t$, from which we may read the reduction formulas shown:

$$\sin\theta = y/r \; (- \text{ in quad. III}), \qquad \sin(180° + t) = -\sin t,$$
$$\cos\theta = x/r \; (- \text{ in quad. III}), \qquad \cos(180° + t) = -\cos t,$$
$$\tan\theta = y/x \; (+ \text{ in quad. III}), \qquad \tan(180° + t) = \tan t.$$

Reduction formulas in general are discussed in Section 3.1.

Computer Problems (Optional)

All right triangles are labeled in standard form (as in Fig. 2.4 or 2.15).

Section 2.2

1. Write a program that will allow you to enter (INPUT) two legs of a right triangle and the output of which will include:
a] the length of the hypotenuse,
b] all six trigonometric functions of α and β from definitions (2.1) and (2.3), rounded off to four decimal places.

Section 2.3

2. Most programming languages (such as BASIC) have built-in functions SIN, COS, TAN, which will accept values only in radians. That is, PRINT SIN(1.5) returns 0.997495, not sin 1.5° = 0.026177. Write a program that will evaluate sin x, cos x, tan x if you enter x in
a] degrees,
b] degrees, minutes, and seconds.

3. Write a program that will evaluate (sin x)/x for a sequence of positive values (in radians, of course) approaching zero.

Is the output consistent with Problem 20 of Exercise 2.3? What happens if the sequence consists of negative numbers approaching zero?

Section 2.4

4. If any two of the quantities *a, b, c, α, β* for a right triangle (Fig. 2.15) are given (where at least one is a length), the remaining three are determined. Most computer languages have only one inverse trigonometric function from which angles may be found. The function is ATN, the

inverse tangent function, and the output is always in radians. For instance, if b and c are given, to find the angle α, we might use $x = \tan \alpha = a/b = \sqrt{c^2 - b^2}/b$, from which we can get $\alpha = \text{ATN}(x)$. Write a program that will accept some specified pair of quantities such as a, α (in radians) and the output of which will be the remaining three quantities. Then modify the program to handle angles in degrees. Different choices of pairs from a, b, c, α, β will naturally require quite different programs.

Sections 2.5 and 2.6

5. Write a program that will accept *two integers* a and b where not both are zero. Suppose point $P(a, b)$ lies on the terminal side of an angle x in standard position.
   a] Write a program that will tell the quadrant in which x is located or identifies x as a quadrantal angle.
   b] Write a program that will give $\sin x$, $\cos x$, and $\tan x$ in *exact form*. Since $r = \sqrt{a^2 + b^2}$ is not always an integer, the program should give the value of r when it is an integer and print $\sqrt{a^2 + b^2}$ using SQR when r is not an integer. For instance, if $a = 3$ and $b = 5$, then for $\sin x$ the output should read "sin x = 5/SQR(34)."
   c] Write a program that will give $\sin x$, $\cos x$, and $\tan x$ rounded off to three decimal places, where a and b can be any real numbers (not necessarily integers).

6. In Problem 40 of Exercise 2.6, a bug travels on a rotating turntable. Write a program that will give its position (x, y coordinates) at five-second intervals from $t = 0$ to $t = 80$. Plot the (x, y) points and draw a curve that shows the bug's path of travel.

Section 2.7

7. Write a program that will accept any real number t and the output of which will give the coordinates (rounded off to three decimal places) of the point on the unit circle at arc length t from point $A(1, 0)$. See Fig. 2.50 and the reference to "generalized arc length."

Review Exercises

1. Make a sketch showing the given angles in standard position (a reasonable approximation is sufficient):
   a] $135°$
   b] $-240°$
   c] $\dfrac{5\pi}{2}$
   d] $-137°$
   e] -2.34
   f] $\dfrac{17\pi}{6}$

2. Determine the quadrant in which the given angles are located:
   a] 235°   b] 4.7   c] −2.47
   d] −640°   e] 841°   f] 30

In Problems 3–10, give the answers in *exact form*.

3. Evaluate the following:
   a] sin 90°   b] tan 30°   c] sec 150°
   d] cos (−240°)   e] tan (−180°)   f] csc 450°
   g] cot (−315°)   h] sin 270°

4. Evaluate the following:

   a] $\cos 3\pi$ b] $\cot(-\pi)$ c] $\sin \dfrac{5\pi}{3}$

   d] $\cos\left(-\dfrac{7\pi}{3}\right)$ e] $\tan \dfrac{7\pi}{6}$ f] $\sec \dfrac{3\pi}{2}$

   g] $\sec\left(\pi - \dfrac{\pi}{6}\right)$ h] $\csc\left(\dfrac{\pi}{3} + \dfrac{5\pi}{6}\right)$

5. If θ is an angle in the third quadrant and $\tan \theta = 4/3$, determine the following:
   a] $\sin \theta$ b] $\sec \theta$ c] $\cos(\theta + \pi)$

   d] $\tan(\theta - \pi)$ e] $\csc\left(\theta - \dfrac{\pi}{2}\right)$ f] $\cos\left(\theta + \dfrac{\pi}{2}\right)$

6. Determine θ from the given information:

   a] $\sin \theta = -\dfrac{\sqrt{2}}{2}$ and $\pi < \theta < \dfrac{3\pi}{2}$

   b] $\cos \theta = -\dfrac{1}{2}$ and $0 < \theta < \pi$

   c] $\tan \theta = -1$ and $-2\pi < \theta < -\pi$

   d] $\sec \theta = -1$ and $0 < \theta < 2\pi$

7. Determine α from the given information:
   a] $\sin \alpha = -1$ and $0° \le \alpha \le 360°$
   b] $\csc \alpha = 2$ and $-90° < \alpha < 90°$

   c] $\cos \alpha = -\dfrac{1}{\sqrt{2}}$ and $0° \le \alpha \le 180°$

   d] $\tan \alpha = -1$ and $-90° \le \alpha \le 90°$

8. If $\alpha = 3\pi/2$, $\beta = \pi/3$, and $\gamma = 5\pi/6$, evaluate the following:
   a] $\sin \alpha$ b] $\tan \gamma$ c] $\cos(\alpha - \beta)$
   d] $\sec(\beta + \gamma)$ e] $\sec(\gamma - \alpha)$ f] $\cos(\alpha + \gamma - \beta)$

9. If $\alpha = 30°$, $\beta = 90°$, and $\gamma = 210°$, evaluate the following:
   a] $\sin(\alpha + \gamma)$ b] $\sin \alpha + \sin \gamma$ c] $\cos(\alpha - \beta)$
   d] $\cos \alpha - \cos \beta$ e] $\tan 2\gamma$ f] $2 \tan \gamma$

10. If $\cos \theta = -0.75$ and $\tan \theta$ is negative, determine the following:
    a] $\sin \theta$ b] $\cot \theta$

    c] $\sec\left(\theta - \dfrac{\pi}{2}\right)$ d] $\tan(\theta + \pi)$

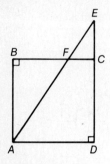

FIGURE 2.58

In Problems 11–16, evaluate the given expressions and give answers rounded off to four decimal places:

11. **a]** $\sin 43°$ **b]** $\tan 154°$ **c]** $\cos 57°16'$
 d] $\cot 48°$ · **e]** $\sec 327°12'$ **f]** $\sin(-231°)$

12. **a]** $\cos 1.43$ **b]** $\sin 3.86$ **c]** $\tan(5\pi/12)$ **d]** $\cot(12/5\pi)$

13. **a]** $\sin(53° + 75°)$ **b]** $\sin 53° + \sin 75°$

14. **a]** $\tan(1.36 + 2.14)$ **b]** $\tan 1.36 + \tan 2.14$

15. **a]** $(\sin 153°)^2 + (\cos 153°)^2$ **b]** $(\sin 1.5)^2 + (\cos 1.5)^2$

16. **a]** $2\left(\sin \dfrac{\pi}{12}\right)\left(\cos \dfrac{\pi}{12}\right)$ **b]** $\left(\cos \dfrac{\pi}{3}\right)^2 - \left(\sin \dfrac{\pi}{3}\right)^2$

17. Determine whether the given statements are true or false:
 a] π and $-\pi$ are coterminal angles

 b] $-\dfrac{3\pi}{2}$ and $-\dfrac{\pi}{2}$ are coterminal angles

 c] $210°$ and $-\dfrac{5\pi}{6}$ are coterminal angles

 d] An angle in standard position with terminal side passing through point $(-1, 2)$ is coterminal with $150°$.

18. The hypotenuse of a right triangle is 37.42 cm, and one angle is 48°12′. Find the lengths of the two sides.

19. If ABC is an isosceles triangle with $|\overline{AB}| = |\overline{AC}| = 4.73$ and the angle opposite AB is 52°10′, find the length of the altitude from A to $\overline{BC}$. Then find the area of the triangle.

20. If the hypotenuse of a right triangle is 24.3 cm and one of the sides is 15.4 cm, find the length of the other side. Determine the angles to the nearest ten minutes.

21. In Fig. 2.58, $ABCD$ is a square with length of side 18.7 cm. If $|\overline{EC}| = 8.43$ cm, find the length of $\overline{AF}$.

22. In Fig. 2.59, $\alpha = 34°$, $\beta = 120°$, and $|\overline{CD}| = 15$ cm. Find the length of $\overline{AB}$.

23. In Fig. 2.60, the center of the circle is O, $\overline{AB}$ is a tangent to the circle at B, and C is a point on the circle and on $\overline{OA}$. If the radius of the circle is 12 cm and the length of arc $\overset{\frown}{BC}$ is 9.0 cm, find the area of the shaded region.

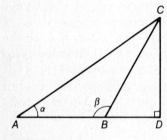

FIGURE 2.59

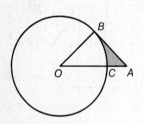

FIGURE 2.60

Graphs of Trigonometric Functions; Inverse Functions

3

In the study of mathematics, graphs are emphasized because of their great utility. Many crucial features of a function become obvious from a reasonable graph. In the first two sections of this chapter we study graphs of the basic trigonometric functions; we delay until Chapter 6 the study of graphs of the general trigonometric functions.

In Sections 3.4 through 3.6 we develop the important topic of inverse trigonometric functions, in which we rely heavily on graphs drawn in the first two sections.

3.1 Periodicity and Graphs of the Sine and Cosine Functions

We want to build on the observation made in Section 2.7 that the coordinates of each point P on the unit circle are expressible in the form $(\cos t, \sin t)$, where t is the radian measure of the central angle or the length of arc from the point $A(1, 0)$ to P.

Consider what happens as P begins at the location $A(1, 0)$ (where $t = 0$) and moves counterclockwise around the unit circle. The y-coordinate of P is the value of $\sin t$, so we can observe how $\sin t$ changes with t. As P moves from its starting point at $A(1, 0)$ around to $B(0, 1)$ (see Fig. 3.1a), the y-coordinate increases steadily from 0 to 1, so $\sin t$ increases from 0 to 1 as t changes from 0 to $\pi/2$.

To help visualize how $\sin t$ changes in relation to t, we show both the unit circle and the graph of the function $f(t) = \sin t$. Figure 3.1(a) shows P with its coordinates and what they represent, with t as the radian measure of the central angle or as the arc length from A to P. Figure 3.1(b) shows the pairs $(t, \sin t)$ belonging to the function $f(t) = \sin t$. We have used the same vertical scale in both parts of

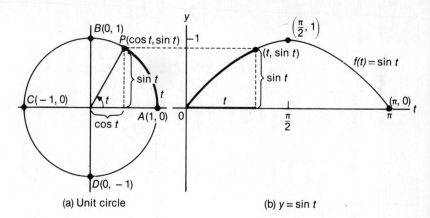

(a) Unit circle   (b) $y = \sin t$

FIGURE 3.1

Fig. 3.1 so that we see how $\sin t$ increases with t in the interval $0 \leq t \leq \pi/2$. The same number t is shown as the arc length in part (a) and as the horizontal distance in part (b), as it would appear if the arc were unrolled along the horizontal axis. The portion of the graph of the sine function is the "arch" traced out on the interval $[0, \pi]$.

It is possible, of course, to graph the sine function without the visual assistance provided by the unit circle. As an exercise, we recommend that the student take values of $\sin t$ from the calculator and carefully plot them on graph paper. The result should look very much like the appropriate portion of Fig. 3.1(b). (See Problem 2 in Exercise 3.1.)

Even though it is possible to get the necessary information about the sine function without using the unit circle, seeing how the coordinates of P change remains the most helpful way we know to visually understand the behavior of $\sin t$ and $\cos t$ as t varies. This is especially true for some of the important properties of *symmetry* and *periodicity*.

To continue with our analysis of the sine function, for example, as P continues around the unit circle from $B(0, 1)$ to $C(-1, 0)$, its y-coordinate ($\sin t$) *decreases* from 1 to 0 *at exactly the same rate* as $\sin t$ increased in the first quarter-revolution from A to B. This means that the portion of the graph of $f(t) = \sin t$ for the interval $\pi/2 \leq t \leq \pi$ is just the reflection of the portion on the interval $[0, \pi/2]$, that is, the right half of the arch in Fig. 3.1(b).

As t increases from π to 2π, the point P traverses the bottom half of the unit circle, from $C(-1, 0)$ through $D(0, -1)$ back to $A(1, 0)$. It should be clear that the y-coordinate goes from 0 to -1 and back to 0 and that the graph of $f(t) = \sin t$ is symmetric about the point $(\pi, 0)$, as shown in Fig. 3.2(a), as if we had picked up the first arch and tipped it upside down for the portion of the graph from $t = \pi$ to $t = 2\pi$.

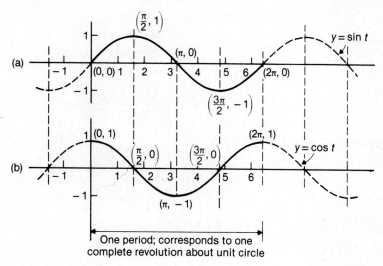

FIGURE 3.2

Having followed P through one complete revolution, we return to the point $A(1,0)$. It should now be obvious that as P continues around the unit circle, the pattern of each of the coordinates $\sin t$ and $\cos t$ *repeats itself exactly*. This is also a consequence of the previously observed fact that *trigonometric functions of coterminal angles are the same*. This is described in functional terms by saying that the functions are *periodic*. Symbolically, for any integer k,

$$\sin(t + 2k\pi) = \sin t \quad \text{and} \quad \cos(t + 2k\pi) = \cos t. \qquad \textbf{[3.1]}$$

It means that *when we know the behavior of the function for one period, we know its behavior everywhere*. Graphically, from the graph of one period (in this case $[0, 2\pi]$) we can repeat that cycle over and over again, in both directions. See Fig. 3.2(a).

In Fig. 3.2 we show the graphs of both the sine and cosine functions, even though we have discussed only the behavior of the sine. The first coordinate of P is $\cos t$. As P begins at $A(1,0)$ and moves around the unit circle, it is clear that $\cos t$ begins at 1 and decreases to 0 in a quarter-revolution (as t increases to $\pi/2$). More important than the specifics, however, is the observation that the *variation of the x-coordinate is just like the variation of the y-coordinate* except for its starting value and the direction of change. The x-coordinate starts at 1 and decreases, just as the y-coordinate does from $B(0,1)$ onwards. If we were to project the x-coordinate downwards, below the unit circle in Fig. 3.1(a), as we have projected the y-coordinate to the right, we would observe exactly the same kind of motion as shown in Fig. 3.1(b), except that the graph begins at the top of the arch.

The result is that the graphs of sin t and cos t, shown in Fig. 3.2, are really the same except for a shift. If we started to follow the cosine from the negative y-axis (where $t = -\pi/2$), the graphs of the sine and cosine would be identical. The relationship may be made precise in the equation

$$\sin(t + \pi/2) = \cos t. \qquad \textbf{[3.2]}$$

It may be helpful to look at Fig. 3.6, where we have superimposed the graphs of the cosine and sine. The period of the cosine function is also 2π; and from the graph of the fundamental period, we can extend the graph in both directions.

The differences between the graphs of the sine function and the cosine function may be recalled in many ways (as, for example, by reference to the unit circle), but a knowledge of the distinctive properties of each graph will answer many, many questions about trigonometric functions.

> It is difficult to overestimate the importance of the graphs in Fig. 3.2. Virtually all the properties of the sine and cosine functions can be deduced from the graphs. *Each student should be able to quickly sketch one full period of each graph.*

Periodic functions

Any function repeating itself over consecutive intervals of fixed length is said to be a periodic function. Many scientific investigations involve phenomena of a cyclic nature that can be described in terms of periodic functions. It is an interesting and important fact that practically all periodic functions can be expressed as combinations of sine and cosine functions.* This is one of the facts that makes trigonometry extremely useful in applications of mathematics to many real-life problems.

DEFINITION   If f is any function with the property that there is a positive number p such that

$$f(x + p) = f(x) \qquad \textbf{[3.3]}$$

for all values of x in the domain of f, then f is said to be a *periodic function*. If p is the *smallest positive number* for which Eq. (3.3) holds, then p is called the *period* of the function.

---

* This is the basis for a broad topic in advanced mathematics called Fourier analysis.

> The sine and cosine functions are periodic, and each has a period of 2π.

Summary

From the graphs in Fig. 3.2 we observe the following important properties of the sine and cosine functions.

Table 3.1

Function	Domain	Range	Period	x-Intercepts
sine	**R**	$-1 \le y \le 1$	2π	$0, \pm\pi, \pm 2\pi, \pm 3\pi, \ldots,$
cosine	**R**	$-1 \le y \le 1$	2π	$\pm\dfrac{\pi}{2}, \pm\dfrac{3\pi}{2}, \pm\dfrac{5\pi}{2}, \ldots$

The symbol **R** denotes the set of real numbers.

Reduction formulas

A reduction formula is one that allows us to reduce the evaluation of a trigonometric function for certain numbers to a simpler or more familiar setting. Equations (3.1) are examples of reduction formulas, consequences of the fact that sine and cosine are periodic functions of period 2π. Equation (3.2) is another.

There are many reduction formulas that appear in different books. Some are more useful than others, but it is difficult to anticipate just which ones will be most needed. We list several below, but no effort should be made to memorize an extensive list. Our feeling is that it is a great deal more important to *understand how the graphs or the unit circle can be used to derive the reduction formulas.*

Reduction formulas from graphs

In each of the graphs there are several obvious kinds of symmetry. Each symmetry may be expressed as a reduction formula. In Fig. 3.3(a) we have taken an arbitrary number t and noted that the height of the graph at t (which is $\sin t$) is the *same* at $\pi - t$ and the *negative of the height* at $-t$, $\pi + t$, and $2\pi - t$. This immediately gives the four reduction formulas listed in Eqs. (3.4) for the sine. In a similar manner we can use the graph in Fig. 3.3(b) to get the reduction formulas listed in Eqs. (3.5).

$$
\begin{aligned}
\sin(-t) &= -\sin t, \\
\sin(\pi - t) &= \sin t, \\
\sin(\pi + t) &= -\sin t, \\
\sin(2\pi - t) &= -\sin t.
\end{aligned}
\qquad \textbf{[3.4]}
$$

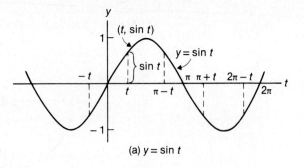

(a) $y = \sin t$

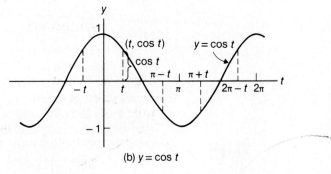

(b) $y = \cos t$

FIGURE 3.3

$$\cos (-t) = \cos t,$$

$$\cos (\pi - t) = -\cos t,$$

$$\cos (\pi + t) = -\cos t,$$

$$\cos (2\pi - t) = \cos t.$$

[3.5]

Reduction formulas from the unit circle

The unit circle may also be used to derive reduction formulas. Perhaps the biggest difficulty is that the circle becomes cluttered if too many points are considered. In Fig. 3.4 we have $P(u, v)$ or $P(\cos t, \sin t)$ associated with the central angle t. If the point M is associated with the angle $-t$, then the coordinates of M are given by $(u, -v)$ or $[\cos (-t), \sin (-t)]$. This is true for a central angle of any size, even though we have used a first quadrant angle in Fig. 3.4. Since P and M are points of the unit circle, we may read

$$\cos (-t) = u = \cos t \qquad \text{and} \qquad \sin (-t) = -v = -\sin t,$$

the same relationships that we had already observed from the graphs in Fig. 3.3.

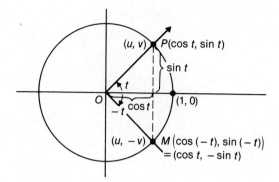

FIGURE 3.4

It should be clear that we can get the same information from either graphs or the unit circle. We recommend that students become familiar with the advantages of each.

Since we can express the other four trigonometric functions in terms of $\sin t$ and $\cos t$ (Eqs. 2.7), we have reduction formulas for all as soon as we get formulas for sine and cosine. For example,

$$\tan(-t) = \frac{\sin(-t)}{\cos(-t)} = \frac{-\sin t}{\cos t} = -\tan t,$$

$$\csc(-t) = \frac{1}{\sin(-t)} = \frac{1}{-\sin t} = -\csc t,$$

$$\sec(-t) = \frac{1}{\cos(-t)} = \frac{1}{\cos t} = \sec t,$$

$$\cot(-t) = \frac{1}{\tan(-t)} = \frac{1}{-\tan t} = -\cot t.$$

[3.6]

Odd and even functions

The trigonometric functions are important examples of what are called *odd* and *even* functions. An *odd function* f is defined by $f(-x) = -f(x)$, and f is *even* if $f(-x) = f(x)$, for all x in the domain of f. It follows that *sine, tangent, cosecant, and cotangent are all odd functions. Cosine and secant are even functions.*

EXAMPLE 1 Show that the period of the tangent function is π, that is, that $\tan(x + \pi) = \tan x$ for all x in the domain of the tangent function.

Solution   We know that $\tan t = (\sin t)/(\cos t)$, from which

$$\tan(x + \pi) = \frac{\sin(x + \pi)}{\cos(x + \pi)} = \frac{-\sin x}{-\cos x} \qquad \text{from Eqs. (3.4) and (3.5)}$$

$$= \frac{\sin x}{\cos x} = \tan x.$$

This holds for all values of x for which $\cos x$ and $\cos(x + \pi)$ are nonzero. From the graph, $\cos x$ and $\cos(x + \pi)$ are both nonzero as long as x is not an odd multiple of $\pi/2$, and these of course are the numbers not in the domain of the tangent function. In the next section we draw the graph of $y = \tan x$ (see Fig. 3.9). From the graph we can see that π is the smallest positive number p such that $\tan(x + p) = \tan x$, from which the period is π.   ■

EXAMPLE 2   Show that $\sin(\pi/2 + s) = \cos s$ for all s.

Solution 1   In a unit circle we may take an arbitrary central angle s with associated point $P(u, v)$. In Fig. 3.5, we show s as a second-quadrant angle, for which $u = \cos s$ is negative and $v = \sin s$ is positive. To get $s + \pi/2$, we add a right angle to get the terminal side $\overline{OQ}$. From the reference triangles (congruent but oriented differently), we can see that the coordinates of Q are $(-v, u)$. From point Q, $\sin(\pi/2 + s) = u$, and from point P, $\cos s = u$. Therefore $\sin(\pi/2 + s) = \cos s$.

Solution 2   In Fig. 3.6 we show the graphs of both $y = \sin x$ (solid) and $y = \cos x$ (broken). It is apparent that the sine graph is shifted $\pi/2$ units to the right of the cosine graph. Accordingly, $\sin(\pi/2 + s) = \cos s$.   ■

FIGURE 3.5

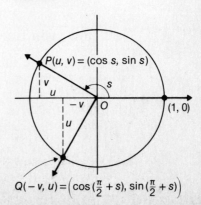

$$Q(-v, u) = \left(\cos\left(\tfrac{\pi}{2} + s\right), \sin\left(\tfrac{\pi}{2} + s\right)\right)$$

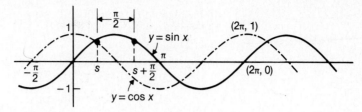

FIGURE 3.6

EXERCISE 3.1

For Problems 1 and 2, use a calculator to evaluate the given function at the values indicated in the table. Give entries to two decimal places. Then make a large-scale (one full page) graph to plot the points from your table, and sketch the graph of the function on the interval.

1. Graph $y = \cos x$ on $0 \le x \le \pi/2$.

x	0	$\frac{\pi}{20}$	$\frac{\pi}{10}$	$\frac{3\pi}{20}$	$\frac{\pi}{5}$	$\frac{\pi}{4}$	$\frac{3\pi}{10}$	$\frac{7\pi}{20}$	$\frac{2\pi}{5}$	$\frac{9\pi}{20}$	$\frac{\pi}{2}$
cos x											

2. Graph $y = \sin x$ for $\pi/2 \le x \le \pi$.

x	$\frac{\pi}{2}$	$\frac{11\pi}{20}$	$\frac{3\pi}{5}$	$\frac{13\pi}{20}$	$\frac{7\pi}{10}$	$\frac{3\pi}{4}$	$\frac{4\pi}{5}$	$\frac{17\pi}{20}$	$\frac{9\pi}{10}$	$\frac{19\pi}{20}$	π
sin x											

3. Note that each of the x-values in the table for Problem 2 is $\pi/2$ greater than the corresponding x-values in the table in Problem 1. Compare your calculated values of y for the two problems or compare the two graphs. What obvious reduction formula can be deduced?

4. From the graph of $y = \cos x$ (Fig. 3.2b), justify the reduction formula $\cos [(\pi/2) - t] = -\cos [(\pi/2) + t]$.

5. From the graph of $y = \sin x$ (Fig. 3.2a), justify the reduction formula $\sin [(\pi/2) - t] = \sin [(\pi/2) + t]$.

6. From Problems 4 and 5, deduce the reduction formula

$$\tan\left(\frac{\pi}{2} - t\right) = -\tan\left(\frac{\pi}{2} + t\right).$$

In each of Problems 7–10, you can fill in the box ☐ with any of several appropriate expressions to produce a valid reduction formula. Use the graphs shown in Fig. 3.2 to help you in finding three such expressions.

7. $\sin(-\pi + t) = \sin(\boxed{})$ **8.** $\sin\left(\frac{3\pi}{2} + t\right) = \sin(\boxed{})$

9. $\cos\left(\frac{3\pi}{2} + t\right) = \cos(\boxed{})$ **10.** $\cos(-t) = \cos(\boxed{})$

In Problems 11–16, use a unit circle diagram similar to that in Example 2 to justify the given reduction formulas. You may wish to take s as a first-quadrant angle. That is, take point $P(\cos s, \sin s)$ in the first quadrant.

11. $\cos\left(\dfrac{\pi}{2} - s\right) = \sin s$ $\qquad$ 12. $\sin\left(s - \dfrac{\pi}{2}\right) = -\cos s$

13. $\cos\left(\dfrac{\pi}{2} + s\right) = -\sin s$ $\qquad$ 14. $\sin\left(\dfrac{3\pi}{2} + s\right) = -\cos s$

15. $\sin(\pi - s) = \sin s$ $\qquad$ 16. $\cos(s - \pi) = -\cos s$

In Problems 17–24, use the graphs in Fig. 3.2 (extended as necessary) to determine which of the given formulas are valid reduction formulas for all values of t.

17. $\cos\left(t + \dfrac{3\pi}{2}\right) = \sin t$ $\qquad$ 18. $\sin\left(t + \dfrac{3\pi}{2}\right) = -\cos t$

19. $\cos\left(\dfrac{3\pi}{2} - t\right) = -\sin t$ $\qquad$ 20. $\sin\left(\dfrac{5\pi}{2} - t\right) = \sin t$

21. $\sin(t + 3\pi) = \sin t$ $\qquad$ 22. $\cos(t + 3\pi) = -\cos t$

23. $\sin(t - 3\pi) = -\sin t$ $\qquad$ 24. $\cos(t - 3\pi) = \cos t$

For Problems 25–30, use the coordinates of the points on the unit circle for $t = \pi/6, \pi/4, \pi/3$ as shown in Fig. 3.7 and whatever reduction formulas are appropriate to evaluate each of the given expressions in exact form.

25. $\sin\dfrac{3\pi}{4}$ $\qquad$ 26. $\cos\dfrac{3\pi}{4}$ $\qquad$ 27. $\cos\dfrac{5\pi}{4}$

28. $\sin\dfrac{8\pi}{3}$ $\qquad$ 29. $\cos\left(-\dfrac{11\pi}{6}\right)$ $\qquad$ 30. $\sin\left(-\dfrac{2\pi}{3}\right)$

31. In Problem 20 of Exercise 2.3 we consider the function $f(x) = (\sin x)/x$ for values of x approaching zero. The limiting value of $f(x)$ is 1, and this suggests that $\sin x$ can be approximated by x. We write:

$$\sin x \approx x \qquad \text{for small } x.$$

On the same system of coordinates, draw large-scale graphs of $y = \sin x$ and $y = x$ on the interval $-0.5 \le x \le 0.5$. Note how closely the graphs approximate each other.

FIGURE 3.7

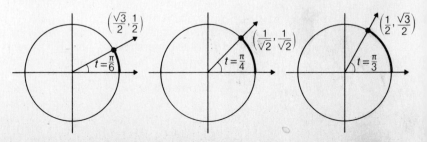

32. Consider the function $f(x) = 2(1 - \cos x)/x^2$. The function is not defined at $x = 0$, but we can examine its behavior for x near zero. Evaluate $f(x)$ at $x = \pm 0.5, \pm 0.1, \pm 0.05, \pm 0.01, \pm 0.005$. Note that $f(x)$ approaches 1, and so this suggests the approximation

$$\frac{2(1 - \cos x)}{x^2} \approx 1.$$

Solve for $\cos x$ and get

$$\cos x \approx 1 - \tfrac{1}{2}x^2 \qquad \text{for small } x.$$

On the same system of coordinates, draw large-scale graphs of $y = \cos x$ and $y = 1 - \tfrac{1}{2}x^2$ (a parabola) on the interval $-0.5 \le x \le 0.5$. Note how closely the graphs approximate each other.

3.2 Graphs of Tangent, Cotangent, Secant, and Cosecant Functions

Graph of the tangent function

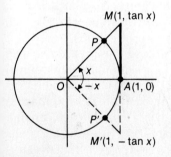

FIGURE 3.8

Heretofore we have used t to represent arc lengths and central angles. Because x is used so much more frequently to denote the independent variable, it is convenient to think of graphing the function $y = \tan x$. For that reason we will change our notation and use x to denote the central angle in Fig. 3.8. Then for most of the remainder of the book our graphs will be of functions of the form $y = f(x)$.

There are two aids that can help in understanding how $\tan x$ varies with x. We use the fact that $\tan x = \sin x / \cos x$ and also identify a segment associated with the unit circle whose length is $\tan x$. We established in Example 1 of the preceding section that $\tan(x + \pi) = \tan x$ for every number x. This means that when we have the graph of $f(x) = \tan x$ for any interval of length π, we may extend it automatically.

Thus consider the unit circle in Fig. 3.8. For first-quadrant angles x, as shown, triangle $\triangle OMA$ is a right triangle with acute angle x. Then $\tan x$ is the ratio of opposite to adjacent sides:

$$\tan x = \frac{|\overline{AM}|}{|\overline{OA}|} = \frac{|\overline{AM}|}{1} = |\overline{AM}| \qquad \text{(see Problem 14 in Exercise 2.7).}$$

This means that the coordinates of M are $(1, \tan x)$, and as x approaches $\pi/2$, the length of the vertical segment $\overline{AM}$, and hence $\tan x$, increases "without bound." We may use the fact that $\tan(-x) = -\tan x$ (Eqs. 3.6) to get the values of $\tan x$ in the fourth quadrant from

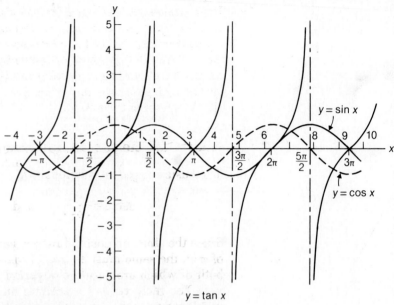

FIGURE 3.9

the first-quadrant values. The same information may be seen from the fact that the point M' in Fig. 3.8 has coordinates $(1, -\tan x)$. This fact that $\tan x$ is an odd function means that the *graph is symmetric with respect to the origin*. The graph of $y = \tan x$ is shown in Fig. 3.9, where we also show the graphs of $y = \sin x$ and $y = \cos x$.

By considering the ratio $\tan x = \sin x / \cos x$ it may easily be seen that $\tan x = 0$ whenever $\sin x = 0$ (at $x = \pm k\pi$) and $\tan x$ is undefined whenever $\cos x = 0$ (at $x = \pm(2k - 1)\pi/2$). Furthermore, when $\cos x$ is near 0, $\sin x$ is a number near 1 or -1. For example, as x approaches $\pi/2$ from the left, the ratio $\sin x / \cos x$ has a smaller and smaller (positive) denominator, so $\tan x$ is very large and positive. For x just slightly greater than $\pi/2$, however, $\cos x$ is a very small negative denominator, and $\tan x$ is very large and negative, as may be seen in the graph near each odd multiple of $\pi/2$. The graph becomes arbitrarily close to the vertical broken lines passing through $x = \pm\pi/2, \pm3\pi/2, \ldots$. These lines are called *vertical asymptotes to the graph*. From the graph in Fig. 3.9 we again observe that the *tangent function has period* π.

Graph of the cotangent function

The key property in consideration of the cotangent function is that it is the reciprocal of the tangent function and is thus also defined in terms of sine and cosine by

$$\cot x = \frac{\cos x}{\sin x}.$$

From the analysis relating tan x to sine and cosine it is readily seen that the *cotangent is periodic with period* π and is defined for all real numbers except those for which sin x = 0 (that is, the integer multiples of π). The graph is very similar to the graph of the tangent function (reflected and shifted) and is shown in Fig. 3.10. The vertical lines given by $x = 0, x = \pm\pi, x = \pm 2\pi, \ldots$ are *vertical asymptotes* for the cotangent function.

Graphs of the secant and cosecant functions _____

The basic reciprocal relations give

$$\sec x = \frac{1}{\cos x} \quad \text{and} \quad \csc x = \frac{1}{\sin x}.$$

Since the sine and cosine functions are the same except for the shift of $\pi/2$, the same must be true of the secant and cosecant functions, both of which are *periodic of period* 2π.

   The reciprocal of a positive number less than 1 is a positive number greater than 1; the reciprocal of a number between −1 and 0 is a negative number less than −1. Since cos x and sin x are always in the interval −1 to 1, the graphs of sec x and csc x are always *outside* the interval $-1 < y < 1$. If we consider one arch of the sine function, for example where $0 \le t \le \pi$, we first note that sin $\pi/2 = 1$ and so csc $\pi/2 = 1/1 = 1$. As x increases from $\pi/2$ toward π, sin x decreases

FIGURE 3.10

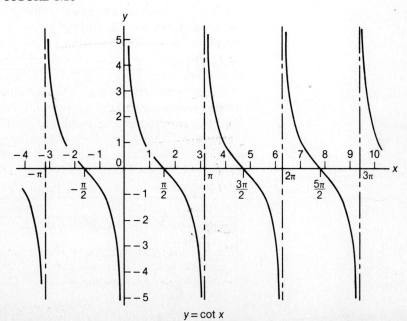

$y = \cot x$

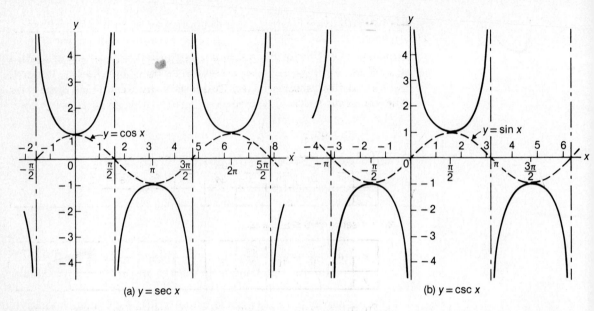

(a) y = sec x   (b) y = csc x

FIGURE 3.11

from 1 to 0, and csc x = 1/sin x *increases* from 1 *without bound*, being undefined at x = π. Because the arch of sin x is symmetric about x = π/2, the corresponding graph for csc x has the same symmetry. The negative arch of sin x on the interval [π, 2π] also gives rise to a symmetric portion of the graph of y = csc x. The graph is shown in Fig. 3.11(b), where y = sin x is also shown. The graph of y = sec x is given in Fig. 3.11(a) in relation to the graph of y = cos x.

Summary

From Figs. 3.9, 3.10, and 3.11 we may readily read the domains, ranges, asymptotes, and periods for the tangent, cotangent, secant, and cosecant functions. These are given in Table 3.2, where k is any integer.

Table 3.2

Function	Domain	Range	Asymptotes	Period
tangent	$x \neq (2k-1)(\pi/2)$	$\mathbb{R}$	$x = (2k-1)(\pi/2)$	π
cotangent	$x \neq k\pi$	$\mathbb{R}$	$x = k\pi$	π
secant	$x \neq (2k-1)(\pi/2)$	$y \leq -1$ or $y \geq 1$	$x = (2k-1)(\pi/2)$	2π
cosecant	$x \neq k\pi$	$y \leq -1$ or $y \geq 1$	$x = k\pi$	2π

EXERCISE 3.2

For Problems 1 and 2, use a calculator to evaluate (to two decimal places) the given function for the values of x indicated in the table, where the function may be undefined at some values. Then make a large-scale graph of the given function on the specified intervals.

1. $y = \tan x$ on $\frac{\pi}{4} \le x \le \frac{3\pi}{4}$.

x.	$\frac{\pi}{4}$	$\frac{3\pi}{10}$	$\frac{7\pi}{20}$	$\frac{2\pi}{5}$	$\frac{9\pi}{20}$	$\frac{\pi}{2}$	$\frac{11\pi}{20}$	$\frac{3\pi}{5}$	$\frac{13\pi}{20}$	$\frac{7\pi}{10}$	$\frac{3\pi}{4}$
y											

2. $y = \sec x$ on $0 \le x \le \pi/2$.

x	0	$\frac{\pi}{20}$	$\frac{\pi}{10}$	$\frac{3\pi}{20}$	$\frac{\pi}{5}$	$\frac{\pi}{4}$	$\frac{3\pi}{10}$	$\frac{7\pi}{20}$	$\frac{2\pi}{5}$	$\frac{9\pi}{20}$	$\frac{\pi}{2}$
y											

3. From the graph of $y = \tan x$ (Fig. 3.9), justify the reduction formulas $\tan(\pi + t) = \tan t$ and $\tan(\pi - t) = -\tan t$.

4. From the graph of $y = \cot x$ (Fig. 3.10), justify the reduction formulas $\cot(\pi + t) = \cot t$ and $\cot(\pi - t) = -\cot t$.

5. From the graph of $y = \sec x$ (Fig. 3.11a), justify the reduction formulas $\sec(\pi - t) = -\sec t$ and $\sec(\pi + t) = -\sec t$.

6. From the graph of $y = \csc x$ (Fig. 3.11b), justify the reduction formulas $\csc(-t) = -\csc t$ and $\csc((\pi/2) + t) = \csc((\pi/2) - t)$.

In each of Problems 7–10 you can fill in the box ☐ with any of several appropriate expressions to produce a valid reduction formula. Use the graphs in this section to help you in finding two such expressions.

7. $\sec\left(\frac{\pi}{2} - t\right) = \sec\left(\boxed{}\right)$

8. $\csc\left(\frac{3\pi}{2} + t\right) = \csc\left(\boxed{}\right)$

9. $\tan(\pi + t) = \tan\left(\boxed{}\right)$

10. $\cot\left(\frac{\pi}{2} + t\right) = \cot\left(\boxed{}\right)$

In Problems 11–16, use appropriate reduction formulas for the sine and cosine functions (along with the relations expressing tan, cot, sec, and csc in terms of sin and cos) to justify the given reduction formula.

11. $\tan\left(\frac{\pi}{2} + t\right) = -\cot t$

12. $\cot(3\pi + t) = \cot t$

13. $\sec\left(\frac{3\pi}{2} - t\right) = -\csc t$

14. $\csc\left(\frac{\pi}{2} + t\right) = \sec t$

15. $\tan\left(\frac{3\pi}{2} - t\right) = \cot t$

16. $\sec(\pi + t) = -\sec t$

In Problems 17–24, use graphs in this section (extended as necessary) to help you determine which of the given formulas are valid reduction formulas.

17. $\tan(3\pi + t) = \tan t$ **18.** $\cot(5\pi - t) = \cot t$

19. $\sec\left(\dfrac{3\pi}{2} + t\right) = \sec t$ **20.** $\csc\left(\dfrac{3\pi}{2} + t\right) = -\sec t$

21. $\tan\left(\dfrac{5\pi}{2} + t\right) = \cot t$ **22.** $\sec(\pi - t) = \sec t$

23. $\cot\left(\dfrac{3\pi}{2} + t\right) = -\tan t$ **24.** $\tan(15\pi + t) = \tan t$

3.3 Review of Inverse Functions From Algebra (Optional)

In Section 2.1 we included a brief review of the basic ideas, notation, and terminology related to functions as they occur in algebra courses. In this section we present a review of inverse functions by considering examples from algebra.

EXAMPLE 1 Suppose function f is given by the rule $f(x) = 2x - 3$ (or $y = 2x - 3$) where the domain of f is the set of real numbers. Discuss the inverse of f.

Solution The rule for f assigns to each number x a corresponding number y. For instance, when f is applied to 4, the corresponding value of y is given by $y = 2(4) - 3 = 5$.

Now suppose we ask an "inverse" question, "What number x gives a corresponding y-value of 7?" We can solve $7 = 2x - 3$ and get $x = 5$ as the answer. In general, suppose we are given any y and we want to determine the x that was used by f to give y. We can solve the equation $y = 2x - 3$ for x in terms of y and get $x = (y + 3)/2$. Hence we have another "rule," which assigns a *unique* number x to each number y. This rule describes a function that is called the inverse of f and denoted by f^{-1}. We write $x = f^{-1}(y) = (y + 3)/2$.* For instance, $f^{-1}(7) = (7 + 3)/2 = 5$.

When the rule for f^{-1} is written as $f^{-1}(y) = (y + 3)/2$, we are assuming that y is the independent variable. It is customary to use the letter x as the independent variable, particularly when we draw graphs, and so the rule for f^{-1} can be written as $f^{-1}(x) = (x + 3)/2$. ∎

---

* The -1 in the symbol f^{-1} is not to be interpreted as a negative exponent. That is, $f^{-1}(y)$ is not equal to $1/f(y)$; f^{-1} is simply notation used to denote the inverse of f.

A technique for finding a formula for $f^{-1}(x)$ is to interchange the x- and y-variables in $y = f(x)$ to get $x = f(y)$ and then solve for y (if possible). This gives $y = f^{-1}(x)$. In the above example the inverse function is found by solving $x = 2y - 3$ for y. Thus $y = (x + 3)/2$ or $f^{-1}(x) = (x + 3)/2$.

EXAMPLE 2 Suppose function g is given by the rule $g(x) = x^2$ (or $y = x^2$) where the domain is the set of real numbers. Discuss the inverse of g.

Solution The rule for g assigns to each real number x the square of x. For instance, if g is applied to 4, it returns 4^2 or 16 as the corresponding value of y.

Now suppose we ask the "inverse" question, "What values of x yield 16 as the corresponding value of y?" Solving the equation $16 = x^2$, we get *two* numbers, 4 and -4, both of which yield 16. In general, if we wish to determine x for any given y (where $y \geq 0$), then solving $y = x^2$ for x gives two solutions, $x = \sqrt{y}$ or $x = -\sqrt{y}$. Thus the "inverse rule" assigns two numbers x for each positive number y, and so it is not a function.

However, it is a relation denoted by g^{-1}, which is given by

$$g^{-1} = \{(y,x) \mid x = \sqrt{y} \quad \text{or} \quad x = -\sqrt{y}\}.$$

The same set of ordered pairs is given by

$$g^{-1} = \{(x,y) \mid y = \sqrt{x} \quad \text{or} \quad y = -\sqrt{x}\}. \quad \blacksquare$$

Inverse functions

In Example 1 the function $f(x) = 2x - 3$ has an inverse that is also a function that is given by the rule $f^{-1}(x) = (x + 3)/2$. However, in Example 2 the function $g(x) = x^2$ has an inverse that is a relation that is not a function.

This leads to an important question: When will a given function, say $y = f(x)$, have an inverse that is also a function? If for each x in the domain of the rule, f assigns *exactly one number* y (then f is a function) and for each y in the range of f there is *precisely one value of* x that yields y, then f has an inverse that is also a function. This essentially says that f is a *one-to-one function*.

How can we determine when a function is one-to-one? Perhaps the most convenient way, if we can draw a graph of $y = f(x)$, is to see that if *each vertical line* and *each horizontal line* intersect the graph in *at most one point*, then the function is one-to-one. Another way of checking is to see if the function is *increasing* everywhere or *decreasing* everywhere on its domain, which may easily be seen from the graph. Compare the graph of $y = x^2$ and the graphs of the functions in Examples 4 and 5 (Figs. 3.12 and 3.13).

Why functions?

Why are we so concerned about dealing with functions rather than the more general concept of relation? In most of mathematics we want results to be unique (only one answer). In the study of calculus, almost every definition or basic concept begins with, "If f is a function, then...."

Technique for finding formulas for inverse functions

Suppose f is a function given by $y = f(x)$. To get the inverse of f, interchange the x- and y-variables to get $x = f(y)$ and solve for y in terms of x (if possible). If we get a unique solution, then f^{-1} is also a function, and we write $y = f^{-1}(x)$. Thus $x = f(y)$ and $y = f^{-1}(x)$ are equivalent.

Properties of inverse functions

Suppose f and its inverse are functions.

1. *Domain–Range.* The domain of f^{-1} is equal to the range of f, and the range of f^{-1} is the domain of f.
2. *Composition Identities.*

$f^{-1}(f(x)) = x$ for each x in the domain of f.
$f(f^{-1}(x)) = x$ for each x in the domain of f^{-1}.   **[3.7]**

These identities essentially say that f and f^{-1} "undo each other."

3. *Graphs.* The graphs of $y = f(x)$ and $y = f^{-1}(x)$ are reflections of each other about the line $y = x$.

EXAMPLE 3   If $f(x) = 2x - 3$,   verify   that $f^{-1}(f(x)) = x$ and $f(f^{-1}(x)) = x$.

Solution   From Example 1 we have $f^{-1}(x) = (x + 3)/2$. Therefore

$$f^{-1}(f(x)) = f^{-1}(2x - 3) = \frac{(2x - 3) + 3}{2} = \frac{2x}{2} = x,$$

$$f(f^{-1}(x)) = f\left(\frac{x + 3}{2}\right) = 2\left(\frac{x + 3}{2}\right) - 3 = (x + 3) - 3 = x. \quad \blacksquare$$

Examples of functions with restricted domains _____

In Example 2 we saw that the inverse of the function $y = x^2$, where the domain is the set of real numbers, is not a function. In the following two examples we consider the same rule, $y = x^2$, with a suitable restriction on the domain so that the inverse is a function. *This is precisely the type of situation we shall encounter in considering inverses of trigonometric functions* in subsequent sections of this chapter.

EXAMPLE 4   Suppose $f(x) = x^2$ where the domain is $\{x \mid x \ge 0\}$. Discuss the inverse of f.

Solution   We first draw a graph of $y = f(x)$ as shown in Fig. 3.12(a). It should be clear from the graph f is a one-to-one function, and so f^{-1} is also a function.

To find a formula for $f^{-1}(x)$, we interchange the x- and y-variables in $y = x^2$ to get $x = y^2$. Solving for y gives $y = \sqrt{x}$. We reject the solution $y = -\sqrt{x}$ because the range of f^{-1} is equal to the domain of f (the set of nonnegative numbers).

Therefore f^{-1} is given by

$$f^{-1}(x) = \sqrt{x} \qquad \text{for } x \ge 0.$$

The graph of the inverse function is shown in Fig. 3.12(b).   ∎

EXAMPLE 5   Suppose function g is given by $g(x) = x^2$, where the domain of g is $\{x \mid x \le 0\}$. Discuss the inverse of g.

Solution   We first draw a graph of $y = g(x)$ as shown in Fig. 3.13(a) and observe from the graph that g is a one-to-one function. Hence the inverse g^{-1} is also a function.

To find a formula for $g^{-1}(x)$, we interchange x and y in $y = x^2$ to get $x = y^2$ and then solve for y. This gives $y = -\sqrt{x}$, where we reject

FIGURE 3.12

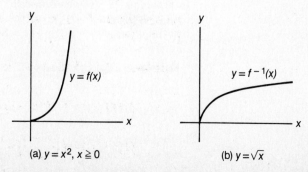

(a) $y = x^2$, $x \ge 0$ (b) $y = \sqrt{x}$

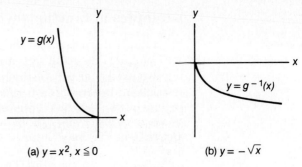

(a) $y = x^2$, $x \leq 0$ (b) $y = -\sqrt{x}$

FIGURE 3.13

the solution $y = \sqrt{x}$, since the range of g^{-1} is equal to the domain of g, the set of negative numbers along with 0.

Therefore g^{-1} is given by

$$g^{-1}(x) = -\sqrt{x}, \qquad x \geq 0.$$

The graph of $y = g^{-1}(x)$ is shown in Fig. 3.13(b).   ∎

EXERCISE 3.3

In Problems 1–10, determine a formula for the inverse function and give it in the form $y = f^{-1}(x)$ along with its domain. Also, evaluate $f^{-1}(0)$, $f^{-1}(1)$, and $f^{-1}(-4)$ for those cases in which these are defined.

1. $f(x) = 3x + 5$
2. $f(x) = 3 - 4x$
3. $f(x) = 2.5x - 4$
4. $f(x) = x + 4$
5. $f(x) = \sqrt{x}$
6. $f(x) = -\sqrt{x}$
7. $f(x) = x^2 + 1$, $x \geq 0$
8. $f(x) = x^2 - 1$, $x \geq 0$
9. $f(x) = x^3$
10. $f(x) = x^3 - 1$

In Problems 11–16, determine the inverse function. Then draw graphs of both $y = f(x)$ and $y = f^{-1}(x)$ on the same system of coordinates. Check to see that the graph of $y = f^{-1}(x)$ is a reflection of the graph of $y = f(x)$ about the line $y = x$.

11. $f(x) = 3 + x$
12. $f(x) = 2x - 4$
13. $f(x) = 1.5x - 3$
14. $f(x) = 4 - 2x$
15. $f(x) = \sqrt{x}$
16. $f(x) = x^2 - 1$, $x \geq 0$

In Problems 17–20, (a) determine a formula for $f^{-1}(x)$ and (b) verify (show work) that $f^{-1}(f(x)) = x$ for each x in the domain of f. (See Example 3.)

17. $f(x) = 2x + 4$ **18.** $f(x) = 3x - 5$

19. $f(x) = x^3 - 1$ **20.** $f(x) = \sqrt{x} + 1$

In Problems 21–24 the inverse of the given function is not a function. Draw a graph and convince yourself that this is the case. Restrict the domain of f in a suitable manner (giving a new function, which we may call g) so that the inverse of g is a function. There are many ways in which the domain can be restricted to give g, and so the answers to these problems will not be unique. Determine $g^{-1}(x)$ for your function and be certain to specify the domains of g and g^{-1}.

21. $f(x) = x^2 - 1$ **22.** $f(x) = |x|$

23. $f(x) = |x - 1|$ **24.** $f(x) = x^2 + 1$

3.4 Inverse Sine and Cosine Functions

In the preceding section we encountered functions for which the inverse is a relation but not a function. However, by a suitable restriction on the domain we were able to get a related function whose inverse is a function (see Examples 2, 4, and 5 in Section 3.3). This is precisely the situation we face in the case of trigonometric functions.

First, let us recall the graph of the sine function discussed in Section 3.1 (Fig. 3.2), which we also show in Fig. 3.14. Note from the graph that for each x there is exactly one value of y (and so $y = \sin x$ represents a function), but for each value of y between -1 and 1 there are infinitely many values of x to which it corresponds. For instance, the values of x to which $y = 1/2$ corresponds are $\pi/6$, $5\pi/6$, ..., $-7\pi/6$, ..., as seen from the graph. Therefore the inverse of $y = \sin x$

FIGURE 3.14

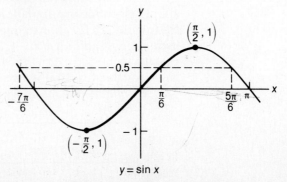

$y = \sin x$

is not a function, and so we shall restrict the domain and get a related function whose inverse is a function. This can be accomplished in many ways, and the choice is ours to make. However, we shall be guided by tradition, and of course we wish to have our inverse function agree with that evaluated by calculators and pocket computers.

Inverse sine function

The domain of the function given by $y = \sin x$ is the set of real numbers (where x is in radians), and the range is the set of numbers for which $-1 \le y \le 1$. That is,

$$\text{Domain} = D(\sin) = \mathbb{R} \quad \text{and} \quad \text{Range} = R(\sin) = \{y \mid -1 \le y \le 1\}.$$

Now let us introduce a new function, denoted by Sin (note the capital letter S), defined by

$$\text{Sin } x = \sin x \quad \text{where } D(\text{Sin}) = \left\{ x \mid -\frac{\pi}{2} \le x \le \frac{\pi}{2} \right\}.$$

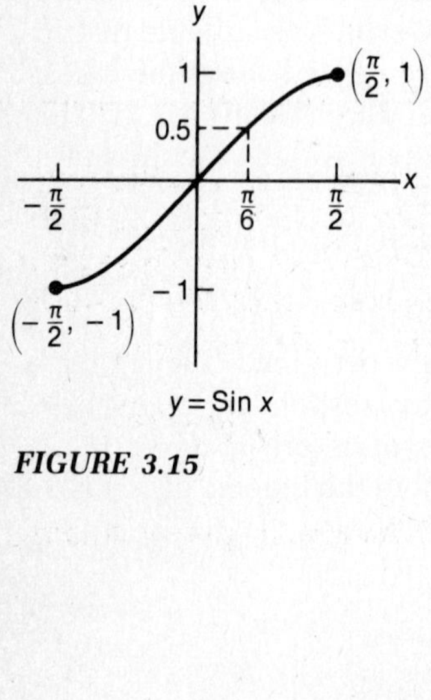

$y = \text{Sin } x$

FIGURE 3.15

Thus the Sine function assumes the same values as the sine function, but its domain is restricted to $-\pi/2 \le x \le \pi/2$. The graph of $y = \text{Sin } x$ (the darker portion of the curve in Fig. 3.14) is shown in Fig. 3.15, from which we see that Sin is an increasing function and conclude that the inverse is also a function. That is, for each y in $-1 \le y \le 1$, there is exactly one value of x to which it corresponds. For instance, if $y = 1/2$, then the only value of x satisfying $1/2 = \text{Sin } x$ is $\pi/6$. From the graph we also observe that the domain of the inverse function is the set of numbers -1 to 1 (the same as the range of Sin) and the range is the set $-\pi/2$ to $\pi/2$ (the same as the domain of Sin).

As was suggested in the preceding section, to find the inverse function with x as the independent variable, we interchange the x- and y-variables and then solve for y. That is, solve the equation $x = \text{Sin } y$ for y. Here the situation is somewhat different from the examples considered in Section 3.3 in that we cannot solve for y by simple algebraic manipulations. However, the important fact is that *the inverse is a function defined by* $x = \text{Sin } y$, *where x is the independent variable* and $-1 \le x \le 1$. Since this is an important function, we give it a name and call it the Sin^{-1} function. Another commonly used name for this function is Arcsin.

We summarize the above discussion in the following definition.

$$\begin{array}{|c|} \hline y = \text{Sin}^{-1} x \quad \text{or} \quad y = \text{Arcsin } x \quad \text{where } -1 \le x \le 1 \\ \text{means } \sin y = x \quad \text{and} \quad -\pi/2 \le y \le \pi/2. \\ \hline \end{array}$$ **[3.8]**

Another way to say this is that the Sin^{-1} function is defined in terms of ordered pairs by

$$\text{Sin}^{-1} = \left\{ (x,y) \mid -1 \le x \le 1, \quad x = \sin y, \quad \text{and} \quad -\frac{\pi}{2} \le y \le \frac{\pi}{2} \right\}.^* \quad \textbf{[3.9]}$$

Note: The definition given in Eq. (3.8) implies that $\text{Sin}^{-1} x$ can be considered as an angle given *in radians* or as a length of arc from $-\pi/2$ to $\pi/2$. This is the context in which the inverse sine function occurs in calculus and also in working with computers. In some situations (such as solving triangles) we will use calculators to get values of y in degrees, $-90° \le y \le 90°$.

Graph of the inverse sine function

We can draw a graph of $y = \text{Sin}^{-1} x$ by simply reflecting the graph of $y = \text{Sin} x$ (Fig. 3.15) about the line $y = x$. Note that $(\pi/2, 1)$ and $(-\pi/2, -1)$ are points on $y = \text{Sin} x$, and so $(1, \pi/2)$ and $(-1, -\pi/2)$ must be points on $y = \text{Sin}^{-1} x$. Likewise, $(-1/\sqrt{2}, -\pi/4)$, $(0,0)$, and $(1/2, \pi/6)$ are also points on $y = \text{Sin}^{-1} x$. Making use of this information, we get the graph of $y = \text{Sin}^{-1} x$ as shown in Fig. 3.16.

Evaluating $\text{Sin}^{-1} x$ in exact form

We need to be able to evaluate $\text{Sin}^{-1} x$ for various values of x between -1 and 1. For instance, if $x = 1/2$, then let $\text{Sin}^{-1}(1/2) = y$. From (3.8) we have $\sin y = 1/2$ and $-\pi/2 \le y \le \pi/2$. Recall that $\sin(\pi/6) = 1/2$, and so we get $y = \pi/6$, or $\text{Sin}^{-1}(1/2) = \pi/6$ (exact form). We can also get a decimal answer by evaluating $\pi/6$, $\text{Sin}^{-1}(1/2) \approx 0.5236$.

EXAMPLE 1 Evaluate

a] $\text{Sin}^{-1}(1/\sqrt{2})$ **b]** $\text{Arcsin}(-\sqrt{3}/2)$.

Give answers in exact form and also as decimal approximations rounded off to three decimal places.

Solution

a] Let $y = \text{Sin}^{-1}(1/\sqrt{2})$; then $\sin y = 1/\sqrt{2}$, and $-\pi/2 \le y \le \pi/2$. This says that y is an angle in the first or fourth quadrant; and since $\sin y$ is positive, it must be in the first. Draw the reference

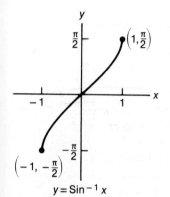

$y = \text{Sin}^{-1} x$

FIGURE 3.16

---

* We use the capital letter S in Sin^{-1} to distinguish the *inverse sine function* from the *inverse sine relation* given by

$$\sin^{-1} = \{ (x,y) \mid -1 \le x \le 1 \quad \text{and} \quad x = \sin y \}.$$

In this case, for each x there are infinitely many corresponding values of y. In some textbooks, $\text{Sin}^{-1} x$ is referred to as the *principal value* of $\sin^{-1} x$.

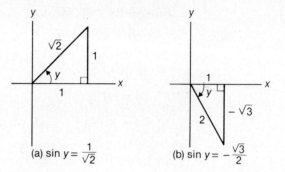

FIGURE 3.17

triangle as shown in Fig. 3.17(a). From it we see that y must be 45°, or in radians $y = \pi/4$. Thus $\mathrm{Sin}^{-1}(1/\sqrt{2}) = \pi/4$ (exact form). Evaluate $\pi/4$ and get $\mathrm{Sin}^{-1}(1/\sqrt{2}) \approx 0.785$.

b] If $y = \mathrm{Arcsin}(-\sqrt{3}/2)$, then $\sin y = -\sqrt{3}/2$ and $-\pi/2 \le y \le \pi/2$. Thus y is an angle in the fourth quadrant as shown in Fig. 3.17(b). The reference triangle shows that $y = -60°$, or in radians $y = -\pi/3$. Hence $\mathrm{Arcsin}(-\sqrt{3}/2) = -\pi/3$ (exact form). Evaluate $-\pi/3$ by calculator to get $\mathrm{Arcsin}(-\sqrt{3}/2) \approx -1.047$. ∎

Composition function identities

As was suggested in the preceding example, it is helpful to think of $\mathrm{Sin}^{-1}x$ as *the angle* θ (between $-\pi/2$ and $\pi/2$) whose sine is x. For instance, $\mathrm{Sin}^{-1}0.5$ is the angle between $-\pi/2$ and $\pi/2$ whose sine is 0.5. There is exactly one such angle, and it is $\pi/6$.

In mathematical notation, "$\mathrm{Sin}^{-1}x$ is that angle whose sine is x" is written as

$$\sin(\mathrm{Sin}^{-1}x) = x \qquad \text{for every } x \text{ in } -1 \le x \le 1. \qquad \textbf{[3.10]}$$

This is precisely the statement given in Eq. (3.7) where the function f is given by $f(x) = \mathrm{Sin}\,x$. The other statement in (3.7) gives

$$\mathrm{Sin}^{-1}(\sin x) = x \qquad \text{for every } x \text{ in } -\frac{\pi}{2} \le x \le \frac{\pi}{2}.* \qquad \textbf{[3.11]}$$

* Here we can see that x must be in the interval $-\pi/2$ to $\pi/2$ because x is equal to $\mathrm{Sin}^{-1}(\mathrm{Sin}\,x)$ and the inverse Sine of any number must be in the interval $-\pi/2$ to $\pi/2$. Note that $\mathrm{Sin}^{-1}(\sin x)$ is not equal to x for all values of x. For instance, if $x = 5\pi/6$, then $\mathrm{Sin}^{-1}(\sin 5\pi/6) = \mathrm{Sin}^{-1}(1/2) = \pi/6$. Thus $\mathrm{Sin}^{-1}(\sin 5\pi/6) \ne 5\pi/6$.

EXAMPLE 2 Evaluate

  a] $\sin\left(\mathrm{Sin}^{-1}(3/4)\right)$ **b]** $\cos\left(\mathrm{Sin}^{-1}(3/4)\right)$.

Give answers in exact form and also in decimal form rounded off to two decimal places.

Solution Let $\theta = \mathrm{Sin}^{-1}(3/4)$; so we want to determine (a) $\sin\theta$ and (b) $\cos\theta$. From Eq. (3.8) we get

$$\sin\theta = \frac{3}{4} \quad \text{and} \quad -\frac{\pi}{2} \le \theta \le \frac{\pi}{2}.$$

The reference triangle for θ is shown in Fig. 3.18.

  a] From Fig. 3.18 we get $\sin\theta = 3/4$. Thus $\sin\left(\mathrm{Sin}^{-1}(3/4)\right) = 3/4 = 0.75$. Note that we could have applied Eq. (3.10) to get the same result.
  b] From Fig. 3.18 we get $\cos\theta = \sqrt{7}/4$, and so $\cos\left(\mathrm{Sin}^{-1}(3/4)\right) = \sqrt{7}/4$ (exact form). Evaluating $\sqrt{7}/4$ by calculator gives $\cos\left(\mathrm{Sin}^{-1}(3/4)\right) \approx 0.66$. ∎

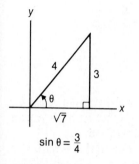

$$\sin\theta = \frac{3}{4}$$

FIGURE 3.18

Evaluating by a calculator
(Approximate decimal form) ――――――――――――――――

We are able to evaluate $\mathrm{Sin}^{-1}x$ in exact form for special values of x, such as $x = 0.5$, in which case we get $\mathrm{Sin}^{-1}0.5 = \pi/6$. However, if $x = 0.6$ and we let $\theta = \mathrm{Sin}^{-1}0.6$, then $\sin\theta = 0.6 = 3/5$. The reference triangle for θ is shown in Fig. 3.19, and we see that θ is not one of the special angles $(30°, 45°, 60°)$. Thus we cannot evaluate $\mathrm{Sin}^{-1}0.6$ in exact form. However, calculators are preprogrammed to give decimal answers to full calculator accuracy. This is illustrated in the following three examples.

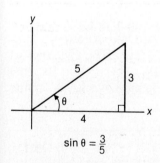

$$\sin\theta = \frac{3}{5}$$

FIGURE 3.19

EXAMPLE 3 Evaluate $\mathrm{Sin}^{-1}0.6$ and give the result rounded off to four decimal places.

Solution Scientific calculators have either a key labeled $\boxed{\text{Sin}^{-1}}$ or keys $\boxed{\text{INV}}$ and $\boxed{\text{Sin}}$. Place the calculator in radian mode, then enter 0.6 into the display, and press $\boxed{\text{Sin}^{-1}}$ or $\boxed{\text{INV}}$, $\boxed{\text{Sin}}$ to get the desired answer, $\mathrm{Sin}^{-1}0.6 \approx 0.6435$. With a pocket computer the sequence $\boxed{\text{ASN}}$, 0.6, $\boxed{\text{EXE}}$ gives the same result.

  If the calculator is in degree mode, then the above evaluations would give the degree equivalent of 0.6435 radians, that is, 36.87° (to two decimal places). This suggests that the calculator deals with two inverse sine functions determined by the mode (radians or degrees). See Section 2.3, in which we referred to the Sin^{R} and $\mathrm{Sin}°$ functions. ∎

EXAMPLE 4 In Example 1 we evaluated $\text{Sin}^{-1}(1/\sqrt{2})$ in exact form and then evaluated the result $(\pi/4)$ by calculator to get the approximate decimal answer, $\text{Sin}^{-1}(1/\sqrt{2}) \approx 0.785$. Evaluate $\text{Sin}^{-1}(1/\sqrt{2})$ directly by calculator.

Solution Place the calculator in radian mode and press the following sequence of keys

Algebraic: 1, ⊙÷, 2, ⊙√x̄, ⊙=, ⊙INV, ⊙Sin

   or   2, ⊙√x̄, ⊙1/x, ⊙INV, ⊙Sin.

RPN: 2, ⊙√x̄, ⊙1/x, ⊙Sin⁻¹.

Pocket Computer: ⊙ASN (1/SQR 2) ⊙EXE.

The display will show 0.78539816. ∎

EXAMPLE 5 Evaluate

a] $\text{Sin}^{-1}(-0.327)$,   **b]** $\text{Arcsin}(\sqrt{5}/3)$,   **c]** $\text{Sin}^{-1}(\sqrt{7}/2)$.

Give answers rounded off to four decimal places.

Solution The definition given in Eq. (3.8) implies that $\text{Sin}^{-1}x$ is an angle in radians, so first be certain that your calculator is in radian mode.

a] First enter -0.327 into the display and press ⊙Sin⁻¹ or ⊙INV ⊙Sin to get $\text{Sin}^{-1}(-0.327) \approx -0.3331$.*
b] First evaluate $\sqrt{5}/3$ and then, with this number in the display, press ⊙Sin⁻¹ or ⊙INV, ⊙Sin to get $\text{Sin}^{-1}(\sqrt{5}/3) \approx 0.8411$.*
c] First evaluate $\sqrt{7}/2$ and then press ⊙Sin⁻¹ or ⊙INV, ⊙Sin. In this case the calculator displays ERROR. The reason is that $\sqrt{7}/2 > 1$, and so $\sqrt{7}/2$ is not in the domain of the Sin^{-1} function. ∎

> In evaluating a function, your calculator will not accept numbers that are not in the domain of the function.

Inverse cosine function

The inverse cosine function can be introduced in a manner similar to that used above for the inverse sine function. First we define the Cos function by having it agree with the cosine function on the interval $0 \le x \le \pi$. That is, we use only the solid portion of the $y = \cos x$ curve

---

* To evaluate $\text{Sin}^{-1}(-0.327)$ with a hand-held computer, first place it in radian mode (use the ⊙MODE key). Then press ⊙ASN $-.327$ ⊙EXE. To evaluate $\text{Sin}^{-1}(\sqrt{5}/3)$, press ⊙ASN (⊙SQR 5/3) ⊙EXE.

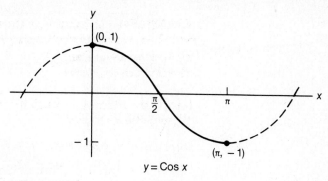

FIGURE 3.20

shown in Fig. 3.20. Thus the function denoted by Cos is defined by

$$\text{Cos}\,x = \cos x \qquad \text{and} \qquad D(\text{Cos}) = \{x \mid 0 \le x \le \pi\}.$$

From the graph in Fig. 3.20 we see that Cos is a decreasing function, and so the inverse of Cos is also a function, which we define as follows.

Definition of the inverse cosine function

The function denoted by Cos^{-1} or by Arccos is defined by

$$\boxed{\begin{array}{l} y = \text{Cos}^{-1}x \quad \text{or} \quad y = \text{Arccos}\,x \quad \text{where } -1 \le x \le 1 \\ \qquad\qquad \text{means} \quad \cos y = x \quad \text{and} \quad 0 \le y \le \pi. \end{array}} \qquad \textbf{[3.12]}$$

Another way to say this is in terms of ordered pairs:

$$\boxed{\text{Cos}^{-1} = \{(x, y) \mid -1 \le x \le 1, \qquad x = \cos y \quad \text{and} \quad 0 \le y \le \pi\}.} \qquad \textbf{[3.13]}$$

FIGURE 3.21

Graph of the inverse cosine function

The graph of $y = \text{Cos}^{-1}x$ can be drawn by reflecting the solid portion of the curve shown in Fig. 3.20 about the line $y = x$. Note that the points $(0, 1)$, $(\pi/2, 0)$, and $(\pi, -1)$ are on the graph of $y = \text{Cos}\,x$, and so $(1, 0)$, $(0, \pi/2)$, and $(-1, \pi)$ are on the $y = \text{Cos}^{-1}x$ graph. The graph of $y = \text{Cos}^{-1}x$ is shown in Fig. 3.21.

Composition identities

Identities analogous to those given in (3.10) and (3.11) are

$$\boxed{\cos\left(\text{Cos}^{-1}x\right) = x \quad \text{for every } x \text{ in} \quad -1 \le x \le 1.} \qquad \textbf{[3.14]}$$

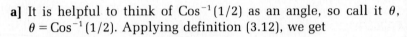

$$\boxed{\text{Cos}^{-1}(\cos x) = x \quad \text{for every } x \text{ in } \quad 0 \le x \le \pi.^*} \qquad \textbf{[3.15]}$$

EXAMPLE 6 Evaluate

a] $\text{Cos}^{-1}(1/2)$ **b]** $\text{Cos}^{-1}(-\sqrt{3}/2)$.

Give results in exact form.

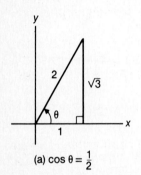

(a) $\cos \theta = \dfrac{1}{2}$

Solution

a] It is helpful to think of $\text{Cos}^{-1}(1/2)$ as an angle, so call it θ, $\theta = \text{Cos}^{-1}(1/2)$. Applying definition (3.12), we get

$$\cos \theta = 1/2 \quad \text{and} \quad 0 \le \theta \le \pi.$$

The reference triangle for θ is shown in Fig. 3.22(a). We see that $\theta = 60°$, or in radians $\theta = \pi/3$. Therefore $\text{Cos}^{-1}(1/2) = \pi/3$.

b] Let $\alpha = \text{Cos}^{-1}(-\sqrt{3}/2)$, and so

$$\cos \alpha = -\sqrt{3}/2 \quad \text{and} \quad 0 \le \alpha \le \pi.$$

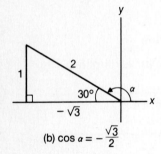

(b) $\cos \alpha = -\dfrac{\sqrt{3}}{2}$

FIGURE 3.22

The angle α satisfying these conditions is shown in Fig. 3.22(b). We note that the reference angle for α is 30°, or in radians $\pi/6$. Therefore $\alpha = \pi - \pi/6 = 5\pi/6$, and so we have $\text{Cos}^{-1}(-\sqrt{3}/2) = 5\pi/6$. ∎

EXAMPLE 7 Evaluate

a] $\text{Cos}^{-1}(0.587)$ **b]** $\text{Cos}^{-1}(-\pi/6)$.

Give answers rounded off to three decimal places.

Solution The definition given in Eq. (3.12) implies that $\text{Cos}^{-1}x$ is an angle in radian measure, so first be certain that your calculator is in *radian mode.* In degree mode the calculator will give an equivalent answer in degrees.

a] After entering 0.587 into the display, press (Cos⁻¹) or (INV), (Cos) to get 0.943448079. Hence $\text{Cos}^{-1}(0.587) \approx 0.943$.

b] First evaluate $\pi/6$ and change sign. Then with the result in the display, press (Cos⁻¹) or (INV), (Cos) to get $\text{Cos}^{-1}(-\pi/6) \approx 2.122.^\dagger$ ∎

EXAMPLE 8 Evaluate

a] $\text{Cos}^{-1}(\sin(-\pi/6))$ **b]** $\text{Cos}^{-1}(\cos(4\pi/3))$.

Express answers in exact form.

---

* Note that $\text{Cos}^{-1}(\cos x)$ is not equal to x for every number x. Equality holds only when $0 \le x \le \pi$. See Example 8(b).

† To evaluate $\text{Cos}^{-1}0.587$ with a hand-held computer, press (ACS) .587 (EXE). For $\text{Cos}^{-1}(-\pi/6)$, press (ACS) $(-\pi/6)$ (EXE).

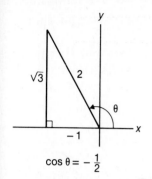

$\cos \theta = -\frac{1}{2}$

FIGURE 3.23

Solution

a] First evaluate $\sin(-\pi/6)$ and get $\sin(-\pi/6) = -1/2$. Thus we want $\text{Cos}^{-1}(-1/2)$. Let $\theta = \text{Cos}^{-1}(-1/2)$, and so $\cos\theta = -1/2$ and $0 \le \theta \le \pi$. Angle θ is shown in Fig. 3.23, and we see that the reference angle for θ is $60°$ and so $\theta = 120°$ or $\theta = 2\pi/3$. Hence $\text{Cos}^{-1}(\sin(-\pi/6)) = 2\pi/3$.

b] Our first temptation is to apply the identity given in Eq. (3.15) and get $4\pi/3$ as the answer. However, that *identity applies only* to $\theta \le x \le \pi$, and $4\pi/3$ is not such a number.

First evaluate $\cos(4\pi/3)$ (or $\cos 240°$) and get $-1/2$. Thus $\text{Cos}^{-1}(\cos(4\pi/3)) = \text{Cos}^{-1}(-1/2)$. In part (a) we saw that $\text{Cos}^{-1}(-1/2) = 2\pi/3$, and so $\text{Cos}^{-1}(\cos(4\pi/3)) = 2\pi/3$. ∎

EXAMPLE 9 Evaluate $\tan(\text{Sin}^{-1}(3/4))$ in exact form.

Solution
Let $\theta = \text{Sin}^{-1}(3/4)$ and we want to determine $\tan\theta$, where $\sin\theta = 3/4$ and $-\pi/2 \le \theta \le \pi/2$. Angle θ is shown in Fig. 3.24. We see that $\tan\theta = 3/\sqrt{7}$, and so $\tan(\text{Sin}^{-1}(3/4)) = 3/\sqrt{7}$. ∎

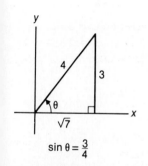

$\sin \theta = \frac{3}{4}$

FIGURE 3.24

EXAMPLE 10 Evaluate

a] $\cos(\pi/2 + \text{Sin}^{-1}(2/5))$ **b]** $\tan(\pi - \text{Cos}^{-1}(3/7))$.

Give answers in exact form.

Solution

a] Let $\theta = \text{Sin}^{-1}(2/5)$ and so $\sin\theta = 2/5$. We want to evaluate $\cos(\pi/2 + \theta)$. Using the reduction formula $\cos(\pi/2 + \theta) = -\sin\theta$, where $\sin\theta = 2/5$, gives $\cos(\pi/2 + \text{Sin}^{-1}(2/5)) = -2/5$.

b] Let $\alpha = \text{Cos}^{-1}(3/7)$. Then $\cos\alpha = 3/7$, and $0 \le \alpha \le \pi$. Angle α is shown in Fig. 3.25. Here we can use the reduction formula $\tan(\pi - \alpha) = -\tan\alpha$. From Fig. 3.25, $\tan\alpha = \sqrt{40}/3$. Therefore

$$\tan\left(\pi - \text{Cos}^{-1}\frac{3}{7}\right) = -\frac{\sqrt{40}}{3} = -\frac{2\sqrt{10}}{3}. \quad ∎$$

FIGURE 3.25

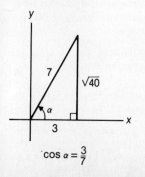

$\cos \alpha = \frac{3}{7}$

EXAMPLE 11 Determine the value of x that will satisfy

a] $\text{Sin}^{-1}x = 1.2$ **b]** $\text{Cos}^{-1}x = -1.4$

Express answers rounded off to two decimal places.

Solution

a] First we note that 1.2 is a number between $-\pi/2$ and $\pi/2$, and so there is a number x satisfying the given equation. Take sine of both sides and then apply Eq. (3.10) to get

$$\sin(\text{Sin}^{-1}x) = \sin 1.2 \qquad x = \sin 1.2$$

Now with the calculator in radian mode, evaluate $\sin 1.2$, $x \approx 0.93$.

b] Since $0 \le \text{Cos}^{-1} x \le \pi$ for every x in $-1 \le x \le 1$ and -1.4 is not between 0 and π, there is no value of x that will satisfy the given equation. We say that the given equation has no solution. ∎

EXERCISE 3.4

There may be some problems in which the given expression is undefined. If a calculator is used, the display will indicate ERROR. Explain what part of the problem causes such a response.

In Problems 1–12, evaluate the given expressions and state answers in exact form (involving π where necessary).

1. $\text{Cos}^{-1} 1$ **2.** $\text{Sin}^{-1} \dfrac{\sqrt{2}}{2}$ **3.** $\text{Cos}^{-1}\left(-\dfrac{1}{\sqrt{2}}\right)$

4. $\text{Sin}^{-1}\left(-\dfrac{1}{2}\right)$ **5.** $\text{Cos}^{-1}\left(\dfrac{\sqrt{3}}{2}\right)$ **6.** $\text{Sin}^{-1}\left(-\dfrac{\sqrt{2}}{2}\right)$

7. $\text{Sin}^{-1}\left(-\dfrac{3}{2\sqrt{3}}\right)$ **8.** $\text{Arccos } 0$ **9.** $\text{Sin}^{-1}(-1)$

10. $\text{Sin}^{-1}\left(\dfrac{2}{\sqrt{3}}\right)$ **11.** $\text{Arcsin}\left(-\dfrac{2}{\sqrt{2}}\right)$ **12.** $\text{Arccos}\left(\dfrac{1}{2}\right)$

In Problems 13–24, evaluate the given expressions and give answers rounded off to two decimal places.

13. $\text{Sin}^{-1} 0.376$ **14.** $\text{Cos}^{-1} 0.573$ **15.** $\text{Arccos } 0.53$

16. $\text{Arcsin } 2.37$ **17.** $\text{Cos}^{-1}(-0.431)$ **18.** $\text{Arcsin}\left(\dfrac{\sqrt{5}}{8}\right)$

19. $\text{Cos}^{-1}(1 - \sqrt{3})$ **20.** $\text{Cos}^{-1}\left(-\dfrac{\pi}{4}\right)$ **21.** $\text{Sin}^{-1}\left(\dfrac{\pi}{6}\right)$

22. $\text{Cos}^{-1}(\sin 1.5)$ **23.** $\text{Sin}^{-1}(\cos 140°)$ **24.** $\text{Sin}^{-1}\left(\cos \dfrac{2}{\pi}\right)$

25. a] Construct a table of x-, y-values satisfying $y = \text{Sin}^{-1} x$ for values of x: $-1.0, -0.8, -0.6, \ldots, 0.8, 1.0$. Round off y-values to two decimal places.

    b] Plot the (x, y) points corresponding to the values given in your table and draw a graph of $y = \text{Sin}^{-1} x$. Compare with the graph shown in Fig. 3.16.

26. Follow a procedure similar to that in Problem 25 and draw a graph of $y = \text{Cos}^{-1} x$. Compare your graph with that given in Fig. 3.21.

In Problems 27–30, for the function described by the given equation, (a) state the domain, (b) draw a graph of the function, and (c) use the graph to determine the range of the function.

27. $y = \sin(\text{Sin}^{-1} x)$ **28.** $y = \text{Sin}^{-1}(\sin x)$

29. $y = \text{Cos}^{-1}(\cos x)$ **30.** $y = \cos(\text{Cos}^{-1} x)$

In Problems 31–40, evaluate the given expressions in exact form.

31. $\sin\left(\text{Sin}^{-1}\frac{2}{7}\right)$ **32.** $\cos\left(\text{Cos}^{-1}0.37\right)$

33. $\text{Cos}^{-1}\left(\cos\left(-\frac{\pi}{3}\right)\right)$ **34.** $\sin\left(\text{Cos}^{-1}\frac{1}{2}\right)$

35. $\tan\left(\text{Sin}^{-1}\left(-\frac{1}{2}\right)\right)$ **36.** $\sec\left(\text{Cos}^{-1}\frac{\sqrt{3}}{2}\right)$

37. $\sin\left(\text{Sin}^{-1}\frac{1}{2}+\text{Cos}^{-1}\frac{1}{2}\right)$ **38.** $\sin\left(2\,\text{Sin}^{-1}0.5\right)$

39. $\tan\left(\text{Sin}^{-1}\frac{1}{\sqrt{2}}+\text{Cos}^{-1}0\right)$ **40.** $\text{Cos}^{-1}\left(\tan 135°\right)$

In Problems 41–46, evaluate the given expressions and give answers in exact form. You may wish to use reduction formulas; see Table 3.5 in the chapter summary.

41. $\sin\left(\frac{\pi}{2}-\text{Cos}^{-1}\frac{3}{7}\right)$ **42.** $\tan\left(\pi+\text{Sin}^{-1}\frac{3}{4}\right)$

43. $\cos\left(\frac{3\pi}{2}-\text{Sin}^{-1}\frac{2}{7}\right)$ **44.** $\sin\left(\frac{\pi}{2}+\text{Sin}^{-1}\frac{3}{4}\right)$

45. $\sin\left(-\text{Cos}^{-1}\frac{4}{5}\right)$ **46.** $\cos\left(-\text{Cos}^{-1}\left(-\frac{2}{3}\right)\right)$

In Problems 47–56, evaluate the given expressions. Give results rounded off to three decimal places.

47. $\sin\left(\text{Cos}^{-1}0.41\right)$ **48.** $\tan\left(\text{Cos}^{-1}0.53\right)$

49. $\sin\left(\text{Sin}^{-1}0.4+\text{Cos}^{-1}0.6\right)$ **50.** $\sin\left(\text{Sin}^{-1}0.4\right)+\sin\left(\text{Cos}^{-1}0.6\right)$

51. $\text{Cos}^{-1}\left(\sin 1.5+\cos 0.3\right)$ **52.** $\sec\left(\text{Sin}^{-1}(-0.75)\right)$

53. $\sin\left(\text{Cos}^{-1}(-0.3)\right)$ **54.** $\tan\left(\text{Sin}^{-1}\frac{\pi}{6}\right)$

55. $\sin\left(\text{Sin}^{-1}0.3+\text{Sin}^{-1}0.7\right)$ **56.** $\text{Sin}^{-1}\left(\tan 20°+\sin 32°\right)$

In Problems 57–61, determine the values of x (if any) that will satisfy the given equations. Express answers rounded off to two decimal places.

57. $\text{Sin}^{-1}x=-0.48$ **58.** $\text{Cos}^{-1}x=1.86$

59. $\text{Cos}^{-1}x=-0.25$ **60.** $\text{Sin}^{-1}x=\frac{\pi}{4}$

61. $\text{Cos}^{-1}x=\frac{5\pi}{6}$

62. Prove that $\text{Sin}^{-1}x+\text{Cos}^{-1}x=\pi/2$ for every x in $-1\le x\le 1$.

63. Prove that $\text{Sin}^{-1}(-x)=-\text{Sin}^{-1}x$ for every x in $-1\le x\le 1$. [Hint: Take sin of both sides and use Eq. (3.10).]

64. Prove that $\text{Cos}^{-1}(-x)=\pi-\text{Cos}^{-1}x$ for every x in $-1\le x\le 1$.

65. Prove that $\text{Sin}^{-1}(-\sin x)=-x$ for every x in $-\pi/2\le x\le \pi/2$. (Hint: Use Problem 63 and Eq. (3.11).)

3.5 Inverse Tangent Function

The graph of the tangent function was discussed in Section 3.2 and is reproduced in Fig. 3.26. From the graph we see that the inverse is not a function. For instance, if $y = 1$, then there are infinitely many values of x to which it corresponds (such as $\pi/4, 5\pi/4, -3\pi/4, \dots$).

Suppose we restrict the domain of the tangent function to that corresponding to the solid portion of the curve $(-\pi/2 < x < \pi/2)$. We get a new function, which we denote by Tan (note the capital letter T), which is defined by

$$\text{Tan}\, x = \tan x \qquad \text{where Domain of Tan} = D(\text{Tan}) = \left\{ x \mid -\frac{\pi}{2} < x < \frac{\pi}{2} \right\}.$$

From the graph in Fig. 3.26 (the solid curve) we see that the inverse of the Tan function is also a function with domain the set of real numbers and range $\{y \mid -\pi/2 < y < \pi/2\}$. This leads to the following definition.

Definition of the inverse tangent function

The function denoted by Tan^{-1} or by Arctan is defined by

$$\boxed{\begin{array}{l} y = \text{Tan}^{-1} x \quad \text{or} \quad y = \text{Arctan}\, x \quad \text{for any } x \text{ in } \mathbb{R} \\[2mm] \qquad \text{means} \quad \tan y = x \quad and \quad -\frac{\pi}{2} < y < \frac{\pi}{2}. \end{array}}$$

[3.16]

FIGURE 3.26

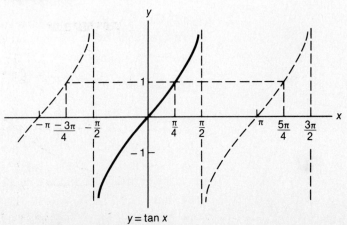

$y = \tan x$

In terms of ordered pairs we can say the same thing as follows

$$\text{Tan}^{-1} = \left\{ (x, y) \mid x = \tan y \quad \text{and} \quad -\frac{\pi}{2} < y < \frac{\pi}{2} \right\}. \qquad [3.17]$$

Graph of the inverse tangent function

We can draw the graph of $y = \text{Tan}^{-1} x$ by simply reflecting the graph of $y = \text{Tan} x$ (the solid curve in Fig. 3.26) about the line $y = x$. It may be helpful to locate a few points. For instance the following points are on the graph of $y = \text{Tan} x$.

$$\left(-\frac{\pi}{3}, -\sqrt{3} \right), \quad \left(-\frac{\pi}{4}, -1 \right), \quad (0, 0), \quad \left(\frac{\pi}{4}, 1 \right), \quad \left(\frac{\pi}{3}, \sqrt{3} \right).$$

The corresponding points on the graph of $y = \text{Tan}^{-1} x$ are

$$\left(-\sqrt{3}, -\frac{\pi}{3} \right), \quad \left(-1, -\frac{\pi}{4} \right), \quad (0, 0), \quad \left(1, \frac{\pi}{4} \right), \quad \left(\sqrt{3}, \frac{\pi}{3} \right).$$

Now plot these points and draw the curve shown in Fig. 3.27.

One reason the inverse tangent function $y = \text{Tan}^{-1} x$ is used more than the inverse sine or cosine is that the *domain*, as noted above and as shown by the graph, *is all real numbers*, whereas the other functions are only defined for $-1 \le x \le 1$. We mention again that higher-level *computer languages include the arctan function (ATN) but not the arcsine or arccosine functions.* See the optional Computer Problems at the end of this chapter.

EXAMPLE 1 Evaluate

   a] $\text{Tan}^{-1} 1$ **b]** $\text{Tan}^{-1}(-\sqrt{3})$.

Give answers in exact form.

FIGURE 3.27

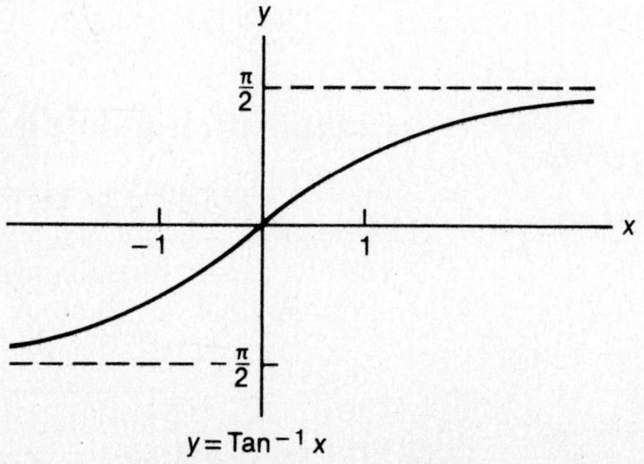

$y = \text{Tan}^{-1} x$

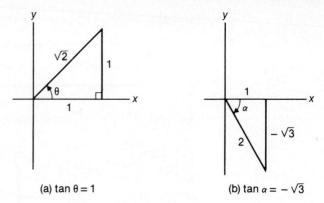

(a) tan θ = 1   (b) tan α = −√3

FIGURE 3.28

Solution

a] It is helpful to think of $\text{Tan}^{-1}1$ as an angle measured in radians (according to the definition in Eq. (3.16)). Thus let $\theta = \text{Tan}^{-1}1$, and so $\tan\theta = 1$ and $-\pi/2 < \theta < \pi/2$. Angle θ is shown in Fig. 3.28(a), and we see that it is $\pi/4$ radians. Therefore $\text{Tan}^{-1}1 = \pi/4$.

b] Let $\alpha = \text{Tan}^{-1}(-\sqrt{3})$. Then $\tan\alpha = -\sqrt{3}$ and $-\pi/2 < \alpha < \pi/2$. Angle α is shown in Fig. 3.28(b) as a fourth-quadrant angle. From the reference triangle for α we see that $\alpha = -60°$, or in radians $\alpha = -\pi/3$. Note that α must be between $-\pi/2$ and $\pi/2$. There is only one angle α satisfying this and $\tan\alpha = -\sqrt{3}$, namely $-\pi/3$.

> *Warning: Do not use the coterminal angle $5\pi/3$ because it is not between $-\pi/2$ and $\pi/2$.*

Thus $\text{Tan}^{-1}(-\sqrt{3}) = -\pi/3$. ∎

Composition identities _____

As was suggested in Example 1, it is helpful to think of $\text{Tan}^{-1}x$ as an angle measured in radians. The definition in Eq. (3.16) says that *$\text{Tan}^{-1}x$ is that angle between $-\pi/2$ and $\pi/2$ whose tangent is x.* In mathematical terminology this translates into

$$\tan(\text{Tan}^{-1}x) = x \quad \text{for every } x \text{ in } \mathbf{R}. \qquad [3.18]$$

Similarly,

$$\text{Tan}^{-1}(\tan x) = x \quad \text{for every } x \text{ in} \quad -\frac{\pi}{2} < x < \frac{\pi}{2}.^* \qquad \text{[3.19]}$$

EXAMPLE 2 Evaluate

a] $\tan(\text{Tan}^{-1}(5/17))$ b] $\cos(\text{Tan}^{-1}(-3/5))$.

Give answers in exact form and also as a decimal approximation rounded off to three places.

Solution

a] Apply Eq. (3.18) to get $\tan(\text{Tan}^{-1}(5/17)) = 5/17$ (exact form). Express $5/17$ as a decimal to get $\tan(\text{Tan}^{-1}(5/17)) \approx 0.294$.

b] Let $\theta = \text{Tan}^{-1}(-3/5)$. Then $\tan\theta = -3/5$, and $-\pi/2 < \theta < \pi/2$. Angle θ is shown in Fig. 3.29, and from the reference triangle we see that $\cos\theta = 5/\sqrt{34}$. Hence $\cos(\text{Tan}^{-1}(-3/5)) = 5/\sqrt{34}$ (exact form). Now evaluate $5/\sqrt{34}$ to get $\cos(\text{Tan}^{-1}(-3/5)) \approx 0.857$. ∎

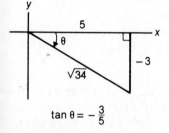

$$\tan\theta = -\frac{3}{5}$$

FIGURE 3.29

Evaluating inverse functions by calculator

In Example 1 we were able to evaluate $\text{Tan}^{-1}x$ in exact form. This is possible for special values of x, such as $0, 1/\sqrt{3}, 1,$ or $\sqrt{3}$, where the corresponding special angles are $0, \pi/6, \pi/4, \pi/3$. If we wish to evaluate say $\text{Tan}^{-1}1.24$, we can rely on a calculator or computer to give the result to several decimal places. Place the calculator in radian mode, then enter 1.24 and press (Tan⁻¹) or (INV), (Tan) keys. This gives $\text{Tan}^{-1}1.24 \approx 0.892\,1338$ (to seven decimal places). Hand-held computers have a function labeled ATN; in radian mode, press (ATN) 1.24, (EXE) to get $\text{Tan}^{-1}1.24 \approx 0.8921338361$.

EXAMPLE 3 Evaluate

a] $\text{Tan}^{-1}(0.876)$ b] $\text{Arctan}(-\sqrt{15}/3)$.

Give answers rounded off to three decimal places.

Solution   First place your calculator in *radian mode*.

a] After entering 0.876 into the display, press (Tan⁻¹) or (INV), (Tan) to get $\text{Tan}^{-1}(0.876) \approx 0.719$. With a hand-held computer (in radian mode), press (ATN) .876 (EXE).

---

* Note that x is restricted to the interval between $-\pi/2$ and $\pi/2$ because x is equal to $\text{Tan}^{-1}(\tan x)$ and Tan^{-1} evaluated at any number must be between $-\pi/2$ and $\pi/2$. Also note that $\text{Tan}^{-1}(\tan x)$ is not equal to x for every number x. Equality holds only when x is between $-\pi/2$ and $\pi/2$. For instance, if $x = 5\pi/4$, then $\text{Tan}^{-1}(\tan 5\pi/4) = \text{Tan}^{-1}(1) = \pi/4$. Thus $\text{Tan}^{-1}(\tan 5\pi/4) = \pi/4$ and not $5\pi/4$.

b] First evaluate $-\sqrt{15}/3$ and, with the result in the display, press
Ⓣⓐⓝ⁻¹ or ⒤ⓝⓥ, Ⓣⓐⓝ to get Arctan$(-\sqrt{15}/3) \approx -0.912$. With a
hand-held computer, press Ⓐⓣⓝ $(-$ Ⓢⓠⓡ $15/3)$ Ⓔⓧⓔ.   ■

EXAMPLE 4   Evaluate

a] $\cos(\mathrm{Sin}^{-1}0.4 + \mathrm{Tan}^{-1}3.54)$ **b]** $\mathrm{Tan}^{-1}(\cos 47°)$.

Express answers rounded off to three decimal places.

Solution

a] Place your calculator in radian mode. First evaluate
$\mathrm{Sin}^{-1}0.4 + \mathrm{Tan}^{-1}3.54$ (you should see 1.7070008, but do
not record this number), then press ⒸⓄⓈ. This gives
$\mathrm{Cos}(\mathrm{Sin}^{-1}0.4 + \mathrm{Tan}^{-1}3.54) \approx -0.136$.

b] Place your calculator in degree mode and press 47, ⒸⓄⓈ; then
in radian mode press Ⓣⓐⓝ⁻¹ or ⒤ⓝⓥ, Ⓣⓐⓝ to get $\mathrm{Tan}^{-1}(\cos 47°) \approx$
0.599.   ■

EXAMPLE 5   Evaluate $\cos(\pi/2 + \mathrm{Tan}^{-1}(5/7))$ in exact form.

Solution   Let $\theta = \mathrm{Tan}^{-1}(5/7)$, then $\tan \theta = 5/7$, and $-\pi/2 < \theta < \pi/2$.
Angle θ is shown in Fig. 3.30. Now we want to evaluate $\cos(\pi/2 + \theta)$.
We can use the reduction formula $\cos(\pi/2 + \theta) = -\sin\theta$, and from
the reference triangle in Fig. 3.30 we see that $\sin\theta = 5/\sqrt{74}$. Therefore $\cos(\pi/2 + \mathrm{Tan}^{-1}(5/7)) = -5/\sqrt{74}$.   ■

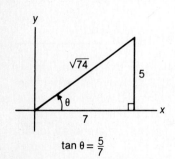

$$\tan\theta = \frac{5}{7}$$

FIGURE 3.30

EXAMPLE 6   Prove that $\mathrm{Sin}^{-1}x = \mathrm{Tan}^{-1}(x/\sqrt{1-x^2})$ for every x in
$-1 < x < 1$.*

Solution   Let $\theta = \mathrm{Sin}^{-1}x$. Then $\sin\theta = x$ and $-\pi/2 < \theta < \pi/2$ for
$-1 < x < 1$. First suppose that $0 < x < 1$; then angle θ is in the first
quadrant as shown in Fig. 3.31. From the reference triangle for θ we
see that $\tan\theta = x/\sqrt{1-x^2}$, and so $\mathrm{Tan}^{-1}(\tan\theta) = \mathrm{Tan}^{-1}(x/\sqrt{1-x^2})$.
Now apply Eq. (3.19) to the left-hand side to get $\theta = \mathrm{Tan}^{-1}(x/\sqrt{1-x^2})$. Therefore $\mathrm{Sin}^{-1}x = \mathrm{Tan}^{-1}(x/\sqrt{1-x^2})$ for $0 < x < 1$.

FIGURE 3.31

For $-1 < x < 0$ we get a similar situation except that θ is an
angle in the fourth quadrant. We leave details to the reader. For $x = 0$,
both sides of the given equation are equal to zero. Thus $\mathrm{Sin}^{-1}x = \mathrm{Tan}^{-1}(x/\sqrt{1-x^2})$ for every $-1 < x < 1$.   ■

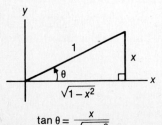

$$\tan\theta = \frac{x}{\sqrt{1-x^2}}$$

EXAMPLE 7   Determine the value of x that will satisfy

a] $\mathrm{Tan}^{-1}x = \dfrac{\pi}{3}$ **b]** $\mathrm{Tan}^{-1}x = -\dfrac{2\pi}{3}$.

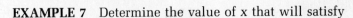

* The identity stated here and the one given in Problem 32 are useful in evaluating $\mathrm{Sin}^{-1}x$
and $\mathrm{Cos}^{-1}x$ using BASIC.

Solution $\mathrm{Tan}^{-1}x$ is an angle between $-\pi/2$ and $\pi/2$. Since $\pi/3$ is such an angle but $-2\pi/3$ is not (that is, $-2\pi/3$ is not in the range of the Tan^{-1} function), then there is a number satisfying the equation in (a) but no solution for that in (b).

a] Taking the tangent of both sides of the given equation and then using the result given in Eq. (3.18) we get

$$\tan\left(\mathrm{Tan}^{-1}x\right) = \tan\frac{\pi}{3}, \qquad x = \tan\frac{\pi}{3} = \sqrt{3}.$$

Therefore $x = \sqrt{3}$.

b] Warning: If we blindly take the tangent of both sides, we get $\tan\left(\mathrm{Tan}^{-1}x\right) = \tan\left(-2\pi/3\right) = \sqrt{3}$. This suggests that $x = \sqrt{3}$. If we check this "solution" in the given equation, we get $\mathrm{Tan}^{-1}\sqrt{3} = \pi/3$ and not $-2\pi/3$. Thus $\sqrt{3}$ is not a solution. ■

EXERCISE 3.5

In Problems 1–6, evaluate the given expressions in exact form.

1. $\mathrm{Tan}^{-1}(-1)$

2. $\mathrm{Tan}^{-1}\sqrt{3}$

3. $\mathrm{Arctan}\,(1/\sqrt{3})$

4. $\mathrm{Arctan}\,(-\sqrt{3}/3)$

5. $\mathrm{Tan}^{-1}1 - \mathrm{Tan}^{-1}(-1)$

6. $\mathrm{Tan}^{-1}(-\sqrt{3}) - \mathrm{Tan}^{-1}(-1/\sqrt{3})$

In Problems 7–12, evaluate the given expressions and give answers rounded off to three decimal places. Your calculator should be in radian mode.

7. $\mathrm{Tan}^{-1}0.738$

8. $\mathrm{Arctan}\,(-1.48)$

9. $\mathrm{Tan}^{-1}\sqrt{5}$

10. $\mathrm{Tan}^{-1}(\pi/2)$

11. $\mathrm{Tan}^{-1}0.3 + \mathrm{Tan}^{-1}2.4$

12. $\mathrm{Arctan}\,(-2\pi/3)$

In Problems 13–22, evaluate the given expressions and give answers in exact form. In Problems 16–21, first use reduction formulas (see Table 3.5 in the Chapter Summary).

13. $\tan\left(\mathrm{Tan}^{-1}\dfrac{4}{3}\right)$

14. $\mathrm{Tan}^{-1}\left(\tan\dfrac{3\pi}{4}\right)$

15. $\cot\left(\mathrm{Tan}^{-1}\dfrac{-5}{7}\right)$

16. $\sin\left(\dfrac{\pi}{2} - \mathrm{Tan}^{-1}\dfrac{4}{3}\right)$

17. $\cos\left(\pi + \mathrm{Tan}^{-1}\dfrac{4}{3}\right)$

18. $\tan\left(\dfrac{\pi}{2} + \mathrm{Tan}^{-1}\dfrac{3}{4}\right)$

19. $\cos\left(\dfrac{3\pi}{2} + \mathrm{Tan}^{-1}\dfrac{5}{12}\right)$

20. $\sin\left(\pi - \mathrm{Tan}^{-1}\dfrac{1}{4}\right)$

21. $\tan\left(\pi + \mathrm{Tan}^{-1}\dfrac{4}{7}\right)$

22. $\mathrm{Tan}^{-1}\left(\cot\dfrac{5\pi}{6}\right)$

23. Sketch a graph of $y = \text{Tan}^{-1} x$ by constructing a table of x, y-values, plotting the corresponding points, and drawing a smooth curve through them. Check to see whether your graph agrees with that shown in Fig. 3.27.

In Problems 24–29, determine the values of x (if any) that will satisfy the given equations. Express answers rounded off to two decimal places.

24. $\text{Tan}^{-1} x = \dfrac{\pi}{6}$ **25.** $\text{Tan}^{-1} x = 2$

26. $\text{Tan}^{-1} x = \tan 0.5$ **27.** $\text{Tan}^{-1} x = -1.3$

28. $\text{Tan}^{-1} x = \cos \dfrac{\pi}{3}$ **29.** $\text{Tan}^{-1} x = \dfrac{2\pi}{3}$

30. Prove that $\text{Tan}^{-1} x + \text{Tan}^{-1} 1/x = \pi/2$ for every $x > 0$. (Hint: Look at the reference triangle for $\theta = \text{Tan}^{-1} x$.)

31. Use the result in Problem 30 to show that $\text{Tan}^{-1} 3/4 + \text{Tan}^{-1} 4/3 = \text{Sin}^{-1} 1$.

32. Use Problem 62 of Exercise 3.4 and Example 6 of Section 3.5 to prove that

$$\text{Cos}^{-1} x = \frac{\pi}{2} - \text{Tan}^{-1} \frac{x}{\sqrt{1 - x^2}} \quad \text{for} \quad -1 < x < 1.$$

3.6 Inverse Cotangent, Secant, and Cosecant Functions (Optional)

The inverse cotangent, secant, and cosecant functions do not occur in applications nearly as often as the other inverse trigonometric functions, and so the material of this section is labeled "optional." Each of these functions can be introduced in a manner analogous to what has been done in the preceding two sections with reference to the graphs of $y = \cot x$, $y = \sec x$, and $y = \csc x$ as shown in Section 3.2. Here we include the essential results and leave details to the interested reader.

Definition of the inverse cotangent function _____

The function denoted by Cot^{-1} or by Arccot is defined by

$$\boxed{\begin{array}{l} y = \text{Cot}^{-1} x \quad \text{for any real number } x \text{ means} \\ \qquad \cot y = x \quad \text{and} \quad 0 < y < \pi. \end{array}} \quad \textbf{[3.20]}$$

The inequality $0 < y < \pi$ tells us that $\text{Cot}^{-1}x$ is an angle in radians in the first or second quadrant.

Graph of the inverse cotangent function

Let $y = \text{Cot}^{-1}x$. The domain is IR, and the range is $\{y \mid 0 < y < \pi\}$. The graph is shown in Fig. 3.32.

EXAMPLE 1   Evaluate $\text{Cot}^{-1}(-\sqrt{3})$ in exact form.

Solution   Let $\theta = \text{Cot}^{-1}(-\sqrt{3})$. Then from the definition given in Eq. (3.20) we have $\cot \theta = -\sqrt{3}$ and $0 < \theta < \pi$. Thus θ is an angle in the second quadrant, as shown in Fig. 3.33. Note that the reference angle for θ is 30°, and so θ is 150° or in radians $\theta = 5\pi/6$. Therefore $\text{Cot}^{-1}(-\sqrt{3}) = 5\pi/6$.   ∎

EXAMPLE 2   Evaluate $\text{Cot}^{-1}(-0.56)$ and give your answer rounded off to three decimal places.

Solution   Your calculator is not preprogrammed to evaluate the Cot^{-1} function directly (it has no Cot⁻¹ or Cot key). Therefore let

$$\theta = \text{Cot}^{-1}(-0.56),$$

and so

$$\cot \theta = -0.56 \quad \text{and} \quad 0 < \theta < \pi.$$

Angle θ is shown in Fig. 3.34(a). If we designate the reference angle for θ by α, then α is a first quadrant angle whose tangent is $\tan \alpha = 1/0.56$ as shown in Fig. 3.34(b). We may evaluate $\alpha = \text{Tan}^{-1}1/0.56$ by calculator. The angle θ is then $\pi - \alpha$:

$$\text{Cot}^{-1}(-0.56) = \theta = \pi - \alpha = \pi - \text{Tan}^{-1}\left(\frac{1}{0.56}\right) \approx 2.081.   ∎$$

FIGURE 3.33

FIGURE 3.32

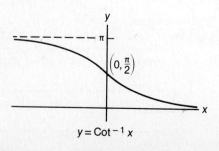

$y = \text{Cot}^{-1}x$

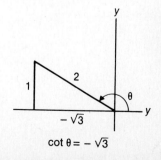

$\cot \theta = -\sqrt{3}$

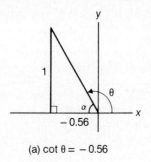

(a) $\cot \theta = -0.56$

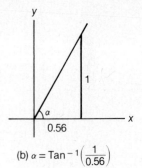

(b) $\alpha = \text{Tan}^{-1}\left(\dfrac{1}{0.56}\right)$

FIGURE 3.34

Definition of inverse secant function* _____

The function denoted by Sec^{-1} or by Arcsec is defined by

$$\boxed{\begin{array}{l} y = \text{Sec}^{-1}x \quad \text{for any } x \le -1 \text{ or } x \ge 1 \quad \text{means} \\[4pt] \sec y = x \quad \text{and} \quad 0 \le y \le \pi, \qquad y \ne \dfrac{\pi}{2}. \end{array}}$$

[3.21]

Note that in place of $x \le -1$ or $x \ge 1$, we could have written $|x| \ge 1$.

Graph of inverse secant function _____

Let $y = \text{Sec}^{-1}x$. The definition stated in Eq. (3.21) gives $\sec y = x$, where for $x \ge 1$, $0 \le y < \pi/2$ and for $x \le -1$, $\pi/2 < y \le \pi$. A few points on the graph are

$$(-1, \pi), \quad \left(-2, \frac{2\pi}{3}\right), \quad (1, 0), \quad \left(2, \frac{\pi}{3}\right).$$

Plotting these and using the fact that the graph is a reflection about the line $y = x$ of the appropriate part of the $y = \sec x$ curve (see Fig. 3.11), we get the curve shown in Fig. 3.35.

Definition of inverse cosecant function _____

The function denoted by Csc^{-1} or by Arccsc is defined by

$$\boxed{\begin{array}{l} y = \text{Csc}^{-1}x \quad \text{for } x \le -1 \text{ or } x \ge 1 \quad \text{means} \\[4pt] x = \csc y \quad \text{and} \quad -\dfrac{\pi}{2} \le y \le \dfrac{\pi}{2}, \qquad y \ne 0. \end{array}}$$

[3.22]

_____

* The definition given here is consistent with that given in most calculus books. However, some textbooks use a slightly different definition.

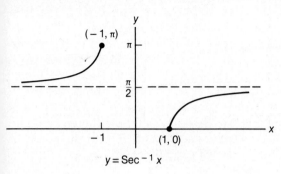

FIGURE 3.35

FIGURE 3.36

Graph of the inverse cosecant function _____

Let $y = \mathrm{Csc}^{-1}x$. From the definition given in Eq. (3.22) we get $x = \csc y$, where for $x \le -1$, $-\pi/2 \le y < 0$ and for $x \ge 1$, $0 < y \le \pi/2$. A few points on the graph are

$$\left(-1, -\frac{\pi}{2}\right), \quad \left(-2, -\frac{\pi}{6}\right), \quad \left(1, \frac{\pi}{2}\right), \quad \left(2, \frac{\pi}{6}\right).$$

Plotting these and using the fact that the graph is a reflection about the line $y = x$ of the appropriate part of the $y = \csc x$ curve (see Fig. 3.11), we get the curve shown in Fig. 3.36.

EXAMPLE 3 Evaluate in exact form

  a] $\mathrm{Sec}^{-1}(-\sqrt{2})$ **b]** $\mathrm{Csc}^{-1}(2/\sqrt{3})$.

FIGURE 3.37

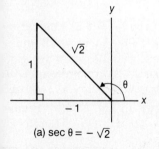

(a) $\sec \theta = -\sqrt{2}$

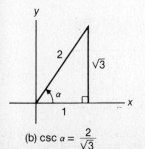

(b) $\csc \alpha = \dfrac{2}{\sqrt{3}}$

Solution

  a] Let $\theta = \mathrm{Sec}^{-1}(-\sqrt{2})$; then $\sec \theta = -\sqrt{2}$ and $0 \le \theta \le \pi$. Thus angle θ is in the second quadrant as shown in Fig. 3.37(a). From the reference triangle we see that $\theta = 3\pi/4$, and so $\mathrm{Sec}^{-1}(-\sqrt{2}) = 3\pi/4$.

  b] Let $\alpha = \mathrm{Csc}^{-1}(2/\sqrt{3})$. Then $\csc \alpha = 2/\sqrt{3}$ and $-\pi/2 \le \alpha \le \pi/2$. Thus angle α is in the first quadrant, as shown in Fig. 3.37(b). From the reference triangle we conclude that $\alpha = \pi/3$, and so $\mathrm{Csc}^{-1}(2/\sqrt{3}) = \pi/3$. ∎

EXAMPLE 4 Evaluate and give answers rounded off to three decimal places.

  a] Arcsec 3.48   **b]** $\mathrm{Csc}^{-1}(-1.53)$

Solution Since calculators are not preprogrammed to evaluate inverse secant and cosecant functions, we formulate our problems to eventually involve the Cos^{-1} and Sin^{-1} functions.

a] Let $\theta = \text{Arcsec}\,3.48$. Then $\sec\theta = 3.48$, and θ is in the first quadrant. Replace $\sec\theta$ by $1/(\cos\theta)$ to get $1/(\cos\theta) = 3.48$, or $\cos\theta = 1/3.48$. Therefore $\theta = \text{Cos}^{-1}1/3.48$. First evaluate $1/3.48$ and with the result in the display, press ⃝Cos⁻¹ or ⃝INV, ⃝cos to get $\theta \approx 1.279$. Hence $\text{Arcsec}\,3.48 \approx 1.279$.

b] Let $\alpha = \text{Csc}^{-1}(-1.53)$. Then $\csc\alpha = -1.53$, and $-\pi/2 \leq \alpha < 0$. Replace $\csc\alpha$ by $1/\sin\alpha$ to get $1/\sin\alpha = -1.53$, or $\sin\alpha = -1/1.53$. Therefore $\alpha = \text{Sin}^{-1}(-1/1.53)$; this will give an angle in $-\pi/2 \leq \alpha < 0$. First evaluate $-1/1.53$ and with the result in the display, press ⃝Sin⁻¹ or ⃝INV, ⃝Sin. This gives $\alpha \approx -0.712$. Therefore $\text{Csc}^{-1}(-1.53) \approx -0.712$. ∎

The solution in Example 4 suggests the following identities, which can be applied when evaluating the inverse secant or inverse cosecant functions by calculator.*

$$\text{Sec}^{-1}x = \text{Cos}^{-1}\frac{1}{x},$$
$$\text{Csc}^{-1}x = \text{Sin}^{-1}\frac{1}{x}.$$

[3.23]

EXAMPLE 5 Evaluate in exact form

a] $\tan(\text{Sec}^{-1}(-2))$ **b]** $\cos\left(\text{Csc}^{-1}\frac{5}{3}\right)$.

Solution

a] Let $\theta = \text{Sec}^{-1}(-2)$. Then $\sec\theta = -2$, and $\pi/2 < \theta \leq \pi$. Angle θ is in the second quadrant as shown in Fig. 3.38(a). Our problem is to determine $\tan\theta$. From the reference triangle for θ we get $\tan\theta = -\sqrt{3}$. Therefore $\tan(\text{Sec}^{-1}(-2)) = -\sqrt{3}$.

b] Let $\alpha = \text{Csc}^{-1}(5/3)$. Then $\csc\alpha = 5/3$, and $-\pi/2 \leq \alpha \leq \pi/2$. Thus angle α is in the first quadrant, as shown in Fig. 3.38(b). From the reference triangle for α we get $\cos\alpha = 4/5$. Therefore $\cos(\text{Csc}^{-1}(5/3)) = 4/5$. ∎

EXAMPLE 6 Find the value of x that satisfies the given equation.

a] $\text{Sec}^{-1}x = \frac{2\pi}{3}$ **b]** $\text{Csc}^{-1}x = \frac{2\pi}{3}$

---

* The formulas given in Eq. (3.23) are suitable for calculator evaluation. However, in using computer BASIC, see identities given in Problems 48 and 49 of Exercise 3.6.

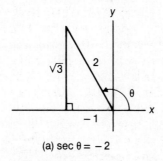

(a) $\sec \theta = -2$ (b) $\csc \alpha = \dfrac{5}{3}$

FIGURE 3.38

Solution

a] From the definition given in Eq. (3.21), $\text{Sec}^{-1} x$ is between 0 and π. Since $2\pi/3$ is between 0 and π, $\text{Sec}^{-1} x = 2\pi/3$ implies that $x = \sec 2\pi/3$. But $\sec 2\pi/3 = -2$, and so $x = -2$.

b] From the definition given in Eq. (3.22) $\text{Csc}^{-1} x$ must be between $-\pi/2$ and $\pi/2$. Since $2\pi/3$ is not in this interval, $\text{Csc}^{-1} x = 2\pi/3$ has no solution. ∎

EXAMPLE 7 Prove that $\text{Sec}^{-1} x = \pi/2 - \text{Tan}^{-1}(1/\sqrt{x^2 - 1})$ for $x > 1$.*

Solution Let $\theta = \text{Sec}^{-1} x$. Then $\sec \theta = x$, and since $x > 1$, $0 < \theta < \pi/2$. Angle θ is shown in Fig. 3.39. From the reference triangle for θ we see that $\tan \beta = 1/\sqrt{x^2 - 1}$, and so $\beta = \text{Tan}^{-1}(1/\sqrt{x^2 - 1})$. But $\theta = \pi/2 - \beta$, and so we have $\text{Sec}^{-1} x = \pi/2 - \text{Tan}^{-1}(1/\sqrt{x^2 - 1})$. ∎

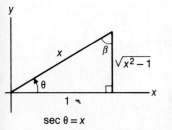

$\sec \theta = x$

FIGURE 3.39

EXERCISE 3.6

There may be some problems in which the given expression is not defined. If a calculator is used, the display will indicate ERROR. Explain what part of the expression causes it to be undefined.

In Problems 1–15, evaluate the given expressions. Give answers in exact form.

1. a] $\text{Cot}^{-1} \sqrt{3}$

  b] $\text{Cot}^{-1}\left(-\dfrac{\sqrt{3}}{3}\right)$

2. a] $\text{Arccot}(-1)$

  b] $\text{Arccot}\left(\dfrac{1}{\sqrt{3}}\right)$

3. a] $\text{Cot}^{-1}(-1) - \text{Cot}^{-1} 1$

  b] $\text{Cot}^{-1} \sqrt{3} + \text{Cot}^{-1}(-\sqrt{3})$

4. a] $\text{Sec}^{-1} 2$

  b] $\text{Csc}^{-1}\left(-\dfrac{2}{\sqrt{3}}\right)$

_____

* This is an example of an identity that is useful for computer BASIC.

5. a] $\operatorname{Sec}^{-1} \sqrt{2}$

  b] $\operatorname{Csc}^{-1}(-\sqrt{2})$

6. a] $\operatorname{Sec}^{-1}\left(-\dfrac{2}{\sqrt{3}}\right)$

  b] $\operatorname{Csc}^{-1} 2$

7. a] $\operatorname{Sec}^{-1}(-1)$

  b] $\operatorname{Csc}^{-1} 1$

8. a] $\operatorname{Sec}^{-1} \dfrac{\pi}{6}$

  b] $\operatorname{Csc}^{-1} \sqrt{2}$

9. a] $\operatorname{Sec}^{-1}\left(\dfrac{2}{\sqrt{3}}\right)$

  b] $\operatorname{Csc}^{-1} \dfrac{\pi}{4}$

10. a] $\sin\left(\operatorname{Csc}^{-1} 5\right)$

  b] $\cos\left(\operatorname{Sec}^{-1}(-3)\right)$

11. a] $\tan\left(\operatorname{Sec}^{-1} \dfrac{5}{4}\right)$

  b] $\sec\left(\operatorname{Csc}^{-1}\left(-\dfrac{3}{2}\right)\right)$

12. a] $\sin\left(\operatorname{Sec}^{-1}(-1)\right)$

  b] $\cos\left(\operatorname{Csc}^{-1}(-1)\right)$

13. a] $\sin\left(2\operatorname{Sec}^{-1} 2\right)$
  b] $\cos\left(2\operatorname{Sec}^{-1} 2\right)$

14. a] $\sec\left(\operatorname{Sec}^{-1} 0.5\right)$
  b] $\csc\left(\operatorname{Sec}^{-1} 2\right)$

15. a] $\sin\left(\operatorname{Sec}^{-1}(-1) + \operatorname{Csc}^{-1} 1\right)$ **b]** $\tan\left(\operatorname{Sin}^{-1} 2 - \operatorname{Csc}^{-1} 2\right)$

In Problems 16–29, evaluate the given expressions and give answers rounded off to three decimal places. Your calculator should be in radian mode.

16. $\operatorname{Cot}^{-1} 0.53$ **17.** $\operatorname{Cot}^{-1}(-2.41)$ **18.** $\operatorname{Arccot} 1.74$

19. $\operatorname{Cot}^{-1} \dfrac{\pi}{2}$ **20.** $\operatorname{Cot}^{-1} \dfrac{\pi}{4}$ **21.** $\operatorname{Cot}^{-1}(-\sqrt{3})$

22. a] $\operatorname{Sec}^{-1} 1.68$ **b]** $\operatorname{Csc}^{-1}(-2.53)$

23. a] $\operatorname{Sec}^{-1}(-2.58)$ **b]** $\operatorname{Csc}^{-1} 4.31$

24. a] $\sin\left(\operatorname{Sec}^{-1} 1.3 + \operatorname{Sec}^{-1} 2.5\right)$ **b]** $\tan\left(-2\operatorname{Sec}^{-1} 3\right)$

25. a] $\operatorname{Sec}^{-1}(\tan 1.3)$ **b]** $\operatorname{Csc}^{-1}(\cot 0.42)$

26. a] $\operatorname{Sec}^{-1}\left(\dfrac{\pi}{2}\right)$ **b]** $\operatorname{Csc}^{-1}\left(\dfrac{3\pi}{2}\right)$

27. a] $\operatorname{Sec}^{-1}(0.84)$ **b]** $\operatorname{Csc}^{-1}\left(-\dfrac{2}{\sqrt{3}}\right)$

28. a] $\tan\left(\operatorname{Sec}^{-1} \sqrt{5}\right)$ **b]** $\cot\left(\operatorname{Csc}^{-1} \sqrt{15}\right)$

29. a] $\sin\left(2\operatorname{Sec}^{-1} 3\right)$ **b]** $\cos\left(2\operatorname{Csc}^{-1} 3\right)$

In Problems 30–38, evaluate the given expressions in exact form. You may wish to apply reduction formulas in Problems 35–38. See Table 3.5 in the summary at the end of this chapter.

30. $\cot\left(\operatorname{Cot}^{-1} \dfrac{3}{7}\right)$ **31.** $\tan\left(\operatorname{Cot}^{-1} \dfrac{2}{7}\right)$ **32.** $\sin\left(\operatorname{Cot}^{-1}\left(-\dfrac{5}{12}\right)\right)$

33. $\cos\left(\operatorname{Cot}^{-1} \dfrac{12}{5}\right)$ **34.** $\sin\left(\operatorname{Cot}^{-1} 1 + \operatorname{Cot}^{-1}(-1)\right)$

35. $\cos\left(\dfrac{\pi}{2} + \operatorname{Cot}^{-1} \dfrac{12}{5}\right)$ **36.** $\sin\left(\dfrac{3\pi}{2} - \operatorname{Cot}^{-1} 3\right)$

37. $\cot\left(\pi + \operatorname{Cot}^{-1} 2\right)$ **38.** $\tan\left(\dfrac{\pi}{2} + \operatorname{Cot}^{-1}(-2.3)\right)$

In Problems 39–43, solve the given equations for x. In some cases there may be no solution. Give answers rounded off to two decimal places.

39. a] $\mathrm{Sec}^{-1}x = 3$ **b]** $\mathrm{Csc}^{-1}x = -2$

40. a] $\mathrm{Sec}^{-1}x = -1.4$ **b]** $\mathrm{Csc}^{-1}x = 1.2$

41. a] $\mathrm{Sec}^{-1}x = \dfrac{\pi}{6}$ **b]** $\mathrm{Sec}^{-1}x = \dfrac{6}{\pi}$

42. a] $\mathrm{Csc}^{-1}x = \dfrac{3\pi}{4}$ **b]** $\mathrm{Csc}^{-1}x = \dfrac{4}{\pi}$

43. a] $\mathrm{Csc}^{-1}x = \tan 1.4$ **b]** $\mathrm{Sec}^{-1}x = \sec\dfrac{\pi}{3}$

44. Prove that $\mathrm{Tan}^{-1}x + \mathrm{Cot}^{-1}x = \pi/2$ for $x > 0$. (Hint: Look at the reference triangle for $\theta = \mathrm{Tan}^{-1}x$. Is this same equality valid for $x < 0$?)

45. Use Problem 44 to evaluate $\mathrm{Cot}^{-1}1.54$ rounded off to three decimal places. Also see Problem 46.

46. a] Prove that $\mathrm{Cot}^{-1}x = \mathrm{Tan}^{-1}1/x$ for $x > 0$. (Hint: Look at the reference triangle for $\theta = \mathrm{Cot}^{-1}x$.)
  b] Use this to calculate $\mathrm{Cot}^{-1}1.54$ rounded off to three decimal places. Also see Problem 45.

47. Prove that $\mathrm{Sec}^{-1}x + \mathrm{Csc}^{-1}x = \pi/2$ for every x for which $|x| \geq 1$.

48. Prove that $\mathrm{Sec}^{-1}x = \mathrm{Tan}^{-1}\sqrt{x^2 - 1}$ for $x \geq 1$.

49. Prove that $\mathrm{Csc}^{-1}x = \mathrm{Tan}^{-1}(1/\sqrt{x^2 - 1})$ for $x > 1$.

50. Prove that $\mathrm{Sec}^{-1}(1/x) = \mathrm{Cos}^{-1}x$ for $0 < |x| \leq 1$.

Summary

Trigonometric functions (see graphs in Sections 3.1 and 3.2)

Table 3.3

Function	Domain	Range	Period	x-Intercepts
$y = \sin x$	x any number	$-1 \leq y \leq 1$	2π	$0, \pm\pi, \pm2\pi,\ldots$
$y = \cos x$	x any number	$-1 \leq y \leq 1$	2π	$\pm\dfrac{\pi}{2}, \pm\dfrac{3\pi}{2}, \pm\dfrac{5\pi}{2},\ldots$
$y = \tan x$	$x \neq \pm\dfrac{\pi}{2}, \pm\dfrac{3\pi}{2},\ldots$	y any number	π	$0, \pm\pi, \pm2\pi,\ldots$
$y = \cot x$	$x \neq 0, \pm\pi, \pm2\pi,\ldots$	y any number	π	$\pm\dfrac{\pi}{2}, \pm\dfrac{3\pi}{2}, \pm\dfrac{5\pi}{2},\ldots$
$y = \sec x$	$x \neq \pm\dfrac{\pi}{2}, \pm\dfrac{3\pi}{2},\ldots$	$y \leq -1$ or $y \geq 1$	2π	none
$y = \csc x$	$x \neq 0, \pm\pi, \pm2\pi,\ldots$	$y \leq -1$ or $y \geq 1$	2π	none

Inverse trigonometric functions (see graphs in Sections 3.4, 3.5, and 3.6)

Table 3.4

Function	Domain	Range
$y = \text{Sin}^{-1} x$	$-1 \le x \le 1$	$-\dfrac{\pi}{2} \le y \le \dfrac{\pi}{2}$
$y = \text{Cos}^{-1} x$	$-1 \le x \le 1$	$0 \le y \le \pi$
$y = \text{Tan}^{-1} x$	x any number	$-\dfrac{\pi}{2} < y < \dfrac{\pi}{2}$
$y = \text{Cot}^{-1} x$	x any number	$0 < y < \pi$
$y = \text{Sec}^{-1} x$	$x \le -1$ or $x \ge 1$	$0 \le y \le \pi,\ \ y \ne \dfrac{\pi}{2}$
$y = \text{Csc}^{-1} x$	$x \le -1$ or $x \ge 1$	$-\dfrac{\pi}{2} \le y \le \dfrac{\pi}{2},\ \ y \ne 0$

Reduction formulas (see graphs in Section 3.1)

Table 3.5

x (rad)	Function						x (deg)
	sin	cos	tan	cot	sec	csc	
$-x$	$-\sin x$	$\cos x$	$-\tan x$	$-\cot x$	$\sec x$	$-\csc x$	$-x$
$\dfrac{\pi}{2} - x$	$\cos x$	$\sin x$	$\cot x$	$\tan x$	$\csc x$	$\sec x$	$90° - x$
$\dfrac{\pi}{2} + x$	$\cos x$	$-\sin x$	$-\cot x$	$-\tan x$	$-\csc x$	$\sec x$	$90° + x$
$\pi - x$	$\sin x$	$-\cos x$	$-\tan x$	$-\cot x$	$-\sec x$	$\csc x$	$180° - x$
$\pi + x$	$-\sin x$	$-\cos x$	$\tan x$	$\cot x$	$-\sec x$	$-\csc x$	$180° + x$
$\dfrac{3\pi}{2} - x$	$-\cos x$	$-\sin x$	$\cot x$	$\tan x$	$-\csc x$	$-\sec x$	$270° - x$
$\dfrac{3\pi}{2} + x$	$-\cos x$	$\sin x$	$-\cot x$	$-\tan x$	$\csc x$	$-\sec x$	$270° + x$
$2\pi - x$	$-\sin x$	$\cos x$	$-\tan x$	$-\cot x$	$\sec x$	$-\csc x$	$360° - x$
$2\pi + x$	$\sin x$	$\cos x$	$\tan x$	$\cot x$	$\sec x$	$\csc x$	$360° + x$

Computer Problems (Optional)

Section 3.1

1. Write a program that will give x, y-values for $y = \sin x$, where $x = 0$, 0.1, 0.2, 0.3, ..., 3.0, 3.2. Have your program round off the y-values to two decimal places. Plot the (x, y) points and draw the corresponding portion of the graph of $y = \sin x$. Compare your graph with that in Fig. 3.1(b).

2. Write a program that will give three columns of numbers x, $\cos(\pi/2 + x)$, and $\sin x$, where x assumes the values -4.0, -3.5, $-3.0, ..., 3.5$, 4.0. Compare the second and third columns and see whether they support the reduction formula $\cos(\pi/2 + x) = -\sin x$.

Section 3.2

3. Write a program that will give two columns of numbers x, y, where $y = \tan x$ and $x = -2.50, -2.25, -2.00, \ldots, 2.50$. Have your program list the y-values rounded off to two significant digits. Use your table of x, y-values to help you draw a graph of $y = \tan x$.

4. Write a program that will give two columns of numbers x, y, where $y = \sec x$ and x assumes values from -3.50 to 3.50 in increments of 0.25. Have your program list the y-values rounded off to two decimal places. Plot the (x, y) points and draw the corresponding portion of the graph of $y = \sec x$. Compare your graph with that shown in Fig. 3.11.

Section 3.4 and 3.5

The following identities are taken from Example 6 of Section 3.5 and Problem 32 of Exercise 3.5 for $-1 < x < 1$,

$$\mathrm{Sin}^{-1} x = \mathrm{Tan}^{-1} \frac{x}{\sqrt{1 - x^2}},$$

$$\mathrm{Cos}^{-1} x = \frac{\pi}{2} - \mathrm{Tan}^{-1} \frac{x}{\sqrt{1 - x^2}}.$$

5. **a]** Write a program that will allow the user to enter x and the output will be $\mathrm{Sin}^{-1} x$. Have your program display two columns of numbers x, $\mathrm{Sin}^{-1} x$, where x assumes values $-0.9, -0.8, -0.7, \ldots, 0.8,$ 0.9 and $\mathrm{Sin}^{-1} x$ is rounded off to three decimal places. Plot a graph of $y = \mathrm{Sin}^{-1} x$.

   b] Follow the instructions in part (a) for $\mathrm{Cos}^{-1} x$.

6. Write a program that will give x-, y-values for $y = \cos(\mathrm{Sin}^{-1} x)$ and for $y = \sqrt{1 - x^2}$, where x assumes the values $-0.9, -0.8, -0.7, \ldots, 0.8,$ 0.9. Compare the corresponding y-values for the two functions and arrive at a conjecture relating $\cos(\mathrm{Sin}^{-1} x)$ and $\sqrt{1 - x^2}$.

Review Exercises

In Problems 1–18, evaluate the given expressions and give answers in exact form.

1. $\mathrm{Sin}^{-1} \dfrac{\sqrt{2}}{2}$

2. $\mathrm{Cos}^{-1} \left(-\dfrac{\sqrt{3}}{2} \right)$

3. $\mathrm{Tan}^{-1} (-\sqrt{3})$

4. $\mathrm{Sin}^{-1} (-1)$

5. $\mathrm{Sin}^{-1} \dfrac{\sqrt{3}}{2} + \mathrm{Cos}^{-1} \dfrac{1}{2}$

6. $3\,\mathrm{Sin}^{-1} \dfrac{1}{2} - \mathrm{Cos}^{-1} \dfrac{\sqrt{3}}{2}$

7. $\tan\left(\mathrm{Sin}^{-1} \dfrac{1}{\sqrt{2}} \right)$

8. $\cos\left(\mathrm{Sin}^{-1} \dfrac{1}{3} \right)$

9. $\tan\left(\mathrm{Cos}^{-1} \left(-\dfrac{2}{5} \right) \right)$

10. $\sin(\mathrm{Tan}^{-1} 3)$

11. $\cos\left(\mathrm{Tan}^{-1}(-2)\right)$ **12.** $\sec\left(\mathrm{Cos}^{-1}(-1)\right)$

13. $\cot\left(\mathrm{Tan}^{-1}\dfrac{2}{7}\right)$ **14.** $\mathrm{Cos}^{-1}\left(\cos\dfrac{4\pi}{3}\right)$

15. $\mathrm{Sin}^{-1}\left(\sin\dfrac{7\pi}{6}\right)$ **16.** $\cos\left(\dfrac{\pi}{2}-\mathrm{Sin}^{-1}\dfrac{1}{4}\right)$

17. $\tan\left(\dfrac{\pi}{2}+\mathrm{Cos}^{-1}\dfrac{1}{3}\right)$ **18.** $\sin\left(\dfrac{3\pi}{2}-\mathrm{Cos}^{-1}\dfrac{1}{5}\right)$

In Problems 19–30, evaluate the given expressions and give answers rounded off to two decimal places.

19. $\mathrm{Arcsin}\,0.34$ **20.** $\mathrm{Arcos}\,(-0.75)$ **21.** $\mathrm{Tan}^{-1}1.48$

22. $\mathrm{Sin}^{-1}\dfrac{\sqrt{3}}{4}$ **23.** $\sin\left(\mathrm{Cos}^{-1}0.8\right)$ **24.** $\mathrm{Cos}^{-1}(\sin 1.24)$

25. $\mathrm{Tan}^{-1}(\cos 2.53)$ **26.** $\mathrm{Sin}^{-1}(\cos 4.2)$ **27.** $\cot\left(\mathrm{Tan}^{-1}2.35\right)$

28. $\sec\left(\mathrm{Cos}^{-1}(-0.3)\right)$ **29.** $\mathrm{Sin}^{-1}0.6-\mathrm{Cos}^{-1}(-0.2)$

30. $\mathrm{Cos}^{-1}(-1)+\mathrm{Tan}^{-1}1.5$

31. a] Complete the following table giving (x,y) pairs for $y=\sin x$ and for $y=2\sin x$. Give y-values to two decimal places.

x (rad)	0	0.2	0.4	0.6	...	2.8	3.0	3.2
sin x					...			
2 sin x					...			

b] Use the (x,y) pairs from the table to draw graphs of $y=\sin x$ and $y=2\sin x$ on the same system of coordinates.

32. Draw graphs of $y=\cos x$ and $y=2\cos x$. Follow a procedure similar to that in Problem 31.

In Problems 33–36, for each of the functions described by the given equations, determine the domain. Then draw a graph of the function. Use your graph to state the range.

33. $y=\cos\left(\mathrm{Cos}^{-1}x\right)$ **34.** $y=\cos\left(\mathrm{Sin}^{-1}x\right)$

35. $y=\sin\left(\mathrm{Cos}^{-1}x\right)$ **36.** $y=\tan\left(\mathrm{Tan}^{-1}x\right)$

In Problems 37–42, evaluate the given expressions in exact form.

37. $\mathrm{Cot}^{-1}(-1)$ **38.** $\mathrm{Sec}^{-1}\left(-\dfrac{2}{\sqrt{3}}\right)$ **39.** $\mathrm{Csc}^{-1}(-1)$

40. $\tan\left(\mathrm{Cot}^{-1}2\right)$ **41.** $\cos\left(\mathrm{Sec}^{-1}3\right)$ **42.** $\sin\left(\mathrm{Csc}^{-1}(-2)\right)$

In Problems 43–48, evaluate the given expressions and give answers rounded off to two decimal places.

43. $\mathrm{Cot}^{-1}0.86$ **44.** $\mathrm{Csc}^{-1}2.3$ **45.** $\mathrm{Sec}^{-1}(-1.24)$

46. $\sin\left(\mathrm{Cot}^{-1}3.52\right)$ **47.** $\cos\left(\mathrm{Sec}^{-1}3.2\right)$ **48.** $\sin\left(\mathrm{Cot}^{-1}0.8\right)$

Identities

4

Every student who has ever studied algebra has "solved" (or found the solution set for) equations. There are many different procedures that may be helpful in finding the solution set for an equation. Some techniques may appear to be "tricks" when they are first encountered, but most approaches to problem solving have developed over time, and only the most fruitful have survived. One purpose of any mathematics course is the sharing of effective techniques and ideas with students.

One of the most useful tools available, for solving equations and for handling trigonometric functions in calculus and applications, is the use of "identities" to change the form of an equation or expression to make it more tractable. Perhaps the most familiar examples of identities in algebra are formulas for factoring, but there are many others.

Identities and conditional equations

Basically, we shall call an equation an *identity* if it yields a true statement for *all* allowable valuables of the variables (the replacement set). In other words, its solution set is the set of *all numbers* or *number pairs* for which *both sides* of the equation are defined. An equation that is true for a restricted set is called a *conditional equation*. Unfortunately, conditional equations are sometimes called simply "equations," so it is sometimes necessary to look carefully to see what is meant.

Consider the equation

$$x^3 - y^3 = (x - y)(x^2 + xy + y^2).$$

The expressions on both sides are defined for all pairs of numbers (x, y), and the statement is *true* for all such pairs, as may be verified by performing the indicated multiplication on the right-hand side. The equation $x^2 - x - 6 = 0$ is just as obviously a conditional equation. Its solution set consists of the two numbers -2 and 3. An equation such as

$$\frac{x^2 - 1}{x - 1} = x + 1$$

needs closer examination. There is obviously no restriction on the right-hand member of the equation, but the left-hand side is not defined if $x = 1$. Nevertheless, the equation is identically true for *all values of x for which both sides are defined*. This may be seen by factoring the numerator and cancelling the factor $(x - 1)$. The cancellation is permissible, since the denominator cannot be zero. Because there is a difference in the admissible values on the two sides of the equation, it is advisable to explicitly write out the omission:

$$\frac{x^2 - 1}{x - 1} = x + 1 \qquad \text{if} \quad x \neq 1.$$

Proving identities

Proving that a particular equation is an identity (called "proving" the identity) is similar to "solving" a conditional equation. In both cases we want to find the solution set, but in general, *more care is required for an identity* because it is not as easy to check our answer.

If in the process of solving an equation we square both sides or multiply both sides by the same expression, we risk changing the solution set, introducing "extraneous" roots. By going back to the original equation, however, we can easily check to see whether we really have the solutions needed. If the apparent solution set is all real numbers, however, there is no easy way to perform the same kind of check. To make it easier to remember some of the questionable steps that must be avoided, we shall include some *warnings* as reminders where extra caution is required. Some examples may clarify the situation.

EXAMPLE 1 Which of the following are identities?

a] $(x + y)^2 = x^2 + y^2$

b] $\sqrt{x^2} = x$

c] $\dfrac{x^2 - y^2}{x - y} = \sqrt{x^2 + 2xy + y^2}$

Solution

a] This is not an identity, although attempting to square a binomial in this way is an error often made by beginning algebra students. The correct identity is, of course, $(x + y)^2 = x^2 + 2xy + y^2$, so the two sides of the equation in part (a) are equal only if $2xy = 0$, which is certainly not true for all allowable x and y.

To *disprove* an identity (to show that a given equation is not an identity), it is sufficient to produce a *single counterexample*, since an identity must hold for *all* allowable values. In the equation in part (a), almost any pair of numbers will provide a counterexample as, for instance, $x = 3$ and $y = 4$.

b] This expression is true for infinitely many examples. It holds for all nonnonegative x, but *any* negative number provides a counterexample. For instance, if $x = -4$, then $\sqrt{x^2} = \sqrt{(-4)^2} = \sqrt{16} = 4$, while the right-hand side of the equation is simply $x = -4$. Thus $\sqrt{x^2} = x$ is not an identity.

c] The analysis of this equation is more involved. When we first look at it, we may see no obvious counterexample. Indeed, trying several number pairs, we may be led to suspect that the equation is an identity. We examine the expressions on both sides of the equation separately. The numerator of the left-hand side obviously factors, allowing us to cancel $(x - y)$ if we make a note that x *and* y *cannot be equal*. Turning to the right-hand side, we may see that the expression under the radical is a perfect square, so the proposed identity may be rewritten in the form

$$x + y = \sqrt{(x + y)^2}, \qquad \text{if} \quad x \neq y. \qquad \text{[4.1]}$$

In this form, we have the same type of expression that we had in the equation in part (b), so *this is not an identity*. A counterexample may be found by taking any pair of numbers whose sum is negative, say $x = 3$ and $y = -5$. In checking, it is always wise to use the original form. ■

If, when we reached the form (4.1), we had squared both sides, we would have had the obviously true statement $(x + y)^2 = (x + y)^2$, even though the original equation is not always true. Squaring both sides of an equation is a standard and useful procedure when solving equations, but it can lead to serious difficulties when working with identities. This leads to our first warning.

> *Warning:* In attempting to prove that an equation is an identity, do not perform the same operation on both sides of the equation.

Operations about which we must be especially careful include *squaring both sides of an equation* and *multiplying or dividing both sides by some expression involving a variable*. These operations must be avoided because they cannot always be reversed and can change the nature of the solution set.

Additive functions

There is another temptation that causes many difficulties for beginning students, especially until they become familiar with functional notation. A function f is said to be *additive* if $f(x + y) = f(x) + f(y)$ for all allowable x and y. This holds so seldom that we state another warning.

> *Warning: Almost no functions are additive. In symbols, $f(x + y) = f(x) + f(y)$ is almost never an identity.*

The only exceptional case encountered in elementary mathematics is a linear function of the form $f(x) = kx$. This means that all of our familiar functions such as the square function or the square root function, polynomial functions, and all trigonometric functions fail to be additive.

Again, we need to keep in mind the distinction between *solving equations* and *proving identities*. We may have an equation that has a very large solution set even though it is not an identity. The equation $(x + y)^2 = x^2 + y^2$ has infinitely many solutions, namely, any pair of real numbers whose product is zero, but it is not an identity because there are infinitely many pairs of numbers for which the equation is false.

EXAMPLE 2

a] Show that $\sqrt{4x + 11} = \sqrt{2x + 5} + \sqrt{2x + 6}$ is not an identity.

b] Find the solution set for the equation $\sqrt{4x + 11} = \sqrt{2x + 5} + \sqrt{2x + 6}$.

Solution

a] We may observe that this would be an identity if the square root function were additive, since $4x + 11$ is the sum of $(2x + 5)$ and $(2x + 6)$. Since the square root is not additive, we should have no difficulty finding a counterexample. If we substitute $x = 0.32$, for example, we get 3.504 for the left-hand side (LHS) and 4.952 for the right-hand side (RHS).

b] To *solve* this conditional equation, we may square both sides, recognizing that in so doing we may introduce extra solutions, but knowing also that we can check all possibilities. If we square both sides and then subtract $4x + 11$ from both sides, we obtain the equation $2\sqrt{(2x + 5)(2x + 6)} = 0$, from which we find the two roots, $x = -3$ and $x = -5/2$. Checking both roots in the original equation, we find that $x = -3$ does not work, while $x = -5/2$ does. The solution set is $\{-5/2\}$. ■

EXERCISE 4.1

In Problems 1–8, show that each of the given equations is *not* an identity.

1. $\sqrt{x - 2y} = \sqrt{x} - \sqrt{2y}$ **2.** $|x| = x$

3. $|x + 3| - x = 3$ **4.** $\sqrt{x^2 + y^2} = x + y$

5. $x^3 - y^3 = (x - y)^3$ **6.** $\sin(x + y) = \sin x + \sin y$

7. $\tan(\alpha - \beta) = \tan\alpha - \tan\beta$ **8.** $\sin\left(x - \dfrac{\pi}{2}\right) = \cos x$

In Problems 9–16, show that the given function is not additive (see our second *Warning*).

9. $f(x) = 3x - 1$ **10.** $g(x) = x + 2$

11. $h(x) = x^2 + 1$ **12.** $F(x) = \sqrt{x^2 + 1}$

13. $G(x) = x^2 + 2x + 1$ **14.** $f(x) = |3x|$

15. $f(x) = \sin x$ **16.** $g(x) = x + \cos x$

In Problems 17–21, determine which are identities.

17. $\sqrt{x^2} = |x|$ **18.** $\dfrac{x^2 - 1}{x + 1} = x - 1$

19. $x^3 - 1 = (x - 1)(x^2 + 2x + 1)$ **20.** $(x + y)^3 = x^3 + y^3$

21. $21x(1 - x)^{4/3} + 9(1 - x)^{7/3} = 21(1 - x)^{4/3} - 12(1 - x)^{7/3}$

In Problems 22–24, (a) show that the equation is not an identity and (b) find the solution set.

22. $\sqrt{x^2 + 4} = x + 2$ **23.** $\sqrt{x^2 + 4} = x - 4$

24. $\sqrt{x + \sqrt{2x - 1}} - \sqrt{x - \sqrt{2x - 1}} = \sqrt{x + |x - 1|}$

(Hint: Let $A = 2x - 1$ so that the left-hand side becomes $\sqrt{x + \sqrt{A}} - \sqrt{x - \sqrt{A}}$, and then square both sides.)

4.2 Basic Trigonometric Identities

For some reason of historical accident, the word "identity" is often associated with trigonometry, even though the concept is common throughout mathematics and the proof of trigonometric identities often involves algebraic identities as well. It is true, however, that there are many interrelationships among the trigonometric functions. The fact that there is frequent need to change the forms of trigonometric expressions by means of identities is also important.

There is literally no end to the number of trigonometric relations that are identities. We have already seen many of them in Chapters 2 and 3. Students sometimes get the impression that the study of identities is artificial, a game devised for the bedevilment of students. Quite to the contrary, we assert that *the use of identities is one of the most important tools that can be gained from a trigonometry course.* The study of periodic (recurring) phenomena, whether arising in economics or engineering, requires trigonometry; and adequate treatment almost always includes changing forms of expressions by the use of identities. Identities are needed throughout the calculus, as changed forms facilitate or make possible the differentiation and integration of otherwise difficult functions.

Because it is not possible to memorize all of the important identities, or even to list all those which may be needed, we shall concentrate on a few that are truly essential and from which the others can be derived as needed.

The same *warnings* we enumerated in the previous section are as important for trigonometric identities as they are for algebraic identities.

Basic identities

A number of the most fundamental trigonometric identities are already familiar to us. All of these were developed in Chapter 2. We repeat them here partly for convenience of reference but also to emphasize that these familiar facts *are identities*, valid for all allowable values of the variable.

These basic identities are important for another reason as well. Perhaps the greatest value of identities is their applicability to situations beyond the trigonometry course. The first seven identities listed below are so fundamental that students will retain (or quickly relearn) them for use when they are needed, in calculus for instance. If a student has learned how to work from these basics to derive a number of important other identities, then this course will be even more valuable.

$$\tan x = \frac{\sin x}{\cos x} \quad \textbf{[I-1]} \qquad \cot x = \frac{1}{\tan x} = \frac{\cos x}{\sin x} \quad \textbf{[I-2]}$$

$$\sec x = \frac{1}{\cos x} \quad \textbf{[I-3]} \qquad \csc x = \frac{1}{\sin x} \quad \textbf{[I-4]}$$

$$\sin^2 x + \cos^2 x = 1 \qquad \text{(Pythagorean Identity)} \quad \textbf{[I-5]}$$

$$\sin(-x) = -\sin x \quad \textbf{[I-6]} \qquad \cos(-x) = \cos x \quad \textbf{[I-7]}$$

$$\tan^2 x + 1 = \sec^2 x \quad \textbf{[I-8]} \qquad 1 + \cot^2 x = \csc^2 x \quad \textbf{[I-9]}$$

The first four identities are consequences of the definitions of the trigonometric functions and should become part of the very fiber of every student of trigonometry. The Pythagorean Identity is almost as fundamental. Identities (I-6) and (I-7) are statements of the crucial facts that sine is an odd function and cosine is even. They can be readily recalled either from the unit circle or from the graphs of sine and cosine (see Fig. 3.4 and Fig. 3.2).

Identities (I-8) and (I-9) are new to us. They may easily be committed to memory, but it is just as easy (and much more valuable in the long run) to learn how to derive them from the Pythagorean Identity. For identity (I-8) we want to go from $\sin^2 x + \cos^2 x = 1$ to an identity involving $\tan^2 x$. Divide both sides of identity (I-5) by $\cos^2 x$ and then use (I-1) and (I-3) to get

$$\frac{\sin^2 x}{\cos^2 x} + \frac{\cos^2 x}{\cos^2 x} = \frac{1}{\cos^2 x},$$
$$\tan^2 x + 1 = \sec^2 x.$$

Does this simple derivation violate the dire *warning* not to divide both sides of an equation by an expression involving the variable? The answer is *no*, and we need to stress the reason. Here we are beginning with a known identity. The Pythagorean Identity is valid for all numbers; its nature is not in question. If two expressions are *known* to be equal, then performing the same operation on both will result in expressions that are *still equal*.

Arguing correctly from a valid premise justifies the conclusion; reaching a correct conclusion from a questionable premise *does not validate the premise.* If we begin with a statement whose truth is not established, we cannot conclude from a correct conclusion that we had a true statement to begin with. This may be illustrated by a trivial example.

Suppose we wish to prove that $1 = 2$. Starting with this statement, our "proof" might proceed by multiplying both sides by zero:

$$1 = 2,$$
$$0 \cdot 1 = 0 \cdot 2,$$

whence

$$0 = 0.$$

From this obvious truth, can we conclude that $1 = 2$? Clearly *not*! What this establishes is the true (but not very enlightening) theorem, "If $1 = 2$, then $0 = 0$." Had we begun with a less trivial expression and included steps less obviously irreversible, we could have gone badly astray, as in Example 1(c) from Section 4.1. Squaring both sides of Eq. (4.1) would have led to the correct conclusion that $(x + y)^2 = (x + y)^2$, but Eq. (4.1) is *not valid for all number pairs.*

Techniques for proving trigonometric identities _____

When we are asked to determine whether or not a given equation is an identity, the obvious place to begin is by checking several values, taking advantage of the calculator for easy computation. All we need for a counterexample is one value that yields a false statement. In searching for a counterexample with trigonometric identities, *it is wise to avoid angles that are special in any way*, such as quadrantal angles.

If several values all satisfy the equation, we may suspect that the equation is an identity. Our options then are to work with one side of the equation and use basic identities to make it look like the other side, or we may work with each side independently to reach a common form. We illustrate these alternatives in examples.

To avoid the temptation of performing the same operation on both sides of an identity, we find it convenient to organize our work in columns. In the left column we have expressions that are equivalent to the expression on the left-hand side (LHS) of the equation, and in the right column we have expressions that are equivalent to the right-hand side (RHS). Our alternatives then may be restated as follows: Work down the left column until it looks like the RHS, work down the right column until it looks like the LHS, or work down both columns independently to get the same expressions on both sides.

EXAMPLE 1 Prove that $\cos x \tan x = \sin x$ is an identity.

Solution Finding a good starting point is always crucial in establishing an identity. In this case the right-hand side can hardly be simpler, so we look at the possibility of making the LHS equal $\sin x$.

$$\cos x \tan x \overset{?}{=} \sin x$$

Apply identity (I-1)

$$(\cos x)\left(\frac{\sin x}{\cos x}\right)$$

Cancel

$$\sin x$$

From this display we could write the proof as a chain of equalities:

$$\cos x \tan x = \cos x \left(\frac{\sin x}{\cos x}\right) = \sin x,$$

but most often we recognize that the column display contains all of the information necessary to establish a complete proof. ■

EXAMPLE 2 Prove that $\dfrac{1 - \sec \theta}{1 + \sec \theta} = \dfrac{\cos \theta - 1}{\cos \theta + 1}$ is an identity.

Solution Since secants and cosines are reciprocals of each other, it would seem reasonable to try identity (I-3) to get the same function on both sides. The work is essentially the same in either direction, but for variety we shall make the RHS look like the LHS.

$$\frac{1 - \sec \theta}{1 + \sec \theta} \overset{?}{=} \frac{\cos \theta - 1}{\cos \theta + 1}$$

Apply identity (I-3)

$$\frac{\dfrac{1}{\sec \theta} - 1}{\dfrac{1}{\sec \theta} + 1}$$

Add fractions

$$\frac{(1 - \sec \theta)/\sec \theta}{(1 + \sec \theta)/\sec \theta}$$

Simplify

$$\frac{1 - \sec \theta}{1 + \sec \theta}$$

Thus the LHS is identically equal to the RHS. ■

EXAMPLE 3 Determine whether or not $(\sin A + \cos A)^2 = \dfrac{\sec A \ \csc A + 2}{\sec A \ \csc A}$ is an identity.

Solution We choose some numbers at which we may evaluate both sides of the equation. If $A = 1.1$ (radians), both sides yield approximately 1.80850. If $A = -0.2$, then we get 0.61058 for both sides, so it seems reasonable to suspect that the equation is an identity. We shall work with both sides independently to try to make them look alike.

$$(\sin A + \cos A)^2 \overset{?}{=} \frac{\sec A \ \csc A + 2}{\sec A \ \csc A}$$

Square	Write as two fractions
$\sin^2 A + 2 \sin A \cos A + \cos^2 A$	$\dfrac{\sec A \ \csc A}{\sec A \ \csc A} + \dfrac{2}{\sec A \ \csc A}$
Group terms	**Simplify**
$(\sin^2 A + \cos^2 A) + 2 \sin A \cos A$	$1 + 2 \left(\dfrac{1}{\sec A}\right)\left(\dfrac{1}{\csc A}\right)$
Use identity (I-5)	**Apply identities (I-3) and (I-4)**
$1 + 2 \sin A \cos A$	$1 + 2 \cos A \sin A$

Since each side is identically equal to $1 + 2 \sin A \cos A$, the two sides are identically equal to each other. The given equation is an identity. ∎

There are almost always a number of good ways to prove any identity. Do not hesitate to try any possibility that occurs to you, and be patient in working toward a common form, working on the two sides alternately.

EXAMPLE 4 Prove that $\dfrac{\sin \theta}{1 + \cos \theta} = \dfrac{1 - \cos \theta}{\sin \theta}$ is an identity.

Solution If we were able to "cross-multiply," we would get $\sin^2 \theta = (1 - \cos \theta)(1 + \cos \theta) = 1 - \cos^2 \theta$, which is a true statement by identity (I-5). However, cross-multiplication is really just multiplying both sides by the same quantity, here $\sin \theta (1 + \cos \theta)$, and cancelling. This is not a valid step, since we do not know that the two sides are equal to begin with. Nonetheless, this observation can be a very useful guide to suggest how to work with the two sides independently. In particular, if we had a numerator of $\sin^2 \theta$ or $1 - \cos^2 \theta$, we could simplify:

$$\frac{\sin\theta}{1+\cos\theta} \overset{?}{=} \frac{1-\cos\theta}{\sin\theta}$$

$$\frac{\sin\theta}{1+\cos\theta} \cdot \frac{\sin\theta}{\sin\theta}$$

Multiply fractions

$$\frac{\sin^2\theta}{(1+\cos\theta)\sin\theta}$$

Use identity (I-5)

$$\frac{1-\cos^2\theta}{(1+\cos\theta)\sin\theta}$$

Factor

$$\frac{(1-\cos\theta)(1+\cos\theta)}{(1+\cos\theta)\sin\theta}$$

Simplify

$$\frac{1-\cos\theta}{\sin\theta}$$

Thus the given equation is an identity.

Alternative Solution Recognizing that we could get to a correct conclusion by cross-multiplying suggests that we try going backwards from the valid conclusion, reversing the steps, checking each step carefully:

$$\sin^2\theta = 1 - \cos^2\theta = (1+\cos\theta)(1-\cos\theta) \qquad \textbf{From identity (I-5)}$$

We want to divide both sides by $\sin\theta\,(1+\cos\theta)$:

$$\frac{\sin^2\theta}{\sin\theta\,(1+\cos\theta)} = \frac{(1+\cos\theta)(1-\cos\theta)}{\sin\theta\,(1+\cos\theta)}, \qquad \text{if } \sin\theta\,(1+\cos\theta) \neq 0.$$

Simplify to get

$$\frac{\sin\theta}{1+\cos\theta} = \frac{1-\cos\theta}{\sin\theta}$$

This is valid if $\sin\theta\,(1+\cos\theta) \neq 0$, which is precisely the condition that must hold for the two sides of the given equation to be defined. Thus the two sides are identically equal *for all values for which both sides are defined.* ■

EXERCISE 4.2

1. Derive identity (I-9) from identity (I-5).

In Problems 2–40, prove that the given equation is an identity.

2. $\sin\theta\cot\theta = \cos\theta$

3. $\dfrac{\tan\theta}{\sin\theta} = \sec\theta$

4. $\cot\theta = \csc\theta\cos\theta$

5. $\cos x\sec x = 1$

6. $\cos x\tan x = \sin x$

7. $1 - \cos^2 x = \cos^2 x\tan^2 x$

8. $\cot x\sec x = \csc x$

9. $\sin^2 x = (1 - \cos x)(1 + \cos x)$

10. $\dfrac{\cot x}{\sec x} = \csc x - \sin x$

11. $\dfrac{\sin x\,\csc x}{\cot x} = \tan x$

12. $\dfrac{\sin(-\theta)}{\cos\theta} = \tan(-\theta)$

13. $\sec\theta\csc\theta = \tan\theta + \cot\theta$

14. $\sec\theta\,(\csc\theta - \sin\theta) = \csc\theta\cos\theta$

15. $\dfrac{1 - \cos x}{1 + \cos x} = (\cot x - \csc x)^2$

16. $\dfrac{\sin\theta}{1 - \cos\theta} = \dfrac{1 + \cos\theta}{\sin\theta}$

17. $\tan x + \cot x = \dfrac{\csc x}{\cos x}$

18. $\dfrac{1 + \tan\theta}{\sec\theta} = \dfrac{1 + \cot\theta}{\csc\theta}$

19. $\cot\alpha\csc\alpha = \dfrac{1}{\sec\alpha - \cos\alpha}$

20. $\dfrac{1}{1 - \sin x} + \dfrac{1}{1 + \sin x} = 2\sec^2 x$

21. $\sec^2 x + \csc^2 x = \sec^2 x\csc^2 x$

22. $\dfrac{\sin\theta}{1 + \cos\theta} + \dfrac{1 + \cos\theta}{\sin\theta} = \dfrac{2}{\sin\theta}$

23. $(\cos x + 1)(\sec x - 1) = \sec x - \cos x$

24. $\sec\theta - \cos\theta = \sin(-\theta)\tan(-\theta)$

25. $\sin^4 x - \cos^4 x = \sin^2 x - \cos^2 x$

26. $1 + \tan^2 x = \tan x\sec x\csc x$

27. $\dfrac{\tan\theta + \sec\theta}{\sin\theta\cot\theta} = \dfrac{1 + \sin\theta}{\cos^2\theta}$

28. $\cot(-x)\cos(-x) = \sin x - \csc x$

29. $\dfrac{\cos\theta}{\sin\theta} + \dfrac{\sin\theta}{\cos\theta} = \sec\theta\csc\theta$

30. $\dfrac{1 - \sin(-x)}{\cos x} = \tan x + \sec x$

31. $\dfrac{1 - \cos x}{1 + \cos x} = \dfrac{\sec x - 1}{\sec x + 1}$

32. $1 - (\sin x - \cos x)^2 = 2\sin x\cos x$

33. $\dfrac{\csc(-x)}{\cot(-x) + \tan(-x)} = \cos x$

34. $\dfrac{\cos x}{1 - \sin x} = \dfrac{1 + \sin x}{\cos x}$

35. $\sec^4 x - \tan^4 x = \sec^2 x\,(\sin^2 x + 1)$

36. $\tan^2 x - \sec^2 x = -1$

37. $\tan^4 x + \tan^2 x = \sec^4 x - \sec^2 x$

38. $\dfrac{1}{\sec\theta - \tan\theta} = \sec\theta + \tan\theta$

39. $\dfrac{\cot x + \tan x}{\sec x\csc x} = 1$

40. $\sin^2 x\tan^2 x + \sin^2 x = \tan^2 x$

4.3 Basic Trigonometric Identities (Continued)

All of the problems of the preceding section are of the form "Prove that the given equation is an identity." We also discussed the possibility of determining whether or not a given equation is an identity (see Example 3 of Section 4.2). For all of the problems in this section we are asked to determine which equations are identities and to verify those which are.

EXAMPLE 1 Determine whether or not $\sin^4 x + \cos^4 x = 1$ is an identity.

Solution This equation illustrates the suggestion in the previous section that in searching for counterexamples it is wise to check numbers that are not special in any way. The reader should verify that the equation is satisfied by $x = 0$, $\pi/2$, and π, all of which are quadrantal. Checking $x = \pi/4$, however, yields $1/4 + 1/4 = 1/2 \neq 1$. Similarly, $x = 1$ gives $(\sin 1)^4 + (\cos 1)^4 \approx 0.5866 \neq 1$. Therefore we have counterexamples to show that $\sin^4 x + \cos^4 x = 1$ is *not* an identity. ∎

EXAMPLE 2 Determine whether or not $\dfrac{1}{1 - \cos x} = \dfrac{1 + \cos x}{\sin^2 x}$ is an identity.

Solution If we try $x = 1$, calculator evaluation of the expressions on the two sides gives 2.17534. At $x = 1/2$, both sides are equal to 8.16877, so it appears that we may have an identity. Comparing this with Example 4 from Section 4.2, we may proceed as in that Alternative Solution, or we may use identity (I-5) to replace $\sin^2 x$ by $1 - \cos^2 x$.

$$\frac{1}{1 - \cos x} \overset{?}{=} \frac{1 + \cos x}{\sin^2 x}$$

Use identity (I-5)

$$\frac{1 + \cos x}{1 - \cos^2 x}$$

Factor

$$\frac{1 + \cos x}{(1 + \cos x)(1 - \cos x)}$$

Simplify

$$\frac{1}{1 - \cos x}$$

So we have an identity. ∎

EXAMPLE 3 Is $\sqrt{\tan^2 x - \sin^2 x} = \sin x \tan x$ an identity?

Solution The two sides agree for $x = 0$, 1, and 1/2 (check this), so it appears that we may have an identity. To prove that the sides are identically equal, we try expressing both sides in terms of $\sin x$ and $\cos x$.

$$\sqrt{\tan^2 x - \sin^2 x} \overset{?}{=} \sin x \tan x$$

Use identity (I-1) and factor	Use identity (I-1)
$\sqrt{(\sin^2 x)\left(\dfrac{1}{\cos^2 x} - 1\right)}$	$(\sin x)\left(\dfrac{\sin x}{\cos x}\right)$
Combine fractions	**Multiply fractions**
$\sqrt{(\sin^2 x)\dfrac{(1 - \cos^2 x)}{\cos^2 x}}$	$\dfrac{\sin^2 x}{\cos x}$
Apply identity (I-5) and simplify	
$\sqrt{\dfrac{\sin^4 x}{\cos^2 x}}$	

The last two expressions are very similar. In fact we have reduced our proposed identity to the form $\sqrt{u^2} = u$, which we know is *not* true for negative values of u. This suggests that we check a value of x for which the RHS is negative, either a second- or third-quadrant angle. For $x = \dfrac{3\pi}{4}$,

LHS: $\sqrt{\tan^2 x - \sin^2 x} = \sqrt{(-1)^2 - (1/\sqrt{2})^2} = \dfrac{1}{\sqrt{2}}$;

RHS: $\tan x \sin x = (-1)(1/\sqrt{2}) = -\dfrac{1}{\sqrt{2}}$.

So the given equation is *not* an identity. Incidentally, since $\sqrt{u^2} = |u|$, we can see that the equation $\sqrt{\tan^2 x - \sin^2 x} = |\tan x \sin x|$ is an identity. ∎

EXERCISE 4.3

Determine which of the given equations are identities. Give a proof or a counterexample to justify your conclusion.

1. $\dfrac{\sin^2 x}{\cos x} = \sec x - \cos x$ 2. $\sin x \cot x = \cos x$

3. $\sin x \tan x = 1 - \cos x$ 4. $(\sin \theta + \cos \theta)^2 = 1$

5. $(\sin x - \cos x)^2 = \sin^2 x - \cos^2 x$ 6. $\dfrac{\sin \theta + \cos \theta}{\cos \theta} = 1 + \tan \theta$

7. $(\cos x + \sin x)(\sec x + \csc x) = 1$ 8. $\cot x \sec x = \cos x$

9. $\sin x \cos x (\sin x \sec x + \cos x \csc x) = 1$

10. $(\cos x - \sin x)^2 = 1 - 2 \sin x \cos x$

11. $(\sin x + \cos x)^3 = \sin^3 x + \cos^3 x$

12. $\sin^4 x - \cos^4 x = 2 \sin^2 x - 1$ 13. $\sqrt{1 - \cos^2 \theta} = \sin \theta$

14. $\sin^3 x - \cos^3 x = (\sin x - \cos x)(1 + \sin x \cos x)$

15. $(\tan \theta + \cot \theta)^2 = \tan^2 \theta + \cot^2 \theta$ 16. $\sqrt{\cot^2 x - \cos^2 x} = \cos x \cot x$

17. $\sqrt{\sin^2 x + \cos^2 x} = |\sin x| + |\cos x|$

18. $\sec^2 x + \csc^2 x = 1$ 19. $(1 + \tan \theta)^2 = \sec^2 \theta + 2 \tan \theta$

20. $\sqrt{1 + \tan^2 \theta} = \sec \theta$ 21. $\tan \theta - \cot \theta = \dfrac{1 - 2 \cos^2 \theta}{\sin \theta \cos \theta}$

22. $\dfrac{1 + \cos \theta}{\sin \theta} - \dfrac{\sin \theta}{1 - \cos \theta} = 0$ 23. $\dfrac{\cos x}{\cos x - \sin x} = \dfrac{\tan x}{1 - \tan x}$

24. $2 \cos x \cot x - 2 \cos x + \cot x = 1$

25. $\cos^2 x - \sin^2 x = 2 \cos x - \sec x$ 26. $\tan^4 x - \sec^4 x = \tan^2 x - \sec^2 x$

27. $\dfrac{\sec^2 A - \sec A \tan A}{\cos^2 A} = \dfrac{\sec^2 A}{1 + \sin A}$ 28. $\cos B \cot B = \csc B - \sin B$

29. $[1 + \tan(-x)]^2 = \sec^2(-x) + 2 \tan(-x)$

30. $\dfrac{1 + \sin(-x)}{\cos(-x)} = \tan x + \sec x$

4.4 Sum and Difference Identities

In Section 4.1 we observed that very few functions are additive. In particular, the trigonometric functions are not, so $\sin(A + B)$ is not identically equal to $\sin A + \sin B$, and $\cos 2t \neq 2 \cos t$ except for some few exceptional values. There are several very important identities that allow us to express functions of sums and differences in simpler terms. Collectively, these are often called the *sum and difference formulas*:

$$\sin(\alpha + \beta) = \sin \alpha \cos \beta + \cos \alpha \sin \beta \qquad \textbf{[I-10]}$$

$$\cos(\alpha + \beta) = \cos \alpha \cos \beta - \sin \alpha \sin \beta \qquad \textbf{[I-11]}$$

$$\sin(\alpha - \beta) = \sin \alpha \cos \beta - \cos \alpha \sin \beta \qquad \textbf{[I-12]}$$

$$\cos(\alpha - \beta) = \cos \alpha \cos \beta + \sin \alpha \sin \beta \qquad \textbf{[I-13]}$$

$$\tan(\alpha + \beta) = \frac{\tan \alpha + \tan \beta}{1 - \tan \alpha \tan \beta} \qquad \textbf{[I-14]}$$

$$\tan(\alpha - \beta) = \frac{\tan \alpha - \tan \beta}{1 + \tan \alpha \tan \beta} \qquad \textbf{[I-15]}$$

These identities find frequent application throughout analytic geometry and calculus. We mentioned earlier that there are a few truly fundamental identities, from which all the others can be derived as needed. Identities (I-1) through (I-7) are in that category. Beyond these, we should include identities (I-10) and (I-11) to our "must know" list. For derivation purposes it is also convenient to be able to refer to

$$\tan(-x) = -\tan x. \qquad \qquad \text{[I-16]}$$

Proofs of the sum and difference formulas _____

Because there are so many identities relating the trigonometric functions, we could prove all of the sum and difference formulas from any one of the six, but we do need to establish one of them independently.

In Problem 43 of Exercise 2.4 we showed how identities (I-10) and (I-11) could be derived for positive angles whose sum is less than $\pi/2$. We used the fact that *in a right triangle with acute angle θ and hypotenuse h the legs of the triangle have lengths $h \sin \theta$ and $h \cos \theta$.*

In Fig. 4.1, then, the legs of triangle ABC have lengths $1 \cdot \cos \alpha = \cos \alpha$ and $1 \cdot \sin \alpha = \sin \alpha$. Each of these, in turn, is the hypotenuse of a triangle with acute angle β, thus giving the sides $|\overline{CF}| = \cos \alpha \sin \beta$, $|\overline{AF}| = \cos \alpha \cos \beta$, $|\overline{BE}| = \sin \alpha \sin \beta$, and $|\overline{CE}| = \sin \alpha \cos \beta$. Triangle ABD has a hypotenuse of length 1 and an acute angle $(\alpha + \beta)$, from which $|\overline{BD}| = \sin(\alpha + \beta)$ and $|\overline{AD}| = \cos(\alpha + \beta)$. Also, $|\overline{BD}| = |\overline{EF}| = |\overline{EC}| + |\overline{CF}|$ and $|\overline{AD}| = |\overline{AF}| - |\overline{DF}| = |\overline{AF}| - |\overline{BE}|$. Replacing each of the sides by its length in terms of the trigonometric functions of α and β gives

$$\sin(\alpha + \beta) = \sin \alpha \cos \beta + \cos \alpha \sin \beta, \qquad \text{[I-10]}$$
$$\cos(\alpha + \beta) = \cos \alpha \cos \beta - \sin \alpha \sin \beta. \qquad \text{[I-11]}$$

Unfortunately, this derivation is valid only for α and β for which $(\alpha + \beta)$ is an acute angle. The identities must hold for *all* numbers, and we need a proof that is not dependent on the size of the numbers.

Proof of identity (I-13) _____

It is convenient for us first to give a proof of identity (I-13): $\cos(\alpha - \beta) = \cos \alpha \cos \beta + \sin \alpha \sin \beta$. We shall use a unit circle and assume that $\alpha > \beta$, but the validity of the argument does not depend on the relative sizes of α and β. In Fig. 4.2 we have central angles α, β, and $\alpha - \beta$ in a unit circle. Triangle AOB has central angle $\alpha - \beta$, and so does triangle COD (in standard position). Since $\overline{OD}, \overline{OC}, \overline{OB}$,

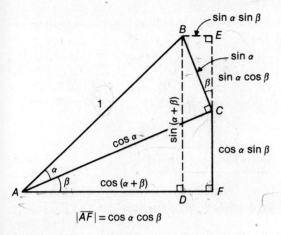

FIGURE 4.1

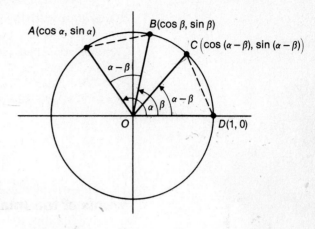

FIGURE 4.2

and $\overline{OA}$ all have unit length, the two triangles are congruent and hence have congruent bases, $|\overline{AB}| = |\overline{CD}|$. We may use the distance formula to calculate the squares of these lengths in terms of their coordinates.

$$|\overline{AB}|^2 = (\cos\alpha - \cos\beta)^2 + (\sin\alpha - \sin\beta)^2$$
$$= \cos^2\alpha - 2\cos\alpha\cos\beta + \cos^2\beta + \sin^2\alpha - 2\sin\alpha\sin\beta + \sin^2\beta$$
$$= (\cos^2\alpha + \sin^2\alpha) + (\cos^2\beta + \sin^2\beta)$$
$$-2(\cos\alpha\cos\beta + \sin\alpha\sin\beta)$$
$$= 2 - 2(\cos\alpha\cos\beta + \sin\alpha\sin\beta) \qquad \textbf{By identity (I-5)}$$

$$|\overline{CD}|^2 = [\cos(\alpha-\beta) - 1]^2 + [\sin(\alpha-\beta) - 0]^2$$
$$= \cos^2(\alpha-\beta) - 2\cos(\alpha-\beta) + 1 + \sin^2(\alpha-\beta)$$
$$= [\cos^2(\alpha-\beta) + \sin^2(\alpha-\beta)] + 1 - 2\cos(\alpha-\beta)$$
$$= 2 - 2\cos(\alpha-\beta) \qquad \textbf{By identity (I-5)}$$

Equating these two results, subtracting 2 from each side, and dividing both sides by (-2), we obtain

$$\cos(\alpha-\beta) = \cos\alpha\cos\beta + \sin\alpha\sin\beta.$$

Proof of identity (I-11)

Having established identity (I-13), we get identity (I-11) by simply changing the sign of β.

$$\cos(\alpha+\beta) = \cos[\alpha - (-\beta)]$$
$$= \cos\alpha\cos(-\beta) + \sin\alpha\sin(-\beta) \qquad \textbf{By identity (I-13)}$$
$$= \cos\alpha\cos\beta + \sin\alpha(-\sin\beta) \qquad \textbf{By identities (I-6) and (I-7)}$$
$$= \cos\alpha\cos\beta - \sin\alpha\sin\beta$$

Proof of identity (I-10)

Any of several reduction formulas from Section 3.1 may be used to derive sine formulas from cosine formulas. We use the facts that $\cos(\pi/2 - t) = \sin t$ and $\sin(\pi/2 - t) = \cos t$ in the following:

$$\sin(\alpha + \beta) = \cos[\pi/2 - (\alpha + \beta)] = \cos[(\pi/2 - \alpha) - \beta]$$
$$= \cos(\pi/2 - \alpha)\cos\beta + \sin(\pi/2 - \alpha)\sin\beta$$
By identity (I-13)
$$= \sin\alpha\cos\beta + \cos\alpha\sin\beta.$$

This establishes identity (I-10). The same sign change we used to derive identity (I-11) from identity (I-13) may be used by the reader to derive identity (I-12) from identity (I-10). (See Problem 1 in Exercise 4.4.)

Proof of identity (I-14)

The derivation of the tangent formulas uses the identities we have just developed for sine and cosine. It seems obvious that we should use

$$\tan(\alpha + \beta) = \frac{\sin(\alpha + \beta)}{\cos(\alpha + \beta)} = \frac{\sin\alpha\cos\beta + \cos\alpha\sin\beta}{\cos\alpha\cos\beta - \sin\alpha\sin\beta},$$

but it is less apparent how we should proceed to translate this last quotient into tangents. If we look at any of the four terms in the quotient, however, we are led to the same division, which makes the whole thing work. Looking at $\sin\alpha\cos\beta$, for example, we can get $\tan\alpha$ by dividing by $\cos\alpha$. The factor of $\cos\beta$ is not readily changed into $\tan\beta$, but we can get rid of it by dividing by $\cos\beta$. This suggests dividing $\sin\alpha\cos\beta$ by $\cos\alpha\cos\beta$; but if one term is divided, *each* term must be divided by the same quantity. Thus dividing each term of the numerator and the denominator by $\cos\alpha\cos\beta$ gives

$$\tan(\alpha + \beta) = \frac{\dfrac{\sin\alpha\cos\beta}{\cos\alpha\cos\beta} + \dfrac{\cos\alpha\sin\beta}{\cos\alpha\cos\beta}}{\dfrac{\cos\alpha\cos\beta}{\cos\alpha\cos\beta} - \dfrac{\sin\alpha\sin\beta}{\cos\alpha\cos\beta}} = \frac{\tan\alpha + \tan\beta}{1 - \tan\alpha\tan\beta},$$

which is identity (I-14). The remaining identity (I-15) may be obtained directly from (I-14) by the use of (I-16) or by using identities (I-12) and (I-13) and proceeding as above. (See Problems 2 and 3 in Exercise 4.4.)

EXAMPLE 1 Prove that $\tan\left(x - \dfrac{\pi}{4}\right) = \dfrac{\sin x - \cos x}{\sin x + \cos x}$ is an identity.

Solution

$$\tan\left(x - \frac{\pi}{4}\right) = \frac{\tan x - \tan(\pi/4)}{1 + \tan x \ \tan(\pi/4)} \qquad \textbf{By identity (I-15)}$$

$$= \frac{\tan x - 1}{1 + \tan x} \qquad \left(\text{since } \tan \frac{\pi}{4} = 1\right)$$

$$= \frac{(\sin x / \cos x) - 1}{1 + (\sin x / \cos x)} \qquad \textbf{By identity (I-1)}$$

$$= \frac{\sin x - \cos x}{\cos x + \sin x} \qquad \textbf{By algebra} \qquad \blacksquare$$

EXAMPLE 2   Evaluate $\sin 75°$ and express the answer in exact form.

Solution

$$\sin 75° = \sin(30° + 45°) = \sin 30° \ \cos 45° + \cos 30° \ \sin 45°$$
$$\textbf{By identity (I-10)}$$

$$= \frac{1}{2} \frac{\sqrt{2}}{2} + \frac{\sqrt{3}}{2} \frac{\sqrt{2}}{2} = \frac{1}{4}(\sqrt{2} + \sqrt{6}). \qquad \blacksquare$$

EXAMPLE 3   Evaluate $\cos(\pi/12)$ and give the answer in exact form.

Solution

$$\cos\frac{\pi}{12} = \cos\left(\frac{\pi}{4} - \frac{\pi}{6}\right) = \cos\frac{\pi}{4} \ \cos\frac{\pi}{6} + \sin\frac{\pi}{4} \ \sin\frac{\pi}{6} \qquad \textbf{By identity (I-13)}$$

$$= \frac{\sqrt{2}}{2} \frac{\sqrt{3}}{2} + \frac{\sqrt{2}}{2} \frac{1}{2} = \frac{1}{4}(\sqrt{6} + \sqrt{2}). \qquad \blacksquare$$

In Section 3.1 we used graphs of the sine and cosine functions (and the unit circle) to establish a number of reduction formulas. Two of these, $\cos(\pi/2 - t) = \sin t$ and $\sin(\pi/2 - t) = \cos t$, were used to derive identity (I-10) from identity (I-13), so logically we should not attempt to use sum and difference formulas to justify reduction formulas. As a practical matter, however, the sum and difference identities are usually easier to remember and may be used to recall desired reduction formulas. We illustrate this logical circularity (but very handy practicality) in the following example.

EXAMPLE 4   Use sum and difference formulas to show that $\cos(\pi/2 - t) = \sin t$.

Solution

$$\cos(\pi/2 - t) = \cos \pi/2 \ \cos t + \sin \pi/2 \ \sin t \qquad \textbf{By identity (I-13)}$$
$$= 0(\cos t) + 1(\sin t) = \sin t. \qquad \blacksquare$$

EXERCISE 4.4

1. Use identities (I-10), (I-6), and (I-7) to prove identity (I-12).
2. Use identities (I-14) and (I-16) to derive identity (I-15).
3. Use identities (I-1), (I-12), and (I-13) to derive identity (I-15).

In Problems 4–9, use sum and difference identities to establish the given reduction formulas.

4. $\cos\left(\frac{\pi}{2}+\theta\right)=-\sin\theta$ 5. $\sin\left(\frac{3\pi}{2}-t\right)=-\cos t$

6. $\tan\left(\frac{\pi}{2}+\theta\right)=-\cot\theta\,^*$ 7. $\csc(180°-A)=\csc A$

8. $\cos(180°-A)=-\cos A$ 9. $\sin(180°+\theta)=-\sin\theta$

In Problems 10–13, prove that the given equations are identities.

10. $\sec\theta-\cos\theta=\sin(-\theta)\tan(-\theta)$
11. $\cot(-x)\cos(-x)=\sin x-\csc x$
12. $\dfrac{1-\sin(-x)}{\cos(-x)}=\tan x+\sec x$ 13. $\dfrac{\csc(-x)}{\cot(-x)+\tan(-x)}=\cos x$

In Problems 14–22, (a) give answers in exact form by expressing the given angle as a sum or difference of special angles (see Examples 2 and 3) and then (b) evaluate your answers by calculator (to two decimal places).

14. $\cos 75°$ 15. $\sin 195°$ 16. $\tan 285°$

17. $\cot 15°$ 18. $\sec 255°$ 19. $\tan\dfrac{7\pi}{12}$

20. $\sec\dfrac{-5\pi}{12}$ 21. $\cot\dfrac{\pi}{12}$ 22. $\sin\dfrac{13\pi}{12}$

23. If $\tan\alpha=3$ and $\tan(\alpha+\beta)=-\dfrac{2}{3}$, find $\tan\beta$.

24. If $x-y=\dfrac{3\pi}{4}$ and $\tan y=3$, find $\tan x$.

25. If $\tan(x-y)=-\dfrac{5}{4}$ and $\tan x=0.4$, find $\tan y$.

In Problems 26–32, determine whether the given equations are identities.

26. $\tan\left(\dfrac{\pi}{4}+x\right)=\dfrac{1+\tan x}{1-\tan x}$ 27. $\sin\left(\dfrac{\pi}{6}-x\right)=\dfrac{1}{2}(\cos x-\sqrt{3}\sin x)$

28. $\dfrac{\cos x-\sin x}{\cos x+\sin x}=\tan\left(\dfrac{\pi}{4}-x\right)$ 29. $\sec(\alpha+\beta)=\sec\alpha+\sec\beta$

---

*Here you have a problem if you try to apply identity (I-14), since $\tan\pi/2$ is not defined. However, try $\tan(\pi/2+\theta)=\sin(\pi/2+\theta)/\cos(\pi/2+\theta)$ and then use identities (I-10) and (I-11).

30. $\csc\left(\dfrac{\pi}{2} - x\right) = \sec x$ **31.** $\sin x + \sin 2x = \sin 3x$

32. $\cos\left(\dfrac{5\pi}{2} + x\right) = -\sin x$

33. Evaluate $\sin 75°$ in exact form by using either $\sin(30° + 45°)$ or $\sin(135° - 60°)$.

34. If $\tan x = 3/4$ and $x + y = \pi/4$, find $\tan y$.

4.5 Product and Factoring Identities

The addition formulas from Section 4.4 can be used to derive formulas that convert products to sums and sums to products. All of these identities find application in calculus and other courses. They are so similar that most people do not attempt to memorize them, but it is important to be aware of them, to know where they are located (for reference), or how to derive them from identities (I-10) through (I-13).

Product to sum identities

$$\sin x \cos y = \tfrac{1}{2}[\sin(x + y) + \sin(x - y)] \qquad \textbf{[I-17]}$$

$$\cos x \sin y = \tfrac{1}{2}[\sin(x + y) - \sin(x - y)] \qquad \textbf{[I-18]}$$

$$\cos x \cos y = \tfrac{1}{2}[\cos(x + y) + \cos(x - y)] \qquad \textbf{[I-19]}$$

$$\sin x \sin y = \tfrac{1}{2}[\cos(x - y) - \cos(x + y)] \qquad \textbf{[I-20]}$$

The proof of the first product formula is given in Example 1. The other derivations are similar and are left to the reader in Problems 1, 2, and 3 of Exercise 4.5.

EXAMPLE 1 Prove that $\sin x \cos y = \tfrac{1}{2}[\sin(x + y) + \sin(x - y)]$.

Solution We have expressions for $\sin(x + y)$ and $\sin(x - y)$ in identities (I-10) and (I-12), which we may add:

$$\sin(x + y) = \sin x \cos y + \cos x \sin y$$
$$\underline{\sin(x - y) = \sin x \cos y - \cos x \sin y}$$
$$\sin(x + y) + \sin(x - y) = 2 \sin x \cos y$$

The desired identity follows when we divide by 2. ∎

Factoring identities (sums to products) ⸻

$$\sin A + \sin B = 2 \sin\left(\frac{A+B}{2}\right) \cos\left(\frac{A-B}{2}\right) \qquad \text{[I-21]}$$

$$\sin A - \sin B = 2 \cos\left(\frac{A+B}{2}\right) \sin\left(\frac{A-B}{2}\right) \qquad \text{[I-22]}$$

$$\cos A + \cos B = 2 \cos\left(\frac{A+B}{2}\right) \cos\left(\frac{A-B}{2}\right) \qquad \text{[I-23]}$$

$$\cos A - \cos B = -2 \sin\left(\frac{A+B}{2}\right) \sin\left(\frac{A-B}{2}\right) \qquad \text{[I-24]}$$

EXAMPLE 2   Prove that $\sin A + \sin B = 2 \sin\left(\frac{A+B}{2}\right) \cos\left(\frac{A-B}{2}\right)$.

Solution   From the last line of Example 1 we have

$$\sin(x+y) + \sin(x-y) = 2 \sin x \cos y.$$

If we set $x + y = A$ and $x - y = B$ and solve for x and y, we find that $x = (A+B)/2$ and $y = (A-B)/2$. Substituting these values for x and y, we obtain the desired identity.   ∎

EXERCISE 4.5

In Problems 1–6, prove that each is an identity. See Example 1 or Example 2.

1. $\cos x \sin y = \frac{1}{2}[\sin(x+y) - \sin(x-y)]$
2. $\cos x \cos y = \frac{1}{2}[\cos(x+y) + \cos(x-y)]$
3. $\sin x \sin y = \frac{1}{2}[\cos(x-y) - \cos(x+y)]$
4. $\sin A - \sin B = 2 \cos\left(\frac{A+B}{2}\right) \sin\left(\frac{A-B}{2}\right)$
5. $\cos A + \cos B = 2 \cos\left(\frac{A+B}{2}\right) \cos\left(\frac{A-B}{2}\right)$
6. $\cos A - \cos B = -2 \sin\left(\frac{A+B}{2}\right) \sin\left(\frac{A-B}{2}\right)$

In Problems 7–12, express each product as a sum or difference.

7. $\sin 3\theta \cos 5\theta$ 8. $\cos 3\theta \cos \theta$ 9. $\sin 2x \sin 4x$
10. $\sin 3x \sin 2x$ 11. $\sin 5x \cos x$ 12. $\cos x \cos 2x$

In Problems 13–16, factor each expression (that is, write as a product).

13. $\sin 2x + \sin 4x$ 14. $\cos x + \cos 2x$
15. $\cos 5x - \cos 3x$ 16. $\sin 3x - \sin x$

In Problems 17–19, simplify the given expression by using the factoring identities.

17. $\dfrac{\sin 3x + \sin x}{\cos 3x + \cos x}$
  18. $\dfrac{\cos x - \cos 4x}{\sin x - \sin 4x}$
  19. $\dfrac{\sin 4x + \sin 2x}{2 \cos x}$

For the remaining problems, use identities from Sections 4.4 and 4.5 as convenient.

In Problems 20–23, write the given expression in terms of $\sin x$ and $\cos x$.

20. $\sin\left(x - \dfrac{\pi}{4}\right)$
  21. $\sin 2x$
  22. $\cos 2x$
  23. $\sin\left(2x - \dfrac{\pi}{3}\right)$

In Problems 24–27, evaluate in exact form.

24. $\sin \dfrac{\pi}{4} \cos \dfrac{\pi}{12} + \sin \dfrac{\pi}{12} \cos \dfrac{\pi}{4}$
  25. $\cos 160° \cos 25° + \sin 160° \sin 25°$

26. $\cos^2 47° + \sin^2 47°$
  27. $\dfrac{\tan 37° - \tan 67°}{1 + \tan 37° \tan 67°}$

28. If A, B, and C are the angles of a triangle, prove that
  a] $\sin C = \sin A \cos B + \cos A \sin B$
  b] $\cos C = \sin A \sin B - \cos A \cos B$

29. A movie marquee on Main Street is 1.5 meters wide with its bottom edge 4 meters above the sidewalk, as shown in Fig. 4.3. A person, with eye level h meters above the sidewalk and x meters from point P directly below the edge of the marquee, is walking along Main Street and observes that the marquee (as measured by angle θ) seems small when viewed from far away (when x is large), but upon getting closer angle θ gets larger until it reaches a maximum, and then it begins to get smaller again until it becomes $0°$ when seen from directly underneath the edge of the marquee. That is, θ is a function of x. Show that this function is given by

$$\theta = \text{Tan}^{-1} \frac{1.5x}{x^2 + (4 - h)(5.5 - h)}.$$

FIGURE 4.3

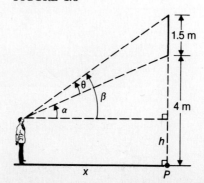

1.5 m

4 m

h

x

P

Hint: Use the two right triangles involving angles α and β and the identity

$$\tan \theta = \tan (\beta - \alpha) = \frac{\tan \beta - \tan \alpha}{1 + \tan \beta \tan \alpha}.$$

30. Suppose the person in Problem 29 is Janet, whose eye level above the sidewalk is 1.5 meters.

   a] Show that her view of the marquee is given by the expression

   $$\theta = \text{Tan}^{-1} \frac{1.5x}{x^2 + 10}.$$

   b] Use your calculator and the result in (a) to complete the following table that gives her view for different values of x in meters. Express angle θ in degrees to two decimal places.

x	40	25	20	10	8	6	5	4	3.5	3.2	3.1	3.0	2.8	2.5	2.0	1.5	1.1	0.5
θ																		

   c] Using the results of (b), make a reasonable estimate of how far from point P Janet should stand to get the best view (that is, the largest value of θ). Refine your estimate by using additional values of x to give an answer correct to two decimal places.

31. Suppose the person in Problem 29 is Preston, whose eye level above the sidewalk is 2 meters.

   a] Show that his view of the marquee is given by the expression

   $$\theta = \text{Tan}^{-1} \frac{1.5x}{x^2 + 7}.$$

   b] Compile a table similar to the one in Problem 30.

   c] How far from point P should Preston stand to get the best view?

32. Using the results found in Problems 30 and 31, answer each of the following:

   a] If Janet is standing at her spot of maximum view, how far behind her should Preston be to get the same view?

   b] When Preston is standing 16 meters from point P, find his view (angle θ) of the marquee from that point.

   c] When Preston is standing 16 meters from P, how far in front of him should Janet be to get the same view he has?

4.6 Double-Angle Formulas

Useful identities can be derived from the addition formulas given in Section 4.4. The following are called *double-angle identities*:

$$\sin 2\theta = 2 \sin \theta \cos \theta$$

[I-25]

$$\boxed{\cos 2\theta = \cos^2 \theta - \sin^2 \theta = 1 - 2\sin^2 \theta = 2\cos^2 \theta - 1} \qquad \text{[I-26]}$$

$$\boxed{\tan 2\theta = \frac{2\tan \theta}{1 - \tan^2 \theta}} \qquad \text{[I-27]}$$

These are special cases of identities (I-10), (I-11), and (I-14) where we take $\alpha = \theta$ and $\beta = \theta$ (see Problem 1 of Exercise 4.6).

Double-angle identities are useful in simplifying certain trigonometric expressions, and the student should become familiar with them. We consider several examples in which the double-angle identities are used along with identities introduced earlier.

EXAMPLE 1 Prove that $\sin 2x = \dfrac{2\tan x}{1 + \tan^2 x}$ is an identity.

Solution

$$\sin 2x \overset{?}{=} \frac{2\tan x}{1 + \tan^2 x}$$

Use identity (I-25)

$$2\sin x \cos x$$

Apply identity (I-8)

$$\frac{2\tan x}{\sec^2 x}$$

Use identities (I-1) and (I-3)

$$\left(\frac{2\sin x}{\cos x}\right)(\cos^2 x)$$

Simplify

$$2\sin x \cos x$$

Thus the given equation is an identity. ∎

EXAMPLE 2 If $\sin \theta = 3/5$ and $\cos \theta$ is negative, evaluate the following:

a] $\sin 2\theta$ **b]** $\cos 2\theta$

FIGURE 4.4

Solution Since $\sin \theta > 0$ and $\cos \theta < 0$, angle θ is in the second quadrant, as shown in Fig. 4.4.

a] To find $\sin 2\theta$, we use Fig. 4.4 to get $\sin \theta$ and $\cos \theta$ and then apply identity (I-25):

$$\sin 2\theta = 2\sin \theta \cos \theta = 2\left(\frac{3}{5}\right)\left(-\frac{4}{5}\right) = -\frac{24}{25}.$$

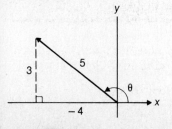

b] To find $\cos 2\theta$ we use identity (I-26):

$$\cos 2\theta = \cos^2 \theta - \sin^2 \theta = \left(-\frac{4}{5}\right)^2 - \left(\frac{3}{5}\right)^2 = \frac{7}{25}. \qquad \blacksquare$$

EXAMPLE 3 Express $\sin 3x$ as a function of $\sin x$.

Solution

$$\sin 3x = \sin (2x + x) = \sin (2x) \cos x + \cos (2x) \sin x$$
$$\textbf{By identity (I-10)}$$
$$= (2 \sin x \cos x) \cos x + (\cos^2 x - \sin^2 x) \sin x$$
$$\textbf{By identities (I-25) and (I-26)}$$
$$= 3 \sin x \cos^2 x - \sin^3 x \qquad \textbf{By algebra}$$
$$= 3 \sin x (1 - \sin^2 x) - \sin^3 x \qquad \textbf{By identity (I-5)}$$
$$= 3 \sin x - 4 \sin^3 x.$$

Therefore $\sin 3x = 3 \sin x - 4 \sin^3 x$ is an identity. $\qquad \blacksquare$

EXAMPLE 4 Find $\sin 22°30'$ in exact form. Using your calculator, evaluate the result and give the answer correct to four decimal places.

Solution We use identity (I-25) in the form $\cos 2\theta = 1 - 2 \sin^2 \theta$, and take $\theta = 22°30'$ (that is, $2\theta = 45°$):

$$\cos 45° = 1 - 2 (\sin 22°30')^2.$$

Solving for $(\sin 22°30')^2$ and using $\cos 45° = \sqrt{2}/2$, we have

$$(\sin 22°30')^2 = \frac{1 - \sqrt{2}/2}{2} = \frac{2 - \sqrt{2}}{4}.$$

Therefore

$$\sin 22°30' = \frac{\sqrt{2 - \sqrt{2}}}{2} \quad \text{(exact form)}.$$

Using a calculator, we evaluate the right-hand side and get

$$\sin 22°30' \approx 0.3827 \quad \text{(approximate decimal form)}. \qquad \blacksquare$$

EXAMPLE 5 Prove that $\sin 4x = 4 \sin x \cos x - 8 \sin^3 x \cos x$ is an identity.

Solution

$$\sin 4x = 2 \sin 2x \cos 2x \qquad \textbf{By identity (I-25)}$$
$$= 2 (2 \sin x \cos x)(1 - 2 \sin^2 x)$$
$$\textbf{By identities (I-25) and (I-26)}$$
$$= 4 \sin x \cos x - 8 \sin^3 x \cos x \qquad \textbf{By algebra}$$

Therefore the given equation is an identity. $\qquad \blacksquare$

EXAMPLE 6 Is $(\sin 6x + \cos 6x)^2 = 1$ an identity?

Solution We first try a few values of x to see if the equation is satisfied:

$$\text{if } x = 0, \text{ then LHS:}\quad (\sin 0 + \cos 0)^2 = (0 + 1)^2 = 1;$$

$$\text{if } x = \frac{\pi}{2}, \text{ then LHS:}\quad (\sin 3\pi + \cos 3\pi)^2 = (0 - 1)^2 = 1;$$

$$\text{if } x = \frac{\pi}{4}, \text{ then LHS:}\quad \left(\sin\frac{3\pi}{2} + \cos\frac{3\pi}{2}\right)^2 = (-1 + 0)^2 = 1.$$

It appears that the equation may represent an identity. However, if we try $x = 1$, we get

$$\text{LHS:}\quad (\sin 6 + \cos 6)^2 \approx 0.46 \quad \text{(to two decimal places).}$$

Therefore $(\sin 6x + \cos 6x)^2 = 1$ is not an identity. ■

EXAMPLE 7 Suppose $\sin\theta = 0.3487$ and $0° < \theta < 90°$. Using a calculator, evaluate each of the following to four decimal places:

a] $\sin 2\theta$ **b]** $\cos 2\theta$ **c]** $\tan 2\theta$

Solution 1 Enter 0.3487 into the display. Then with the calculator in either degree or radian mode, press keys (INV) and (sin) (or (sin⁻¹) key), which gives θ in the display, multiply the result by 2, and store it with the (STO) key. Using the (RCL) key as needed, we get:

a] $\sin 2\theta \approx 0.6536$ **b]** $\cos 2\theta \approx 0.7568$ **c]** $\tan 2\theta \approx 0.8637$

(Note: On some calculators the store and recall keys may be labeled differently from (STO) and (RCL).)

Solution 2 With the identities we now have at our disposal, there are many alternative approaches to this problem. Solution 1 uses the calculator to find θ (by evaluating $\mathrm{Sin}^{-1} 0.3487$). Identity (I-26) gives us $\cos 2\theta = 1 - 2\sin^2\theta$, so we may find $\cos 2\theta = 1 - 2(0.3487)^2 \approx 0.7568$ without evaluating θ. The other values may be obtained, for example, from $\sin 2\theta = \sqrt{1 - \cos^2 2\theta} \approx 0.6536$ and $\tan 2\theta = (\sin 2\theta)/(\cos 2\theta) \approx 0.8637$. ■

EXERCISE 4.6

1. Give details of the proof that identities (I-25), (I-26), and (I-27) are special cases of identities (I-10), (I-11), and (I-14), respectively.

In Problems 2–24, prove that the given equations are identities.

2. $(\sin\theta + \cos\theta)^2 = 1 + \sin 2\theta$ 3. $\dfrac{1}{\csc 2\theta} = 2\sin\theta\cos\theta$

4. $\sin 2\theta \sec \theta = 2 \sin \theta$

5. $(\cos x + \sin x)(\cos x - \sin x) = \cos 2x$

6. $\cos 2x \tan 2x = \sin 2x$　　　　　**7.** $\sin 2x \tan x = 2 \sin^2 x$

8. $(\cos x - \sin x) \sec 2x = \dfrac{1}{\cos x + \sin x}$

9. $(1 + \tan x) \tan 2x = \dfrac{2 \tan x}{1 - \tan x}$

10. $\tan \theta \sin 2\theta = 1 - \cos 2\theta$　　　**11.** $\sin 2\theta \sec^2 \theta = 2 \tan \theta$

12. $\cot x - \tan x = 2 \cot 2x$　　　**13.** $2 \csc 2x = \tan x + \cot x$

14. $\dfrac{2}{1 + \cos 2\theta} = \sec^2 \theta$　　　**15.** $\cot 2\theta = \dfrac{\cot^2 \theta - 1}{2 \cot \theta}$

16. $\cos^4 x - \sin^4 x = \cos 2x$　　　**17.** $\dfrac{1 - \tan x}{1 + \tan x} = \sec 2x - \tan 2x$

18. $\dfrac{\sin 2x}{1 + \cos 2x} = \tan x$　　　**19.** $(\cot x - \tan x) \tan 2x = 2$

20. $2 \tan \theta \csc 2\theta = 1 + \tan^2 \theta$　　　**21.** $\dfrac{1 + \tan^2 \theta}{\tan \theta} = 2 \csc 2\theta$

22. $\dfrac{1 - \cos 2x}{1 + \cos 2x} = \tan^2 x$　　　**23.** $\cos 3x = 4 \cos^3 x - 3 \cos x$

24. $\cos 4x = \cos^4 x - 6 \sin^2 x \cos^2 x + \sin^4 x$

25. If $\cos \theta = -12/13$ and θ is in the second quadrant, find in exact form:

　　a] $\sin 2\theta$　　　　**b]** $\cos 2\theta$　　　　**c]** $\tan 2\theta$

26. If $\sin \theta = -5/13$ and $\cos \theta = 12/13$, find in exact form:

　　a] $\sin 2\theta$　　　　**b]** $\cos 2\theta$　　　　**c]** $\tan 2\theta$

27. Suppose $\cos \theta = 0.5873$ and $0° < \theta < 90°$. Using a calculator, evaluate the following to four decimal places:

　　a] $\sin 2\theta$　　　　**b]** $\cos 2\theta$　　　　**c]** $\tan 2\theta$

28. Suppose $\sin \theta = 0.4385$ and $0 < \theta < \pi/2$. Using a calculator, evaluate to four decimal places:

　　a] $\sin 2\theta$　　　　**b]** $\cos 3\theta$　　　　**c]** $\cot 3\theta$

29. Evaluate the following and give answers in exact form:

　　a] $\sin 15° \cos 15°$　　**b]** $\sin^2 105° - \cos^2 105°$　**c]** $1 - 2 \sin^2 \dfrac{5\pi}{12}$

In Problems 30–39, determine which of the given equations are identities.

30. $\sec 2x = \dfrac{1}{2 \cos x}$　　　　　**31.** $\sin 4x = 2 \sin 2x \cos 2x$

32. $\sin 2x + \sin 3x = \sin 5x$　　　　**33.** $\sin^2 2x = 1 - \cos^2 2x$

34. $2 \cot 2x = \cot x - \tan x$　　　　**35.** $2 \csc 2x = \sec x \csc x$

36. $\sin 3x \sin 2x = \sin 6x$　　　　**37.** $(\sin 2x + \cos 2x)^2 = 1$

38. $(\sin 4x + \cos 4x)^2 = 1$ **39.** $\sec 2x + \tan 2x = \tan\left(\dfrac{\pi}{4} + x\right)$

40. In Example 6 it was seen that $(\sin 6x + \cos 6x)^2 = 1$ is not an identity, and in Problem 37 you can show that $(\sin 2x + \cos 2x)^2 = 1$ is not an identity either. It is obvious that $(\sin 0x + \cos 0x)^2 = 1^2 = 1$ is an identity. Is there any nonzero k such that $(\sin kx + \cos kx)^2 = 1$ is an identity?

41. If α is an acute angle, then the double-angle formulas can be derived by using Fig. 4.5, where triangle ABC is inscribed in a semicircle of unit radius with center Q. Let α be the angle at A and D be the foot of the perpendicular from C. Then

   a] show that the labels given to the angles in the diagram are justified (recall from Section 1.1 that angle ACB is a right angle);

   b] using the triangles shown in the diagram, derive identities (I-25) and (I-26).

In each of Problems 42–45 there is a number c such that the given equation becomes an identity for that value of c.

   a] Determine c. (Hint: An identity is satisfied by all allowable values of x.)

   b] Substitute your number c into the given equation and prove that the resulting equation actually is an identity.

42. $4 \sin^4 x = \cos^2 2x - 2 \cos 2x + c$

43. $\sin^4 x = \cos^4 x - 2 \cos^2 x + c$

44. $\sec^4 x \sin^2 2x = c \tan^2 x$ **45.** $4 \sin^6 x + 4 \cos^6 x = 3 \cos^2 2x + c$

46. The slope of a nonvertical line l is given by $m = \tan \theta$, where the angle of inclination θ is the angle that l makes with the x-axis. Suppose l_1 is an angle bisector line with angle of inclination $\theta/2$ and let its slope be given by $x = \tan (\theta/2)$.

   a] Use identity (I-27) to show that for given m, the value of x can be found by solving the quadratic equation $mx^2 + 2x - m = 0$.

   b] In general, if m is a rational number, say $m = a/b$ where a and b are integers, then x is an irrational number. However, if $a^2 + b^2 = c^2$, where c is also an integer, prove that x is also a rational number.

   c] Prove the converse: If x is a rational number, then m is also a rational number, say $m = a/b$ such that $a^2 + b^2$ is a perfect square.

47. Verify the result in Problem 46(b) for the lines l having slope $3/4, 5/12, 20/21$. That is, find the corresponding slope of l_1 for each of these.

FIGURE 4.5

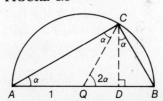

4.7 Half-Angle Formulas

If we write identity (I-26) in the form $\cos 2x = 1 - 2 \sin^2 x$ and then replace x by $\theta/2$, we get $\cos \theta = 1 - 2 \sin^2 (\theta/2)$. Solving for $\sin (\theta/2)$ gives

$$\sin \frac{\theta}{2} = \sqrt{\frac{1 - \cos \theta}{2}} \qquad \text{when } \sin \frac{\theta}{2} \geq 0,$$

$$\sin \frac{\theta}{2} = -\sqrt{\frac{1 - \cos \theta}{2}} \qquad \text{when } \sin \frac{\theta}{2} < 0.$$

These two equations are ordinarily written as

$$\sin \frac{\theta}{2} = \pm \sqrt{\frac{1 - \cos \theta}{2}}, \qquad \textbf{[I-28]}$$

where the "$\pm$" sign does not mean that we get two values for $\sin (\theta/2)$ but that we select the sign that is consistent with the sign of $\sin (\theta/2)$ (depending upon the quadrant in which $\theta/2$ is located).

In a similar manner, if we replace the angle θ by $\theta/2$ in the form $\cos 2\theta = 2 \cos^2 \theta - 1$ of identity (I-26), we get

$$\cos \frac{\theta}{2} = \pm \sqrt{\frac{1 + \cos \theta}{2}}, \qquad \textbf{[I-29]}$$

where again we select the $+$ or $-$ that agrees with the sign of $\cos (\theta/2)$.

We can now get an identity for $\tan (\theta/2)$ by using identities (I-28) and (I-29) along with identity (I-1):

$$\tan \frac{\theta}{2} = \pm \sqrt{\frac{1 - \cos \theta}{1 + \cos \theta}}. \qquad \textbf{[I-30]}$$

Identity (I-30) can be expressed in a more desirable form not involving the "$\pm$" sign. Rather than manipulating identity (I-30) directly, we can proceed as follows. When θ is replaced by $\theta/2$, identities (I-25) and (I-26) can be written in the form

$$\sin \theta = 2 \sin \frac{\theta}{2} \cos \frac{\theta}{2} \qquad \text{and} \qquad 1 + \cos \theta = 2 \cos^2 \frac{\theta}{2},$$

respectively. Dividing these two equations, we get

$$\frac{\sin\theta}{1+\cos\theta}=\frac{2\sin(\theta/2)\cos(\theta/2)}{2\cos^2(\theta/2)}=\frac{\sin(\theta/2)}{\cos(\theta/2)}=\tan\frac{\theta}{2}.$$

Thus

$$\tan\frac{\theta}{2}=\frac{\sin\theta}{1+\cos\theta}.$$

An alternative form of this equation is (see Example 4 in Section 4.2)

$$\tan\frac{\theta}{2}=\frac{1-\cos\theta}{\sin\theta}.$$

Therefore we have the following identities for $\tan(\theta/2)$:

$$\boxed{\tan\frac{\theta}{2}=\frac{\sin\theta}{1+\cos\theta}=\frac{1-\cos\theta}{\sin\theta}.}\qquad\text{[I-31]}$$

EXAMPLE 1 Evaluate each of the following and express the answer in exact form:

    a] $\sin 22°30'$ **b]** $\cos 112.5°$ **c]** $\tan\dfrac{7\pi}{12}$

Solution

    a] $\sin 22°30'=\sin\left(\dfrac{45}{2}\right)^{\!\circ}=\sqrt{\dfrac{1-\cos 45°}{2}}=\dfrac{1}{2}\sqrt{2-\sqrt{2}}\,;$

    b] $\cos 112.5°=\cos\left(\dfrac{225}{2}\right)^{\!\circ}=-\sqrt{\dfrac{1+\cos 225°}{2}}=-\dfrac{1}{2}\sqrt{2-\sqrt{2}}\,;$

    c] $\tan\dfrac{7\pi}{12}=\tan\dfrac{7\pi}{2\cdot 6}=\dfrac{1-\cos(7\pi/6)}{\sin(7\pi/6)}=\dfrac{1-(-\sqrt{3}/2)}{-1/2}$

             $=-(2+\sqrt{3}).$ ■

EXAMPLE 2 If $\cos\theta=-3/5$ and $180°<\theta<270°$ (Fig. 4.6), evaluate the following in exact form:

    a] $\sin\dfrac{\theta}{2}$ **b]** $\cos\dfrac{\theta}{2}$ **c]** $\tan\dfrac{\theta}{2}$

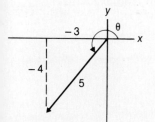

FIGURE 4.6

Solution We first note that $90°<\theta/2<135°$, and so $\sin(\theta/2)$ is positive and $\cos(\theta/2)$ is negative.

    a] $\sin\dfrac{\theta}{2}=\sqrt{\dfrac{1-\cos\theta}{2}}=\sqrt{\dfrac{1-(-3/5)}{2}}=\dfrac{2\sqrt{5}}{5}\,;$

    b] $\cos\dfrac{\theta}{2}=-\sqrt{\dfrac{1+\cos\theta}{2}}=-\sqrt{\dfrac{1+(-3/5)}{2}}=-\dfrac{\sqrt{5}}{5}\,;$

    c] $\tan\dfrac{\theta}{2}=\dfrac{\sin\theta}{1+\cos\theta}=\dfrac{-4/5}{1+(-3/5)}=-2.$ ■

EXAMPLE 3 Evaluate $\sin 15°$ in exact form in two ways:

a] by using identity (I-12)   **b]** by using identity (I-28)

Solution

a] $\sin 15° = \sin(45° - 30°) = \sin 45° \cos 30° - \cos 45° \sin 30°$

$$= \frac{\sqrt{6} - \sqrt{2}}{4};$$

therefore

$$\sin 15° = \frac{\sqrt{6} - \sqrt{2}}{4}.$$

b] $\sin 15° = \sin\left(\frac{30}{2}\right)° = \sqrt{\frac{1 - \cos 30°}{2}} = \frac{1}{2}\sqrt{2 - \sqrt{3}};$

therefore

$$\sin 15° = \frac{1}{2}\sqrt{2 - \sqrt{3}}.$$

It appears that we get two different answers for $\sin 15°$. We leave it for the student to evaluate each with a calculator to see if they might both represent the same number. (See Problem 26 of Exercise 4.7). ■

EXAMPLE 4   Suppose $\sin(\theta/2) = 0.6843$ and $0° < \theta < 180°$. Use a calculator to evaluate each of the following to four decimal places:

a] $\sin \theta$ **b]** $\cos 2\theta$ **c]** $\tan \dfrac{\theta}{4}$

Solution   Enter 0.6843 into the display. Then with the calculator in either degree or radian mode, press (INV) and (sin) keys (or (sin⁻¹) key); then multiply by 2 (this gives θ) and store into memory with the (STO) key. Using the (RCL) key as needed, we get

a] $\sin \theta \approx 0.9980$ **b]** $\cos 2\theta \approx -0.9919$

c] $\tan \dfrac{\theta}{4} \approx 0.3957.$ ■

EXERCISE 4.7

In Problems 1–4, give answers in exact form; evaluate these results to four decimal places and then check by evaluating directly with a calculator.

1. a] $\sin 67°30'$ **b]** $\cos(-22.5°)$ **c]** $\sin 105°$ **d]** $\cos 105°$

2. a] $\tan 165°$ **b]** $\cos(247.5°)$ **c]** $\tan(-195°)$ **d]** $\cos 285°$

3. **a]** $\sin \dfrac{\pi}{12}$ **b]** $\cos \dfrac{5\pi}{8}$ **c]** $\sin \dfrac{11\pi}{8}$ **d]** $\tan \dfrac{13\pi}{12}$

4. **a]** $\cos \dfrac{19\pi}{8}$ **b]** $\sin\left(-\dfrac{7\pi}{8}\right)$ **c]** $\sin \dfrac{21\pi}{8}$ **d]** $\tan\left(-\dfrac{5\pi}{12}\right)$

In Problems 5–12 express answers in exact form.

5. If $\cos \theta = -\dfrac{5}{13}$ and $90° < \theta < 180°$, evaluate

    a] $\sin \dfrac{\theta}{2}$ **b]** $\cos \dfrac{\theta}{2}$ **c]** $\tan \dfrac{\theta}{2}$ **d]** $\sec \dfrac{\theta}{2}$

6. If $\tan \theta = -\dfrac{3}{4}$ and $-\dfrac{\pi}{2} < \theta < 0$, find

    a] $\sin \dfrac{\theta}{2}$ **b]** $\cot \dfrac{\theta}{2}$ **c]** $\sec \dfrac{\theta}{2}$ **d]** $\csc \dfrac{\theta}{2}$

7. If $\sin \theta = \dfrac{1}{2}$ and $360° < \theta < 450°$, find $\cos \dfrac{\theta}{2}$ and $\tan \dfrac{\theta}{2}$.

8. If $\cos \theta = -\dfrac{3}{4}$ and $0° < \theta < 180°$, evaluate

    a] $\sin \dfrac{\theta}{2}$ **b]** $\cos \dfrac{\theta}{2}$ **c]** $\sin 2\theta$ **d]** $\cos 2\theta$

9. If $\tan \alpha = 5$ and $\pi < \alpha < \dfrac{3\pi}{2}$, determine

    a] $\sin \alpha$ **b]** $\sin \dfrac{\alpha}{2}$ **c]** $\sin 2\alpha$ **d]** $\tan \dfrac{\alpha}{2}$

10. If $\cos \beta = -\dfrac{1}{\sqrt{2}}$ and $180° < \beta < 360°$, find

    a] $\cos \dfrac{\beta}{2}$ **b]** $\tan \dfrac{\beta}{2}$ **c]** $\tan 2\beta$ **d]** $\cos 2\beta$

11. If $\sin \dfrac{\theta}{2} = -\dfrac{3}{4}$, find $\cos \theta$.   **12.** If $\cos \dfrac{\theta}{2} = \dfrac{1}{\sqrt{2}}$, find $\cos \theta$.

13. Suppose $\sin \theta = 0.5486$ and $0° < \theta < 90°$. Use a calculator to evaluate each of the following to four decimal places:

    a] $\sin \dfrac{\theta}{2}$ **b]** $\cos \dfrac{\theta}{2}$ **c]** $\tan \dfrac{\theta}{2}$

14. Suppose $\cos \theta/2 = 0.6431$ and $0 < \theta < \pi$. Use a calculator to evaluate each of the following to four decimal places:

    a] $\sin \theta$ **b]** $\cos 2\theta$ **c]** $\tan \dfrac{\theta}{4}$

In Problems 15–22, prove that the given equations are identities.

15. $\tan \dfrac{\theta}{2} = \csc \theta - \cot \theta$ **16.** $\left(\sin \dfrac{\theta}{2} + \cos \dfrac{\theta}{2}\right)^2 = 1 + \sin \theta$

17. $\cos^2 \dfrac{x}{2} - \sin^2 \dfrac{x}{2} = \cos x$ **18.** $2 \sin^2 \dfrac{x}{2} = \dfrac{\sec x - 1}{\sec x}$

19. $\tan \dfrac{x}{2} = \dfrac{\sec x - 1}{\sin x \sec x}$ **20.** $2 \sin^2 \dfrac{x}{2} = \sin x \tan \dfrac{x}{2}$

21. $2 \cos^2 \dfrac{x}{2} = \dfrac{\sin x + \tan x}{\tan x}$ **22.** $\tan \dfrac{3\theta}{2} = \dfrac{\cos 2\theta - \cos \theta}{\sin \theta - \sin 2\theta}$

23. Follow Example 3 of this section and evaluate $\cos 165°$ by two different methods. Check to see that the two answers actually represent the same number.

24. Follow the instructions of Problem 23 for $\cos 15°$.

25. If $\cos \theta = -3/5$ and $90° < \theta < 180°$, find each of the following in exact form:

a] $\cos \dfrac{\theta}{2}$ **b]** $\cos \dfrac{\theta}{4}$

26. In Example 3 of this section we concluded that the two numbers

$$\dfrac{\sqrt{6} - \sqrt{2}}{4} \quad \text{and} \quad \dfrac{1}{2} \sqrt{2 - \sqrt{3}}$$

are equal. Use your calculator to "check" this conclusion; then prove that they are equal without using a calculator. (Hint: Show that the squares of these two numbers are equal. Why isn't this fact sufficient to show that the numbers are equal? What if we know that both are positive?)

27. Derive identity (I-31) directly from (I-30). (Hint: Multiply numerator and denominator of $(1 - \cos \theta)/(1 + \cos \theta)$ by $(1 + \cos \theta)$ and get

$$\sqrt{\dfrac{1 - \cos \theta}{1 + \cos \theta}} = \dfrac{|\sin \theta|}{1 + \cos \theta}.$$

Now show that $\tan \theta/2$ and $\sin \theta$ always agree in sign.)

28. Show that

$$\cos \dfrac{\pi}{4} = \dfrac{1}{2} \sqrt{2}, \quad \cos \dfrac{\pi}{8} = \dfrac{1}{2} \sqrt{2 + \sqrt{2}}, \quad \cos \dfrac{\pi}{16} = \dfrac{1}{2} \sqrt{2 + \sqrt{2 + \sqrt{2}}}.$$

Generalize by telling how many "nested square roots" $\cos (\pi/2^n)$ will have. As n approaches infinity, what can be said about $\sqrt{2 + \sqrt{2 + \sqrt{2 + \dots}}}$?

4.8 Using Identities beyond Trigonometry (Optional)

At this point we have identified 31 identities by number in this chapter. They are listed on the back endpaper of the book for convenience of reference. Including the reduction formulas in Chapter 2 and the identities in the problems in the exercise sets, we have seen many more. Some of the problems in the exercise sets are for practice, intended to reinforce an understanding of the essential ideas and to give some degree of mastery through repetition. However, many of the problems are precisely the kind encountered in applications of trigonometry in subsequent courses.

If there is time and opportunity to check a reference source, then a list of identities such as that found at the end of this book is entirely sufficient. Our experience suggests that a student who has learned how to work from the most fundamental identities has a substantial advantage. A rough analogy may be drawn from the quadratic formula in algebra. Surely it is helpful to know that the formula exists and will permit the solution of any quadratic equation, but the student who must go and check an algebra book for the details of the quadratic formula is handicapped in comparison to the student who can write down the solution immediately.

In classifying identities we consider the reciprocal and quotient identities (I-1) through (I-4) to be part of the basic fabric of definitions; they should not even have to be listed. Identities (I-6) and (I-7), which indicate that the sine is an odd function while the cosine is even, must be remembered, but their content is included in the graphs of the sine and cosine functions. As we stressed in Chapter 2, every student should be able to make a quick sketch of both graphs whenever the trigonometric functions are used. Assuming then, that these facts are automatic, the truly key identities needed beyond are identities (I-5), (I-10), and (I-11).

$$\sin^2 t + \cos^2 t = 1 \qquad\qquad \textbf{[I-5]}$$

$$\sin(\alpha + \beta) = \sin\alpha\,\cos\beta + \cos\alpha\,\sin\beta \qquad\qquad \textbf{[I-10]}$$

$$\cos(\alpha + \beta) = \cos\alpha\,\cos\beta - \sin\alpha\,\sin\beta \qquad\qquad \textbf{[I-11]}$$

It is probably impossible to overemphasize how fundamental identity (I-5) is throughout mathematics and its applications. Any student who takes more mathematics courses will use it often enough to thoroughly absorb it. The other two identities are probably more important as convenient means to get the double-angle and half-angle identities and all the reduction formulas, although they are used in precisely the forms above in calculus in derivation of formulas for the derivatives of the sine and cosine functions.

Exercise 4.8 includes a number of derivations. Several of these have been done already in this chapter. They are repeated here because we feel strongly that the ability to work from the basics to derive needed identities is beneficial. All of the other problems are similar to exercises in previous sections, but *all are taken directly from a first course in calculus and analytic geometry.*

EXERCISE 4.8

Derivations

1. From identities (I-10) and (I-11), derive identities (I-12) and (I-13).

2. From identities (I-10) and (I-11), derive identity (I-14).

3. From identities (I-10) and (I-11), derive identities (I-25) and (I-26).

4. From the double-angle formulas for $\cos 2\theta$, express $\cos^2 \theta$ in terms of $\cos 2\theta$.

5. From the double-angle formulas for $\cos 2\theta$, express $\sin^2 \theta$ in terms of $\cos 2\theta$.

6. Use identity (I-5) to derive an identity involving $\tan^2 t$ and 1.

7. Use identity (I-5) to derive an identity involving $\cot^2 t$ and 1.

8. Having expressed $\cos^2 \theta$ in terms of $\cos 2\theta$ (Problem 4), replace 2θ by A (and θ by $A/2$) to derive a half-angle formula.

9. Follow a procedure similar to that in Problem 8 to derive a half-angle formula for $\sin (A/2)$.

10. Use the fact that each point on the unit circle has coordinates of the form $(\cos t, \sin t)$ to derive identity (I-5).

Reduction formulas

In Problems 11–22, use identities (I-10) and (I-11) to express each given function in simpler form as a function of x alone.

11. $\sin (x + \pi)$	12. $\tan (x - \pi)$	13. $\cos (x + \pi)$
14. $\sec (x + \pi/2)$	15. $\cot (-x)$	16. $\csc (x - \pi/2)$
17. $\sin (\pi/2 + x)$	18. $\cot (x + \pi/2)$	19. $\tan (x + 3\pi/4)$
20. $\cos (3\pi/2 - x)$	21. $\sec (x + \pi)$	22. $\sin (x + \pi/4)$

In the following we give a preview of some of the types of problems from analytic geometry and calculus in which identities are applied.

Identities in analytic geometry

23. *Angle between two lines*: Suppose L_1 and L_2 are two lines that have angles of inclination θ_1 and θ_2, respectively (see Fig. 4.7). Let $m_1 = \tan \theta_1$ and $m_2 = \tan \theta_2$ be the corresponding slopes and θ the angle from L_1 to L_2 as shown.

a] Show that $\theta = \theta_2 - \theta_1$.   b] Show that $\tan \theta = \dfrac{m_2 - m_1}{1 + m_1 m_2}$

24. Use the result of Problem 23 to find the acute angle in degree measure (to two decimal places) between the two lines given by $y = 2x - 3$ and $2y = x + 3$.

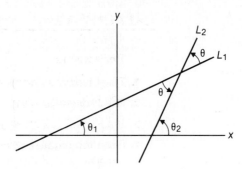

FIGURE 4.7

Rotation of axes

In identifying and graphing equations of the form $Ax^2 + Bxy + Cy^2 + Dx + Ey + F = 0$, a simplifying substitution (rotation of axes) will "remove" the xy-term:

$$x = x' \cos \theta - y' \sin \theta, \qquad y = x' \sin \theta + y' \cos \theta,$$

where $\tan 2\theta = B/(A - C)$. In Problems 25–27, *find the substitution* that will remove the xy-term. That is, identify $\tan 2\theta$ and then use appropriate identities to get $\sin \theta$ and $\cos \theta$ in exact form. It may be assumed that $0 < \theta < \pi/2$.

25. $73x^2 - 72xy + 52y^2 + 30x + 40y - 75 = 0$

26. $x^2 + 4xy + 4y^2 - 3x + 6 = 0$

27. $16x^2 + 24xy + 9y^2 + 100x - 50y = 0$

Identities in calculus

The study of elementary calculus involves two types of *operations on functions*: differentiation and integration. To perform these operations on functions in which trigonometric functions occur, it is frequently helpful to replace the given function by an equivalent one (using identities) to which established calculus formulas can be applied. Each of Problems 28–32 has been extracted from the solution of actual problems in calculus. Do not concern yourself with understanding the concepts or terminology of calculus (this will come later), but concentrate on the identity transformations required.

28. In the function $f(x) = \sin^2 x$ the first and second derivatives are given by the functions $g(x) = 2 \sin x \cos x$ and $h(x) = 2 \cos^2 x - 2 \sin^2 x$, respectively. Simplify the expressions for $g(x)$ and $h(x)$ by expressing each in terms of double angle, $2x$.

29. There is no obvious simple formula that will allow you to integrate the function $f(x) = 1/(1 + \sin x)$. However, there are formulas for the integration of $\sec^2 x$ and $\sec x \tan x$. Express $1/(1 + \sin x)$ in terms of $\sec^2 x$ and $\sec x \tan x$. (Hint: Multiply the numerator and denominator by $1 - \sin x$.)

30. There is no obvious simple formula for the integration of the function $\sin^2 x$. However, by applying identity (I-26), $\sin^2 x$ can be written in terms of $\cos 2x$. Standard formulas can be used to integrate the resulting expression. Express $\sin^2 x$ in terms of $\cos 2x$.

31. Follow a procedure similar to that in Problem 30 for the function $\cos^2 x$. That is, express $\cos^2 x$ in terms of $\cos 2x$.

32. The formula for differentiating a function that is the product of two functions is more complicated than that for differentiating a sum. Thus it is helpful to express a product as a sum if possible. This is one reason that the "product to sum" identities were introduced (Section 4.5). Express each of the following as a sum of sines and cosines.

    a] $f(x) = \sin 2x \cos 4x$　　　　**b]** $f(x) = \sin 3x \sin 5x$
    c] $f(x) = \cos x \cos 3x$

33. A quotient of two functions can frequently be simplified by factoring and "cancelling" common factors. Use the "sum to product" identities given in Section 4.5 to simplify the following:

    a] $\dfrac{\sin 2x + \sin 4x}{\cos 2x + \cos 4x}$　　　　**b]** $\dfrac{\sin x + \sin 2x}{\cos x - \cos 2x}$

34. Follow the instructions given in Problem 33 to simplify the following:

    a] $\dfrac{2 \sin 3x}{\cos 2x - \cos 4x}$　　　　**b]** $\dfrac{\cos 3x + \cos x}{\sin 4x + \sin 2x}$

35. Prove that $\sqrt{1 + \sin x} - \sqrt{1 - \sin x} = 2\sqrt{(1 - \cos x)/2}$ for x in the first quadrant.

36. Prove that $\sqrt{1 + \sin x} - \sqrt{1 - \sin x} = 2\sqrt{(1 + \cos x)/2}$ for x in the second quadrant.

Summary

Warning: In attempting to prove that an equation is an identity, do not perform the same operation on both sides of the equation.

Warning: Very few functions are additive. In particular

$$f(x + y) = f(x) + f(y)$$

is *not an identity* for *any* trigonometric function.

The identities most often needed are grouped for reference:

Group I: Basic identities

$$\tan x = \frac{\sin x}{\cos x}, \qquad \cot x = \frac{1}{\tan x} = \frac{\cos x}{\sin x}, \qquad \sec x = \frac{1}{\cos x},$$

$$\csc x = \frac{1}{\sin x}, \qquad \sin^2 x + \cos^2 x = 1, \qquad \tan^2 x + 1 = \sec^2 x,$$

$$1 + \cot^2 x = \csc^2 x$$

Group II: Odd–even and cofunction identities (reduction formulas)

$$\sin(-x) = -\sin x, \qquad \cos(-x) = \cos x, \qquad \tan(-x) = -\tan x,$$

$$\sin\left(\frac{\pi}{2} - x\right) = \cos x, \qquad \cos\left(\frac{\pi}{2} - x\right) = \sin x, \qquad \tan\left(\frac{\pi}{2} - x\right) = \cot x$$

Group III: Addition identities

$$\sin(\alpha + \beta) = \sin\alpha\,\cos\beta + \cos\alpha\,\sin\beta,$$

$$\sin(\alpha - \beta) = \sin\alpha\,\cos\beta - \cos\alpha\,\sin\beta,$$

$$\cos(\alpha + \beta) = \cos\alpha\,\cos\beta - \sin\alpha\,\sin\beta,$$

$$\cos(\alpha - \beta) = \cos\alpha\,\cos\beta + \sin\alpha\,\sin\beta,$$

$$\tan(\alpha + \beta) = \frac{\tan\alpha + \tan\beta}{1 - \tan\alpha\,\tan\beta},$$

$$\tan(\alpha - \beta) = \frac{\tan\alpha - \tan\beta}{1 + \tan\alpha\,\tan\beta}$$

Group IV: Double-angle and half-angle identities

$$\sin 2x = 2\sin x\,\cos x, \qquad \cos 2x = \cos^2 x - \sin^2 x$$
$$= 2\cos^2 x - 1$$
$$= 1 - 2\sin^2 x,$$

$$\tan 2x = \frac{2\tan x}{1 - \tan^2 x},$$

$$\sin\frac{x}{2} = \pm\sqrt{\frac{1 - \cos x}{2}}, \qquad \cos\frac{x}{2} = \pm\sqrt{\frac{1 + \cos x}{2}}$$

$$\tan\frac{x}{2} = \pm\sqrt{\frac{1 - \cos x}{1 + \cos x}} = \frac{\sin x}{1 + \cos x} = \frac{1 - \cos x}{\sin x},$$

Group V: Product to sum and sum to product (factoring) identities

$$2\sin x\,\cos y = \sin(x + y) + \sin(x - y)$$

$$2\cos x\,\cos y = \cos(x + y) + \cos(x - y)$$

$$2\cos x\,\sin y = \sin(x + y) - \sin(x - y)$$

$$2\sin x\,\sin y = \cos(x - y) - \cos(x + y)$$

$$\sin x + \sin y = 2\sin\left(\frac{x + y}{2}\right)\cos\left(\frac{x - y}{2}\right)$$

$$\cos x + \cos y = 2\cos\left(\frac{x + y}{2}\right)\cos\left(\frac{x - y}{2}\right)$$

$$\sin x - \sin y = 2\cos\left(\frac{x + y}{2}\right)\sin\left(\frac{x - y}{2}\right)$$

$$\cos x - \cos y = -2\sin\left(\frac{x + y}{2}\right)\sin\left(\frac{x - y}{2}\right)$$

Computer Problems (Optional!)

Section 4.3

1. In Exercise 4.3 we were given several equations and asked to determine which are identities. For any of the given equations, write a program that will evaluate the LHS and the RHS for several values of x. Use this approach to help you arrive at decisions for several of the problems in this exercise set.

Section 4.4

2. Follow a procedure similar to that in Problem 1 to help you in solving Problems 26–32 of Exercise 4.4.

Section 4.5

3. Write programs that will help you in solving Problems 30, 31, and 32 of Exercise 4.5.

Section 4.6

4. Follow a procedure similar to that in Problem 1 above to help you arrive at conclusions for Problems 30–39 of Exercise 4.6.

Section 4.7

5. Identity (I-28) is given by

$$\sin \frac{\theta}{2} = \pm \sqrt{\frac{1 - \cos \theta}{2}}.$$

Write a program that will first determine whether the + or − sign should be used for any given angle θ. Then continue the program so that it will evaluate the LHS and the RHS of identity (I-28) for any given angle θ. Check to see whether the results support the claim that (I-28) is an identity. Follow a similar procedure for identity (I-29).

6. In reference to Problem 28 of Exercise 4.7, write a program that will evaluate successive terms of the sequence

$$a_1 = \sqrt{2}, \quad a_2 = \sqrt{2 + \sqrt{2}}, \quad a_3 = \sqrt{2 + \sqrt{2 + \sqrt{2}}}, \quad \ldots$$

That is, $a_1 = \sqrt{2}$ and $a_{n+1} = \sqrt{2 + a_n}$ for $n = 1, 2, 3, \ldots$. Have your program give several terms of this sequence. See whether the sequence is approaching a fixed number.

Review Exercises

In Problems 1–25, prove that the given equations are identities:

1. $\cos x \tan x = \sin x$
2. $\sec (90° - \theta) \tan \theta = \sec \theta$
3. $\csc \theta \sin 2\theta = 2 \cos \theta$
4. $\cos (90° - 2\theta) = 2 \sin \theta \cos \theta$

5. $\tan\left(\theta + \dfrac{3\pi}{4}\right) = \dfrac{\sin\theta - \cos\theta}{\cos\theta + \sin\theta}$

6. $(\sin x + \cos x)^2 = 1 + \sin 2x$

7. $(1 - \sin 2x)(1 + \sin 2x) = \cos^2 2x$

8. $2\csc x \sin^2 \dfrac{x}{2} = \dfrac{\sin x}{1 + \cos x}$

9. $\cos\left(\dfrac{\pi}{2} + x\right)\cot(-x) = \cos x$

10. $\sin\theta \tan\dfrac{\theta}{2} = 1 - \cos\theta$

11. $\left(\sin\dfrac{\theta}{2} - \cos\dfrac{\theta}{2}\right)^2 = 1 - \sin\theta$

12. $\sin^2\dfrac{\theta}{2}\cos^2\dfrac{\theta}{2} = \dfrac{\sin^2\theta}{4}$

13. $\csc x \tan x = \sec x$

14. $\cot\dfrac{x}{2} - \tan\dfrac{x}{2} = 2\cot x$

15. $2\sin\left(\theta + \dfrac{\pi}{6}\right) = \sqrt{3}\sin\theta + \cos\theta$

16. $\sqrt{2}\cos\left(\theta - \dfrac{3\pi}{4}\right) = \sin\theta - \cos\theta$

17. $\tan 2x \csc 2x = \sec 2x$

18. $\left(1 - \cos\dfrac{x}{2}\right)\left(1 + \cos\dfrac{x}{2}\right) = \sin^2\dfrac{x}{2}$

19. $\cos^4\dfrac{x}{2} - \sin^4\dfrac{x}{2} = \cos x$

20. $\cos 2x \tan 2x = 2\sin x \cos x$

21. $(\sec\theta + 1)(\sec\theta - 1) = \tan^2\theta$

22. $(1 + \sin\theta)(1 - \csc\theta) = \sin\theta - \csc\theta$

23. $(1 - \tan\theta)\tan 2\theta = \dfrac{2\tan\theta}{1 + \tan\theta}$

24. $\cos^2\dfrac{x}{2} - \sin^2\dfrac{x}{2} = \cos x$

25. $\cos\theta\,(1 + \sec\theta) = 2\cos^2\dfrac{\theta}{2}$

In Problems 26–32, determine whether or not the given equations are identities. Give good reasons or proofs for your answers.

26. $\sin x = 2\sin\dfrac{x}{2}\cos\dfrac{x}{2}$

27. $\tan x \cot x - \sin^2 x = \cos^2 x$

28. $\sqrt{\sec^2\theta - \tan^2\theta} = 1$

29. $(\cos\theta - \sin\theta)^2 = \cos^2\theta - \sin^2\theta$

30. $\sin 2\theta + (\sin\theta - \cos\theta)^2 = 1$

31. $\sin x + \sin 2x = \sin 3x$

32. $\left(\cos\dfrac{x}{2} + \sin\dfrac{x}{2}\right)\left(\cos\dfrac{x}{2} - \sin\dfrac{x}{2}\right) = \cos x$

In Problems 33–50, evaluate the given expressions in exact form if angles α, β, and γ satisfy the following conditions.

$$\sin\alpha = \frac{3}{5} \quad \text{and} \quad \frac{\pi}{2} \leq \alpha \leq \pi,$$

$$\tan\beta = -\frac{5}{12} \quad \text{and} \quad -\frac{\pi}{2} < \beta < \frac{\pi}{2},$$

$$\cos\gamma = \frac{4}{5} \quad \text{and} \quad 0 \leq \gamma \leq \pi.$$

33. $\cos\alpha$

34. $\sin 2\alpha$

35. $\sin\dfrac{\gamma}{2}$

36. $\sin(\alpha + \beta)$

37. $\tan(\beta - \gamma)$

38. $\cos\dfrac{\beta}{2}$

39. $\cos 2\beta$

40. $\tan 2\gamma$

41. $\cos(\alpha + 2\beta)$

42. $\tan(2\alpha - \gamma)$

43. $1 - \cos^2 \alpha$

44. $\cos^2 \dfrac{\gamma}{2} - \sin^2 \dfrac{\gamma}{2}$

45. $\dfrac{\sin \beta}{\cos \beta}$

46. $\sec^2 \beta - \tan^2 \beta$

47. $\sin\left(\alpha - \dfrac{3\pi}{2}\right)$

48. $\tan\left(\beta + \dfrac{\pi}{4}\right)$

49. $\sin 2(\alpha + \beta)$

50. $\cos\left(\dfrac{\alpha + \beta}{2}\right)$

Trigonometric Equations
5

The reader already has some experience in solving algebraic equations. For example, the equation $x^2 - x - 12 = 0$ is satisfied by $x = 4$ and $x = -3$. That is, if x is replaced in the equation by 4, we get $4^2 - 4 - 12 = 0$, which is a true statement. Similarly, if $x = -3$, we get $(-3)^2 - (-3) - 12 = 0$, another true statement.

The set of possible replacement values for the variable is called the *replacement set* for that equation. In general, unless otherwise specified, the replacement set will be the largest subset of the set of real numbers for which the expressions on the two sides of the given equation are defined (as real numbers). The *solution set* for a given equation is the subset of all numbers from the replacement set that satisfy the equation.

We call an equation an *identity* if the solution set is the entire replacement set; otherwise the equation is called a *conditional equation*. In Chapter 4 we encountered a large number of identity equations involving trigonometric functions. In this chapter we consider conditional equations involving trigonometric functions with the primary goal of developing techniques for finding the solution sets for such equations.

Many of our problems will begin with an equation that we will not solve directly in its given form. We shall make use of the identities in Chapter 4 to replace the given equation by an equivalent one that we can solve. However, before we begin developing techniques for solving equations involving trigonometric functions, we first review some methods introduced in algebra courses.

5.1 Review of Techniques for Solving Equations from Algebra (Optional)

Linear equations

An equation that can be written in the form $ax + b = 0$, where a and b are given numbers and $a \neq 0$, is called a *linear equation*. Such an equation has one solution which is given by $x = -b/a$. For instance, the equation $2x - 6 = 0$ has one *root*, which is given by $x = 6/2 = 3$. We say that the *solution set* is $\{3\}$.

Quadratic equations

Any equation that can be written in the form

$$au^2 + bu + c = 0, \qquad \text{[5.1]}$$

where a, b, and c are given numbers and $a \neq 0$, is called a *quadratic equation* in u. For instance, each of the following is a quadratic equation:

$$2x^2 - 3x - 1 = 0, \qquad (x - 4)^2 = 5, \qquad (\sin x)^2 - 2(\sin x) - 3 = 0.$$

The first two are quadratic in x, and the third is quadratic in $\sin x$ (that is, $u = \sin x$ in Eq. (5.1)).

In algebra courses we study two techniques for solving quadratic equations: by factoring or by applying the quadratic formula.

Quadratic formula

The roots of $au^2 + bu + c = 0$, where $a \neq 0$, are given by

$$u = \frac{-b \pm \sqrt{b^2 - 4ac}}{2a}. \qquad \text{[5.2]}$$

EXAMPLE 1 Solve the equation $x^2 - 2x - 3 = 0$.

Solution By factoring: Since $x^2 - 2x - 3 = (x + 1)(x - 3)$ is an identity, we can replace the left-hand side of the given equation by $(x + 1)(x - 3)$ and get an equivalent equation $(x + 1)(x - 3) = 0$. In this form we can solve our problem by using the fundamental property of numbers that the product of two numbers is zero if and only if one of them is zero. Thus we get $x + 1 = 0$ or $x - 3 = 0$, and our problem is reduced to solving two linear equations, giving $x = -1$ or $x = 3$. The solution set is $\{-1, 3\}$. ∎

EXAMPLE 2 Find the solution set for the equation $2x^2 - 2x - 1 = 0$. Express answers in exact form and also in decimal form rounded off to two decimal places.

Solution The left-hand side of the given equation cannot be factored (in terms of whole numbers), and so we apply the quadratic formula. Here $a = 2$, $b = -2$, and $c = -1$. Substituting into Eq. (5.2) gives

$$x = \frac{-(-2) \pm \sqrt{(-2)^2 - 4(2)(-1)}}{2(2)} = \frac{2 \pm \sqrt{12}}{4}$$

$$= \frac{2(1 \pm \sqrt{3})}{4} = \frac{1 \pm \sqrt{3}}{2}.$$

Thus we have two roots, given by $x_1 = (1 + \sqrt{3})/2$ and $x_2 = (1 - \sqrt{3})/2$. The solution set for the given equation is $\{(1 + \sqrt{3})/2, (1 - \sqrt{3})/2\}$ (exact form), or $\{1.37, -0.37\}$ (approximate decimal form). ∎

EXAMPLE 3 Solve the equation $x^2 - x + 2 = 0$.

Solution Apply the quadratic formula with $a = 1$, $b = -1$, $c = 2$ to get

$$x = \frac{1 \pm \sqrt{1 - 8}}{2} = \frac{1 \pm \sqrt{-7}}{2}.$$

We see that the roots are not real numbers (since $\sqrt{-7}$ is an imaginary number), and thus there is no solution in the system of real numbers. If we were working in the context of the set of complex numbers, the solution set would contain two complex numbers, $(1 \pm \sqrt{7}i)/2$; since we are working with real numbers only, the solution set is the empty set. ∎

Equations involving absolute value

Recall that the *absolute value* of any real number u is defined by

$$|u| = \begin{cases} u & \text{if } u \geq 0, \\ -u & \text{if } u < 0. \end{cases}$$

EXAMPLE 4 Solve the equation $|x - 2| - x = 3$.

Solution Consider two cases.

1. $x - 2 \geq 0$; that is, $x \geq 2$. For $x \geq 2$ the given equation can be written as $(x - 2) - x = 3$, or $0 \cdot x = 5$. There is no real number x satisfying $0 \cdot x = 5$. Thus the given equation has no solutions for which $x \geq 2$.

2. $x - 2 < 0$; that is, $x < 2$. In this case the given equation can be written as $-(x - 2) - x = 3$, or $-2x = 1$. Thus $x = -1/2$. Since $-1/2 < 2$, we get one solution for the given equation, namely, $-1/2$. ∎

EXAMPLE 5 Solve the equation $\sqrt{x^2} - 2x - 3 = 0$.

Solution First, $\sqrt{x^2} = |x|$ is an identity, and so we replace $\sqrt{x^2}$ by $|x|$ in the given equation to get $|x| - 2x - 3 = 0$. Now consider two cases.

1. For $x \geq 0$ the given equation is equivalent to $x - 2x - 3 = 0$, the solution of which is $x = -3$. But -3 is not an x for which $x \geq 0$, so there are no solutions for this case.

2. For $x < 0$ the given equation is equivalent to $-x - 2x - 3 = 0$, which has a solution given by $x = -1$. Since $-1 < 0$, we have a solution given by $x = -1$. ∎

EXAMPLE 6 Solve the equation $\sqrt{x + 2} - x = 0$.

Solution The given equation can be written as $\sqrt{x + 2} = x$. Now square both sides to get $x + 2 = x^2$, or $x^2 - x - 2 = 0$. Hence we have a quadratic equation that can be solved by factoring: $(x + 1)(x - 2) = 0$. The solutions are given by $x = -1$, $x = 2$.

When we square both sides of an equation, the resulting equation may have extra roots that are not solutions to the given equation, and so it is necessary to check our solutions to see whether they *satisfy the original equation.* Substituting -1 for x gives $\sqrt{-1 + 2} - (-1) = 0$, or $2 = 0$, which is not a true statement, and so -1 is not a solution. To check $x = 2$, we have $\sqrt{2 + 2} - 2 = 0$, or $\sqrt{4} - 2 = 0$, which is a true statement.

Our conclusion is that the given equation has one root which is given by $x = 2$. ∎

EXERCISE 5.1

In Problems 1–12, solve the given equations. Express answers in exact form (involving square roots in some cases) and simplify when possible.

1. $3x - 2 = 4$
2. $4x + 3 = 2 - x$
3. $2(x - 3) + 4(2 - x) = 0$
4. $x^2 - 4x - 5 = 0$
5. $2x^2 + 5x - 3 = 0$
6. $2x^2 - 4x - 3 = 0$
7. $x^2 - 2x + 3 = 0$
8. $2x^2 + x + 1 = 0$
9. $\sqrt{x^2} - 2|x| + 3 = 0$
10. $|x + 2| - 3 = 0$
11. $\sqrt{x^2} - 2x - 3 = 0$
12. $(x + 2)(x - 3) = 6$

In Problems 13–20, find the solution set for the given equations. Give answers in exact form and also rounded off to two decimal places.

13. $\sqrt{3}\,x - 2 = x$

14. $x - 2 = 2x + \sqrt{3}$

15. $\sqrt{3}\,(x + 2) = x - 4$

16. $2 - \sqrt{5}\,x = \sqrt{5}$

17. $2x^2 - 3x - 1 = 0$

18. $x^2 + \sqrt{3}\,x - 2 = 0$

19. $(x - \sqrt{3})(x + \sqrt{12}) = 1$

20. $(x - 1 + \sqrt{2})(x - 1 - \sqrt{2}) = 4$

5.2 Linear Trigonometric Equations

In this section we consider techniques for solving "linear" equations that can be reduced to $af(x) + b = 0$, where $f(x)$ is any one of the six trigonometric functions, and a and b are given numbers, $a \neq 0$. This can best be done by illustration with several examples.

EXAMPLE 1 Find all solutions to the equation $2 \sin x - 1 = 0$. Express answers in degree measure and in radian measure.

Solution First solve for $\sin x$, $\sin x = 1/2$. Now we have a problem of the type we saw in Chapter 2. We can draw reference triangles for angles x whose sine is $1/2$ as shown in Fig. 5.1. We observe that two solutions are given by $x_1 = 30°$ and $x_2 = 150°$. If we add or subtract any multiple of $360°$ to either of these, we get corresponding coterminal angles that also are solutions. Thus all solutions are represented by

$$x = 30° + k \cdot 360° \quad \text{or} \quad x = 150° + k \cdot 360°,$$

where k is any integer (positive, negative, or zero). In radian measure the solutions are given by

$$x = \frac{\pi}{6} + k \cdot 2\pi \quad \text{or} \quad x = \frac{5\pi}{6} + k \cdot 2\pi. \quad \blacksquare$$

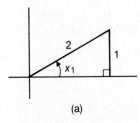

(a)

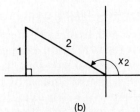

(b)

FIGURE 5.1

EXAMPLE 2 Find the solutions to $\sin(2x - \pi/4) - 1 = 0$, where the replacement set is $\{x \mid 0 \le x \le 2\pi\}$.

Solution The given equation can be written as $\sin(2x - \pi/4) = 1$. Thus $2x - \pi/4$ must be an angle such as $\pi/2$, $5\pi/2, \ldots$ or $-3\pi/2$, $-7\pi/2, \ldots$. That is, $2x - \pi/4 = \pi/2 + k \cdot 2\pi$, where k is any integer. Now solve for x:

$$2x = \frac{\pi}{4} + \frac{\pi}{2} + k \cdot 2\pi, \qquad 2x = \frac{3\pi}{4} + k \cdot 2\pi, \qquad x = \frac{3\pi}{8} + k\pi.$$

We need to select values of k that will give $0 \le x \le 2\pi$. For $k = 0$, $x_1 = 3\pi/8$, and for $k = 1$, $x_2 = 11\pi/8$. Thus there are two solutions, $3\pi/8$ and $11\pi/8$. $\blacksquare$

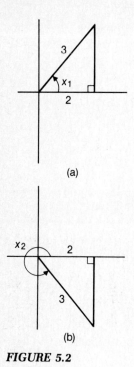

(a)

(b)

FIGURE 5.2

EXAMPLE 3   Find the solution set for $3 \cos x - 2 = 0$, where the replacement set is given by $0 \leq x \leq 2\pi$. Give your answers in exact form and also rounded off to two decimal places.

Solution   The given equation can be written as $\cos x = 2/3$, and so x is an angle in the first or fourth quadrant as shown in Fig. 5.2. In diagram (a) we see that $x_1 = \text{Cos}^{-1}(2/3)$, and from (b) the reference angle for x_2 is $\text{Cos}^{-1}(2/3)$, and so we get $x_2 = 2\pi - \text{Cos}^{-1}(2/3)$. The solution set in exact form is $\{\text{Cos}^{-1}(2/3), 2\pi - \text{Cos}^{-1}(2/3)\}$. With the calculator in radian mode we can evaluate to get the approximate decimal form $\{0.84, 5.44\}$.   ∎

EXAMPLE 4   Solve the equation $3 \sec x + 5 = 0$, where the replacement set is given by $-\pi \leq x \leq \pi$. Give answers rounded off to two decimal places.

Solution   Since $\sec x = 1/(\cos x)$ is an identity, we can replace $\sec x$ in the given equation by $1/(\cos x)$ and then solve for $\cos x$. This gives $\cos x = -3/5$. Since $\cos x$ is negative, x is an angle in the second or third quadrant as shown in Fig. 5.3. Note that we have drawn x_1 and x_2 so that they are angles in the replacement set; x_1 is given by $\text{Cos}^{-1}(-3/5)$ and x_2 is $-\text{Cos}^{-1}(-3/5)$. With the calculator in radian mode, evaluate $\text{Cos}^{-1}(-3/5)$ to get

$$x_1 \approx 2.21 \quad \text{and} \quad x_2 \approx -2.21. \quad \blacksquare$$

EXAMPLE 5   Find the solution set for $\sin x + \sqrt{3} \cos x = 0$. Give answers in radians and in exact form.

Solution   Since the replacement set is not stated explicitly, it is understood to be the set of real numbers. The given equation can be written as $\sin x = -\sqrt{3} \cos x$; dividing by $\cos x$, we get $(\sin x)/(\cos x) = -\sqrt{3}$. But $(\sin x)/(\cos x) = \tan x$ is an identity, and so we

FIGURE 5.3

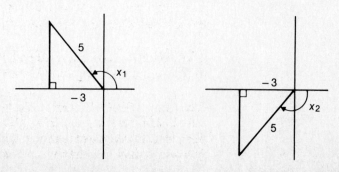

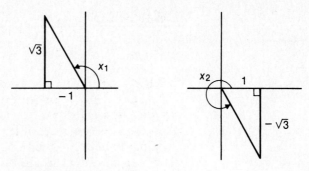

FIGURE 5.4

have $\tan x = -\sqrt{3}$. Therefore the given equation is equivalent to $\tan x = -\sqrt{3}$, and this is an equation we can solve.

Since $\tan x$ is negative, we get two solutions, as shown in Fig. 5.4. We see that $x_1 = 2\pi/3$ and $x_2 = 5\pi/3$. Any angle coterminal with either of these is also a solution. Note that $x_2 = x_1 + \pi$. Hence all solutions are given by $x = (2\pi/3) + k \cdot \pi$, and the solution set is $\{x \mid x = (2\pi/3) + k\pi\}$. ∎

EXAMPLE 6   Solve the equation $2 \sin x - 3 = 0$.

Solution   The given equation can be written as $\sin x = 3/2$. Since $-1 \le \sin x \le 1$ for every x and $3/2 > 1$, the equation $\sin x = 3/2$ has no solutions. Thus the solution set for the given equation is the empty set. ∎

EXAMPLE 7   Find the solution set for $2 \sin x \cos x + \cos x = 0$, where the replacement set is given by $0 \le x \le 2\pi$.

Solution   The given equation can be written as $\cos x (2 \sin x + 1) = 0$. A product is zero if and only if one of the factors is zero, and so we solve two equations: $\cos x = 0$ and $2 \sin x + 1 = 0$. For $\cos x = 0$ there are two solutions in $0 \le x \le 2\pi$, namely, $\pi/2$ and $3\pi/2$. For $2 \sin x + 1 = 0$, or $\sin x = -1/2$, we also get two solutions, which are $7\pi/6$ and $11\pi/6$. Therefore the solution set is $\{\pi/2, 3\pi/2, 7\pi/6, 11\pi/6\}$. ∎

EXAMPLE 8   Determine all values of x at which the function $f(x) = 3 \sin x$ assumes a maximum value.

Solution   Since $-1 \le \sin x \le 1$, the maximum value of $3 \sin x$ will occur when $\sin x = 1$. Thus our problem is to find all roots of $\sin x = 1$. These are given by $x = \pi/2 + k \cdot 2\pi$, where k is any integer (zero, positive, or negative). ∎

EXERCISE 5.2

In Problems 1–9, solve the given equations, where the replacement set is $\{x \mid 0° \leq x \leq 360°\}$.

1. $2 \cos x + 1 = 0$ **2.** $2 \sin x + \sqrt{3} = 0$

3. $\sqrt{3} \tan x - 1 = 0$ **4.** $2 \cos x - \sqrt{3} = 0$

5. $\sin x - 1 = 0$ **6.** $\sec x + 2 = 0$

7. $\cos x + \sqrt{3} = 0$ **8.** $\sqrt{3} \csc x - 2 = 0$

9. $\sin x + \cos x = 0$

In Problems 10–15, solve the given equations, where the replacement set is $\{x \mid 0 \leq x \leq 2\pi\}$. Give answers in exact form.

10. $\sqrt{3} \sin x - \cos x = 0$ **11.** $\sqrt{2} \sin x - 1 = 0$

12. $2 \sin x \cos x - 3 \sin x = 0$ **13.** $\sqrt{2} \sin x \cos x - \cos x = 0$

14. $\cos^2 x - \cos x \sin x = 0$ **15.** $\sin^2 x + \sqrt{3} \sin x \cos x = 0$

In Problems 16–21, find the solution sets for the given equations, where the replacement set is given by $-\pi \leq x \leq \pi$. Give answers in exact form.

16. $2 \cos x + 1 = 0$ **17.** $\sqrt{3} \sin x + \cos x = 0$

18. $\cos^2 x + \sin x \cos x = 0$ **19.** $\sin x \cos x + 3 \cos x = 0$

20. $\tan x + 1 = 0$ **21.** $\cos x \tan x - \cos x = 0$

In Problems 22–30, find the solution sets for the given equations, where the replacement set is given by $0 \leq x \leq 2\pi$. Express answers in set notation with numbers rounded off to two decimal places.

22. $4 \sin x - 3 = 0$ **23.** $\sqrt{3} \cos x - 2 = 0$

24. $2 \tan x - 3 = 0$ **25.** $4 \sec x - 5 = 0$

26. $3 \sin x - 5 \cos x = 0$ **27.** $2 \sin x + 7 \cos x = 0$

28. $3 \cos^2 x - \cos x = 0$ **29.** $\sin x - \sqrt{5} \cos x = 0$

30. $3 \csc x + 4 = 0$

In Problems 31–36, find the solution set for each of the given equations. Here the replacement set is not stated explicitly. Express answers in set notation with numbers (radians) given in exact form.

31. $2 \cos x + 1 = 0$ **32.** $\sqrt{3} \tan x + 1 = 0$

33. $2 \sin x \cos x - \sin x = 0$ **34.** $\cos x \tan x - \sqrt{3} \cos x = 0$

35. $\sqrt{3} \sec x - 2 = 0$ **36.** $\sqrt{2} \sin x \tan x - \tan x = 0$

In Problems 37–40, find all values of x at which the given function will assume a maximum value.

37. $f(x) = 3 \cos x$ **38.** $f(x) = -2 \sin x$

39. $f(x) = 2 - 3 \sin x$ **40.** $f(x) = 2 + 3 \sin x$

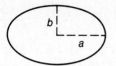

FIGURE 5.5

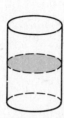

FIGURE 5.6

In Problems 41–44, find all values of x at which the given function will assume a minimum value.

41. $f(x) = 3 \sin x$ **42.** $f(x) = 4 + \cos x$

43. $f(x) = 3 - \cos x$ **44.** $f(x) = 4 - \sin x$

45. The area of the region bounded by an ellipse is given by Area $= \pi ab$, where a and b are lengths of semi-axes as shown in Fig. 5.5. In the special case of $b = a$ we get a circle, and the area formula becomes a familiar one, Area $= \pi a^2$.

A cylindrical can with circular cross-section of radius 5 cm is 16 cm tall and is 3/8 full of water. When it stands upright, the water surface is circular; but when it is inclined, the water surface is elliptical.

a] Find the area K of the elliptical surface when the can is inclined so that its axis makes an angle θ with the vertical direction (see Fig. 5.6).

b] The formula derived in (a) gives K as a function of θ, $K = f(\theta)$. What is the domain of f? That is, for what values of θ is the water surface elliptic?

5.3 Quadratic Trigonometric Equations

In this section we consider problems that can be reduced to solving quadratic equations of the type $a[f(x)]^2 + bf(x) + c = 0$, where $f(x)$ is one of the six trigonometric functions and a, b, and c are given numbers, $a \neq 0$. We illustrate solution techniques by considering several examples.

EXAMPLE 1   Solve $2 \sin^2 x - \sin x = 0$, where the replacement set is

$$\{x \mid 0° \leq x \leq 360°\}.$$

Solution   We first express the given equation in factored form:

$$\sin x (2 \sin x - 1) = 0$$

Now use the basic property of numbers that if the product of two numbers is zero, then at least one of the numbers must be zero. Therefore the given equation is equivalent to

$$\sin x = 0 \quad \text{or} \quad (2 \sin x - 1) = 0;$$

$\sin x = 0$ gives $x = 0°, 180°, 360°$ as solutions, while $2 \sin x - 1 = 0$, or $\sin x = \frac{1}{2}$, gives $x = 30°, 150°$ as solutions. Thus the solution set is

$$\{0°, 30°, 150°, 180°, 360°\}. \quad \blacksquare$$

EXAMPLE 2 Find the solution set for $3 \cos^2 x + \sin^2 x + 3 \cos x - 3 = 0$, where the replacement set is given by $0 \le x \le 2\pi$. Give answers in exact form.

Solution Since $\sin^2 x + \cos^2 x = 1$ is an identity, we can replace $\sin^2 x$ by $1 - \cos^2 x$ in the given equation; and after simplifying, we get $2 \cos^2 x + 3 \cos x - 2 = 0$. We can solve this quadratic equation by factoring $(2 \cos x - 1)(\cos x + 2) = 0$. Therefore our problem reduces to solving two equations of the type appearing in the preceding section, $2 \cos x - 1 = 0$ and $\cos x + 2 = 0$, or $\cos x = \frac{1}{2}$ and $\cos x = -2$. Solutions to the first are given by $x_1 = \pi/3$, $x_2 = 5\pi/3$, while there are no solutions to the second. The solution set is $\{\pi/3, 5\pi/3\}$. $\blacksquare$

EXAMPLE 3 Find all solutions to $\tan^2 x - 1 = 0$. Express answers in exact form.

Solution The given equation can be written in factored form as $(\tan x + 1)(\tan x - 1) = 0$, and so we have two problems of the type studied in the preceding section, $\tan x + 1 = 0$, $\tan x - 1 = 0$. For $\tan x = -1$, two solutions are shown in Fig. 5.7(a), where $x_1 = 3\pi/4$, $x_2 = 7\pi/4$. For $\tan x = 1$, two solutions are shown in Fig. 5.7(b), namely, $x_3 = \pi/4$, $x_4 = 5\pi/4$. Any angle that is coterminal with any one of these is also a solution. The set of all solutions is given by $x = (\pi/4) + k \cdot (\pi/2)$, where k is any integer. $\blacksquare$

FIGURE 5.7

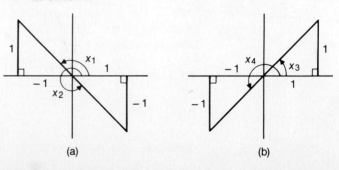

(a)   (b)

EXAMPLE 4 Find the solution set for $2\cos^2 x - 6\cos x + 1 = 0$, where $0 \le x \le 2\pi$. Give answers rounded off to two decimal places.

Solution Since the left-hand side of the given equation does not factor in a simple manner, we use the quadratic formula

$$\cos x = \frac{6 \pm \sqrt{36 - 4 \cdot 2 \cdot 1}}{4} = \frac{3 \pm \sqrt{7}}{2}.$$

Therefore the given equation is equivalent to

$$\cos x = \frac{3 + \sqrt{7}}{2} \quad \text{or} \quad \cos x = \frac{3 - \sqrt{7}}{2}.$$

There is no solution for $\cos x = (3 + \sqrt{7})/2 \approx 2.8229$. For

$$\cos x = \frac{3 - \sqrt{7}}{2} \approx 0.1771,$$

we see that x is in the first or fourth quadrant, as shown in Fig. 5.8. By using a calculator we find

$$x_1 = \text{Cos}^{-1}\frac{3 - \sqrt{7}}{2} \approx 1.3927.$$

For the second solution the reference angle is x_1 (see Fig. 5.8), and so x_2 is given by

$$x_2 = 2\pi - x_1 \approx 2\pi - 1.3927 \approx 4.8905.$$

Therefore the solution set is $\{1.39, 4.89\}$. ∎

EXAMPLE 5 Solve $\sin^2 x - 2\sin x + 2 = 0$.

Solution Using the quadratic formula, we get

$$\sin x = \frac{2 \pm \sqrt{4 - 8}}{2} = 1 \pm \sqrt{-1}.$$

Since $1 \pm \sqrt{-1}$ are not real numbers, there is no value of x that will satisfy the given equation, and so the solution set is the empty set. ∎

FIGURE 5.8

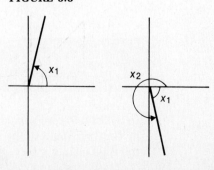

EXERCISE 5.3

In Problems 1–8, find the solution set for the given equations, where the replacement set is $\{x \mid 0° \leq x \leq 360°\}$.

1. $2 \sin^2 x + \sin x - 1 = 0$

2. $\sin^2 x - \cos^2 x = 0$

3. $\cos^2 x + 2 \cos x + 1 = 0$

4. $\tan^2 x - 1 = 0$

5. $1 - 4 \sin^2 x = 0$

6. $3 \sec^2 x + 2 \sec x - 1 = 0$

7. $\cos^2 x - \sin^2 x + 3 \cos x - 1 = 0$

8. $2 \sin^2 x + 5 \sin x - 3 = 0$

In Problems 9–24, find the solution sets for the given equations, where the replacement set is given by $0 \leq x \leq 2\pi$. Give answers in exact form when possible (for example, where the corresponding reference angles are 0, $\pi/6$, $\pi/4$, $\pi/3$, or $\pi/2$). Otherwise, give answers rounded off to two decimal places.

9. $2 \sin^2 x - 5 \sin x - 3 = 0$

10. $3 \cos^2 x - \sin^2 x = 0$

11. $3 \cos^2 x + \cos x - 2 = 0$

12. $4 - \tan^2 x = 0$

13. $\tan^2 x + 2 \tan x + 1 = 0$

14. $2 \sec^2 x - 3 \sec x - 2 = 0$

15. $\cos^2 x + 3 \cos x - 2 = 0$

16. $4 \sin^2 x + 3 \sin x - 1 = 0$

17. $4 \sin^2 x + 3 \cos^2 x - 4 = 0$

18. $\cos^2 x - 3 \cos x - 2 = 0$

19. $\sin^2 x + 2 \sin x + 1 = 0$

20. $2 \tan^2 x - 4 \tan x + 1 = 0$

21. $\sin^2 x + 2 \sin x + \cos^2 x = 0$

22. $\sin^2 x - \sin x + 2 = 0$

23. $3 \cos^2 x + 4 \cos x + 2 = 0$

24. $2 \sin^2 x + 2 \sin x - 1 = 0$

In Problems 25–30, find the solution sets for the given equations. Here the replacement set is not specified, so find *all* solutions. Give answers in degrees.

25. $2 \sin^2 x + \sin x = 0$

26. $4 \cos^2 x - 3 = 0$

27. $2 \cos^2 x + 5 \cos x - 3 = 0$

28. $\tan^2 x - 3 = 0$

29. $\cot^2 x - (\sqrt{3} + 1) \cot x + \sqrt{3} = 0$

30. $\sec^2 x + \sec x - 2 = 0$

5.4 Equations of the Form $a \sin x + b \cos x = c$

In Section 5.2 we solved an equation of the type $a \sin x + b \cos x = 0$ by first writing it in the form $\tan x = -b/a$. If the given equation is of the type $a \sin x + b \cos x = c$, where $c \neq 0$, then we must use a different approach. Suppose we divide both sides of this equation by $\sqrt{a^2+b^2}$:

$$\frac{a}{\sqrt{a^2+b^2}} \sin x + \frac{b}{\sqrt{a^2+b^2}} \cos x = \frac{c}{\sqrt{a^2+b^2}}. \qquad \textbf{[5.3]}$$

Let us introduce an angle α defined by

$$\cos \alpha = \frac{a}{\sqrt{a^2+b^2}} \quad \text{and} \quad \sin \alpha = \frac{b}{\sqrt{a^2+b^2}}. \qquad \textbf{[5.4]}$$

Angle α can be shown in standard position by plotting the point (a, b) and having the terminal side of α pass through it as illustrated in Fig. 5.9 (where, in the diagram shown, a is negative and b is positive). Equation (5.3) can now be written as

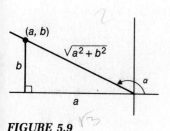

FIGURE 5.9

$$\sin x \, \cos \alpha + \cos x \, \sin \alpha = \frac{c}{\sqrt{a^2+b^2}}.$$

The left-hand side of this equation is identically equal to $\sin(x + \alpha)$, and so the given equation is equivalent to

$$\sin(x + \alpha) = \frac{c}{\sqrt{a^2+b^2}}. \qquad \textbf{[5.5]}$$

The problem is now essentially reduced to solving an equation of the type studied in Section 5.2.

EXAMPLE 1 Solve the equation $\sin x - \sqrt{3} \cos x = 1$, where the replacement set is given by $0° \le x \le 360°$.

Solution We shall first illustrate the procedure described above (Method 1) and then also look at a slight variation (Method 2). First divide both sides of the given equation by $\sqrt{1^2 + (-\sqrt{3})^2} = \sqrt{1+3} = \sqrt{4} = 2$ and get

$$\frac{1}{2} \sin x - \frac{\sqrt{3}}{2} \cos x = \frac{1}{2}. \qquad \textbf{[5.6]}$$

Method 1

Let angle α be determined by

$$\cos \alpha = \frac{1}{2} \quad \text{and} \quad \sin \alpha = -\frac{\sqrt{3}}{2}$$

FIGURE 5.10

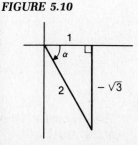

and draw a diagram showing α (Fig. 5.10). We see from the reference triangle for α that $\alpha = -60°$. Equation (5.6) can now be written as $\cos(-60°) \sin x + \sin(-60°) \cos x = 1/2$. Using identity (I-10) gives

$$\sin(x - 60°) = \frac{1}{2}. \qquad \textbf{[5.7]}$$

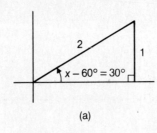 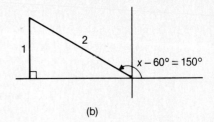

(a)   (b)

FIGURE 5.11

Hence angle $x - 60°$ is an angle of the type shown in Fig. 5.11. The general solution to Eq. (5.7) is given by $x - 60° = 30° + k \cdot 360°$ (see Fig. 5.11a), or $x - 60° = 150° + k \cdot 360°$ (see Fig. 5.11b). Solving for x and simplifying, we get

$$x = 90° + k \cdot 360° \quad \text{or} \quad x = 210° + k \cdot 360°.$$

Now we must choose values of k that give x in $0° \le x \le 360°$. In both cases, take $k = 0$ to get $x_1 = 90°$ and $x_2 = 210°$. Therefore there are two solutions for the given problem, namely, $90°$ and $210°$.

Method 2

Write Eq. (5.6) as follows:

$$\frac{\sqrt{3}}{2} \cos x - \frac{1}{2} \sin x = -\frac{1}{2}. \qquad \text{[5.8]}$$

Now let angle β be determined by

$$\cos \beta = \frac{\sqrt{3}}{2} \quad \text{and} \quad \sin \beta = \frac{1}{2}.$$

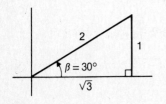

FIGURE 5.12

From Fig. 5.12 we observe that $\beta = 30°$.
   Eq. (5.8) now becomes $\cos x \, \cos 30° - \sin x \, \sin 30° = -1/2$, which by identity (I-11) can be written as

$$\cos (x + 30°) = -\frac{1}{2}.$$

The general solution for this equation is given by (look at Fig. 5.13)

$$x + 30° = 120° + k \cdot 360° \quad \text{or} \quad x + 30° = 240° + k \cdot 360°.$$

Solve for x and simplify to get

$$x = 90° + k \cdot 360° \quad \text{or} \quad x = 210° + k \cdot 360°.$$

To get solutions in the replacement set, take $k = 0$ in each case. This gives $90°$ and $210°$ as solutions. ∎

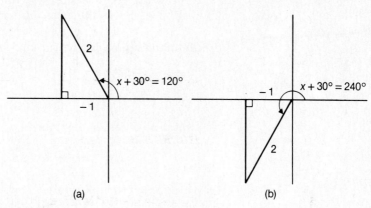

FIGURE 5.13

EXAMPLE 2 Find the solution set for $3 \sin x + 4 \cos x = 2$, where the replacement set is $\{x \mid 0 \le x \le 2\pi\}$. Give answers rounded off to two decimal places.

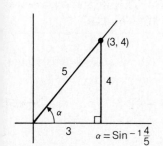

FIGURE 5.14

FIGURE 5.15

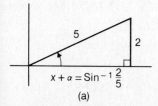

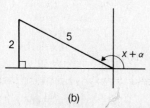

Solution Divide both sides of the given equation by $\sqrt{3^2 + 4^2} = \sqrt{25} = 5$ and get

$$\frac{3}{5} \sin x + \frac{4}{5} \cos x = \frac{2}{5}. \qquad \textbf{[5.9]}$$

Plot the point $(3, 4)$ and let α be an angle in standard position with terminal side through $(3, 4)$ as shown in Fig. 5.14. Then α is given by $\alpha = \text{Sin}^{-1}\,(4/5)$ (or by $\alpha = \text{Cos}^{-1}\,(3/5)$), $\alpha \approx 0.927$.

Since $\cos \alpha = 3/5$ and $\sin \alpha = 4/5$, Eq. (5.9) can be written as $\sin x \cos \alpha + \cos x \sin \alpha = 2/5$. By identity (I-10), this is equivalent to

$$\sin (x + \alpha) = \frac{2}{5}.$$

Solving for $(x + \alpha)$, we get the general solution (see Fig. 5.15)

$$x + \alpha = \text{Sin}^{-1}\frac{2}{5} + k \cdot 2\pi \qquad \text{or} \qquad x + \alpha = \left(\pi - \text{Sin}^{-1}\frac{2}{5}\right) + m \cdot 2\pi.$$

Solve for x:

$$x = \text{Sin}^{-1}\frac{2}{5} - \alpha + k \cdot 2\pi \qquad \text{or} \qquad x = \left(\pi - \text{Sin}^{-1}\frac{2}{5}\right) - \alpha + m \cdot 2\pi.$$

Since $\text{Sin}^{-1}\,2/5 \approx 0.411$ and $\alpha \approx 0.927$, we have

$$x \approx -0.516 + k \cdot 2\pi \qquad \text{or} \qquad x \approx 1.803 + m \cdot 2\pi.$$

Since we want values of x in the replacement set, take $k = 1$ and $m = 0$ to get the desired solutions, namely, 5.77 and 1.80. ■

EXAMPLE 3   Solve the equation $\sin x + 2 \cos x = 3$.

Solution   Divide both sides of the given equation by $\sqrt{1^2 + 2^2} = \sqrt{5}$ to get

$$\frac{1}{\sqrt{5}} \sin x + \frac{2}{\sqrt{5}} \cos x = \frac{3}{\sqrt{5}}. \qquad \textbf{[5.10]}$$

Now let α be an angle described by

$$\cos \alpha = \frac{1}{\sqrt{5}} \quad \text{and} \quad \sin \alpha = \frac{2}{\sqrt{5}}.$$

Using identity (I-10), Eq. (5.10) can be written as

$$\sin (x + \alpha) = \frac{3}{\sqrt{5}}.$$

Since $3/\sqrt{5} > 1$ and $\sin (x + \alpha) \le 1$ for every value of x, we see that there is no solution. That is, there is no value of x that will satisfy the given equation.   ■

EXERCISE 5.4

In Problems 1–10, find the solution set for each of the given equations, where the replacement set is given by $0° \le x \le 360°$.

1. $\sin x + \cos x = 1$
2. $\sin x - \cos x = 1$
3. $\sqrt{3} \sin x + \cos x = 1$
4. $\sin x + \sqrt{3} \cos x = -1$
5. $\sin x + \sqrt{3} \cos x = 0$
6. $\sin x + \cos x = 0$
7. $\sin^2 x + \sin x - \sin x \cos x = 0$
8. $\cos^2 x - \cos x + \sqrt{3} \sin x \cos x = 0$
9. $\sin x + \sqrt{3} \cos x = 2$
10. $3 \sin x + \cos x = 4$

In Problems 11–20, find the *solution set* for each of the given equations, where the replacement set is given by $0 \le x \le 2\pi$. Give answers in set notation with numbers rounded off to two decimal places.

11. $\sqrt{3} \sin x - \cos x = 2$
12. $4 \sin x - 3 \cos x = 5$
13. $3 \sin x + 4 \cos x = 5$
14. $4 \sin x - 3 \cos x = 3$
15. $5 \sin x - 12 \cos x = 5$
16. $\sin x + 2 \cos x = 1$
17. $2 \sin x - 2 \cos x = 5$
18. $\sin^2 x - 2 \sin x \cos x = \sqrt{5} \sin x$
19. $\cos^2 x + 2 \sin x \cos x = \cos x$
20. $1.2 \sin x - 2.4 \cos x = 3.6$

In Problems 21–24, find *all solutions* for the given equations. Here the replacement set is the set of real numbers. Give answers in exact form.

21. $\sin x + \cos x = -1$
22. $\sqrt{3} \sin x - \cos x = -1$
23. $\sin x - \sqrt{3} \cos x = 2$
24. $\cos^2 x - \cos x + \sin x \cos x = 0$

5.5 Using Identities in Solving Equations

In this section we consider a variety of examples in which we see the usefulness of the identities studied in Chapter 4. In general, most problem-solving techniques in mathematics involve a sequence of steps in which the given problem is reduced to an equivalent problem that can be solved by methods learned earlier. Here we are interested in solving equations that can be reduced by applying trigonometric identities to a solution of equations of the type studied in the preceding sections of this chapter.

EXAMPLE 1 Solve $2 \cos (x + 90°) = 1$. Give answers in degrees.

Solution From Section 2.7 we have the reduction identity $\cos (x + 90°) = -\sin x$. Therefore replace $\cos (x + 90°)$ by $-\sin x$ in the given equation and get $-2 \sin x = 1$. Our problem is now reduced to the type discussed in Section 5.2. Thus $\sin x = -1/2$, and so x is an angle in the third or fourth quadrant as shown in Fig. 5.16. From diagram (a) we see that $x_1 = 210°$, and from (b) we get $x_2 = 330°$. Thus all solutions are given by $x = 210° + k \cdot 360°$ or $x = 330° + k \cdot 360°$. ∎

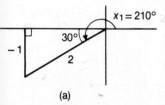

(a)

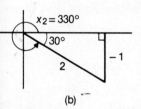

(b)

FIGURE 5.16

EXAMPLE 2 Solve the equation $\sin 2x + 2 \cos^2 x = 0$, where the replacement set is $\{x \mid 0 \le x \le 2\pi\}$.

Solution $\sin 2x = 2 \sin x \cos x$ is an identity (see identity (I-25) of Section 4.6). Therefore the given equation is equivalent to

$$2 \sin x \cos x + 2 \cos^2 x = 0.$$

This can be written in factored form:

$$2 \cos x (\sin x + \cos x) = 0.$$

Here we have a familiar situation of a product equal to zero, and so $\cos x = 0$ or $\sin x + \cos x = 0$. We need to solve two equations of the type encountered in Section 5.2: $\cos x = 0$ gives $\pi/2$ and $3\pi/2$ as solutions; $\sin x + \cos x = 0$ or $\tan x = -1$ gives $3\pi/4$ and $7\pi/4$ as solutions. Therefore the roots of the given equation are $\pi/2$, $3\pi/4$, $3\pi/2$, and $7\pi/4$. ∎

EXAMPLE 3 Solve $\sin^2 x - \cos^2 x - \sin x + 1 = 0$, where the replacement set is given by $-180° \le x \le 180°$.

Solution Replace $\cos^2 x$ by $1 - \sin^2 x$ in the given equation to get

$$\sin^2 x - (1 - \sin^2 x) - \sin x + 1 = 0.$$

This simplifies to $2 \sin^2 x - \sin x = 0$. By factoring, $\sin x (2 \sin x - 1) = 0$, and so $\sin x = 0$ or $2 \sin x - 1 = 0$. $\sin x = 0$ gives $-180°$, $0°$, and $180°$ as solutions; $2 \sin x - 1 = 0$ or $\sin x = 1/2$ gives $30°$ and $150°$ as solutions. Therefore the roots of the given equation are: $-180°$, $0°$, $30°$, $150°$, and $180°$. ■

EXAMPLE 4 Solve $\sin 2x - 3 \cos 2x = 0$, where the replacement set is given by $-\pi \le x \le \pi$. Give answers rounded off to two decimal places.

Solution The given equation can be written as $\sin 2x = 3 \cos 2x$. Divide by $\cos 2x$ and use the identity $(\sin 2x)/(\cos 2x) = \tan 2x$ to get $\tan 2x = 3$. Therefore $2x$ is an angle as shown in Fig. 5.17. From the diagram we see that all solutions are given by $2x = \mathrm{Tan}^{-1}\, 3 + k \cdot \pi$ where k is any integer. Divide both sides by 2 to get

$$x = \frac{1}{2}\, \mathrm{Tan}^{-1}\, 3 + k \cdot \frac{\pi}{2}.$$

The values of k giving the desired values of x between $-\pi$ and π are $k = -2$, $k = -1$, $k = 0$, and $k = 1$.

$$k = -2 \text{ gives } x_1 = \frac{1}{2}\, \mathrm{Tan}^{-1}\, 3 - \pi \approx -2.52;$$

$$k = -1 \text{ gives } x_2 = \frac{1}{2}\, \mathrm{Tan}^{-1}\, 3 - \frac{\pi}{2} \approx -0.95;$$

$$k = 0 \text{ gives } x_3 = \frac{1}{2}\, \mathrm{Tan}^{-1}\, 3 \approx 0.62;$$

$$k = 1 \text{ gives } x_4 = \frac{1}{2}\, \mathrm{Tan}^{-1}\, 3 + \frac{\pi}{2} \approx 2.20. \quad ■$$

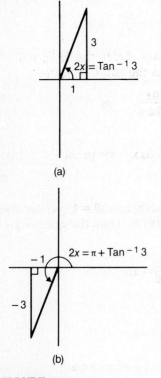

(a)

(b)

FIGURE 5.17

EXAMPLE 5 Find the solution set for $\sin x \tan (x/2) = 0.5$, where the replacement set is given by $0° \le x \le 360°$.

Solution $\tan (x/2) = (1 - \cos x)/(\sin x)$ is an identity (see identity (I-31) in Section 4.7). Thus the given equation can be written as $(\sin x)(1 - \cos x)/(\sin x) = 0.5$. Simplifying gives $1 - \cos x = 0.5$, or $\cos x = 1/2$. Solutions of $\cos x = 1/2$ are $60°$ and $300°$. Therefore the solution set for the given equation is $\{60°, 300°\}$. ■

EXAMPLE 6 Find the solution set for $2 (\sin x + \cos x)^2 = 1$, where $0 \le x \le 2\pi$.

Solution Each of the following equations is equivalent to the given equation:

$$2\left(\sin^2 x + 2\sin x \cos x + \cos^2 x\right) = 1 \qquad \textbf{By algebra}$$

$$2\left(1 + 2\sin x \cos x\right) = 1 \qquad \textbf{By identity I-5}$$

$$2\left(1 + \sin 2x\right) = 1 \qquad \textbf{By identity I-25}$$

$$\sin 2x = -\frac{1}{2} \qquad \textbf{By algebra}$$

Therefore

$$2x = -\frac{\pi}{6} + k \cdot 2\pi \quad \text{or} \quad 2x = -\frac{5\pi}{6} + k \cdot 2\pi,$$

where k is any integer. Then solving these equations for x gives

$$x = -\frac{\pi}{12} + k\pi \quad \text{or} \quad x = -\frac{5\pi}{12} + k\pi.$$

We select those values of k that give values of x in the replacement set. In both cases we use $k = 1$ or 2. Thus the solution set is

$$\left\{\frac{11\pi}{12}, \frac{23\pi}{12}, \frac{7\pi}{12}, \frac{19\pi}{12}\right\}. \qquad \blacksquare$$

EXAMPLE 7   Find the solution set for $\cos x - \sin(x/2) = 1$, where $-180° \leq x \leq 180°$.

Solution   We use the double-angle identity $\cos 2\theta = 1 - 2\sin^2 \theta$ to replace $\cos x = \cos[2(x/2)]$ by $1 - 2\sin^2(x/2)$. Then the given equation is equivalent to

$$1 - 2\sin^2 \frac{x}{2} - \sin \frac{x}{2} = 1.$$

Simplifying and factoring, we get

$$\left(\sin \frac{x}{2}\right)\left(2\sin \frac{x}{2} + 1\right) = 0.$$

Now set each factor on the left-hand side equal to zero.

$$\sin \frac{x}{2} = 0 \quad \text{or} \quad \sin \frac{x}{2} = -\frac{1}{2}.$$

From $\sin(x/2) = 0$ we get $x/2 = 0°$ as the only solution that gives x in the interval $-180° \leq x \leq 180°$. Thus $x = 0°$. From $\sin(x/2) = -1/2$ we see that angle $x/2$ is in the third or fourth quadrant, and so the only angle that gives x in the interval $-180° \leq x \leq 180°$ is $x/2 = -30°$. That is, $x = -60°$. Therefore the solution set is $\{0°, -60°\}$. $\blacksquare$

EXAMPLE 8   Find the solution set for $\sin 3x + \sin x - \cos x = 0$, where the replacement set is given by $0 \leq x \leq \pi$.

Solution $\sin A + \sin B = 2\,\sin((A+B)/2)\,\cos((A-B)/2)$ is an identity (see identity (I-21) in Section 4.5). If we take the first two terms on the left-hand side of the given equation and apply this identity we get

$$\sin 3x + \sin x = 2\sin\left(\frac{3x+x}{2}\right)\cos\left(\frac{3x-x}{2}\right) = 2\,\sin 2x\,\cos x.$$

Thus we can replace $\sin 3x + \sin x$ by $2\sin 2x \cos x$, and the given equation becomes

$$2\,\sin 2x\,\cos x - \cos x = 0$$
$$\cos x\,(2\,\sin 2x - 1) = 0$$
$$\cos x = 0 \quad\text{or}\quad 2\,\sin 2x - 1 = 0.$$

Thus we have reduced our problem to the solution of two simple equations where we want roots in $0 \le x \le \pi$.

$$\cos x = 0 \text{ gives } x_1 = \frac{\pi}{2};$$

$$2\sin 2x - 1 = 0, \text{ or } \sin 2x = \frac{1}{2}, \text{ gives } 2x_2 = \frac{\pi}{6} \text{ or } 2x_3 = \frac{5\pi}{6};$$

and so

$$x_2 = \frac{\pi}{12}, \qquad x_3 = \frac{5\pi}{12}.$$

Therefore the solution set for the given equation is $\{\pi/12, 5\pi/12, \pi/2\}$.   ■

EXAMPLE 9   Find the solution set for the equation
$$\cos^3 x + \sin^2 x \cos x - \cos x = 0, \qquad \text{where } 0 \le x \le 2\pi.$$

Solution   The given equation is equivalent to each of the following.
$$\cos x\,(\cos^2 x + \sin^2 x) - \cos x = 0,$$
$$\cos x - \cos x = 0.$$

In this form we have an equation that is satisfied by all values of x. Therefore the solution set is equal to the replacement set $\{x \mid 0 \le x \le 2\pi\}$. Thus the given equation is an identity.   ■

EXERCISE 5.5

In each of the following, find the *solution set* for the given equation where the replacement set is specified. Use set notation and express answers in exact form unless otherwise specified.

In Problems 1–8 the replacement set is $\{x \mid 0° \le x \le 360°\}$.

1. $\cos x \tan x + \sin x = 1$ **2.** $\sin x \tan x - \cos x = 0$

3. $2 \sin (x + 180°) = 1$ **4.** $\sqrt{3} \cot (x + 90°) = 1$

5. $\sin 2x = \cos x$ **6.** $\sin 2x = \sin x$

7. $\sin 2x - \cos 2x = 0$ **8.** $(\sin x - \cos x)^2 = 0.5$

In Problems 9–16 the replacement set is $\{x \mid 0 \le x \le 2\pi\}$.

9. $\sin x \cot x + \cos x = 1$ **10.** $\cos x \cot x - \sin x = 0$

11. $\cos 2x + 2 \sin x = 1$ **12.** $(1 - \tan x) \tan \left(x + \dfrac{\pi}{4}\right) = 2$

13. $\sin 2x - 4 \sin x \cos x = 1$ **14.** $\cos 2x - \cos x + 1 = 0$

15. $(\sin x + \cos x)^2 = 2$ **16.** $\cos \left(x + \dfrac{\pi}{3}\right) + \cos \left(x - \dfrac{\pi}{3}\right) = 1$

In Problems 17–20 the replacement set is given by $0 \le x \le \pi$. Give answers rounded off to two decimal places.

17. $2 \sin 2x + \sin x = 0$ **18.** $\cos 2x + 2 \cos x = 0$

19. $\cos 2x = 2 \sin x$ **20.** $4 \cos x \tan x - \sin x = 1$

In Problems 21–24 the replacement set is $\{x \mid -90° \le x \le 90°\}$.

21. $\cos 2x - \cos x + 1 = 0$ **22.** $(\sin x + \cos x)^2 - 2 \sin 2x = 0$

23. $\sin 2x + \cos x = 0$ **24.** $\cos x \tan x - 3 \sin x = 1$

In Problems 25–30 the replacement set is given by $0 \le x \le \pi$.

25. $2 \cos^2 x = 1 + \sin x$ **26.** $\cos^2 x - \sin^2 x - \sin 2x = 0$

27. $\sin \left(\dfrac{\pi}{2} + x\right) - \sin \left(\dfrac{\pi}{2} - x\right) = 1$ **28.** $(\sin x + \cos x) \tan \left(x - \dfrac{\pi}{4}\right) = 1$

29. $\sin 3x + \sin x + \cos x = 0$ **30.** $\sin 3x + \sin x - \sin 2x = 0$
(Hint: See Example 8.)   (Hint: See Example 8.)

In Problems 31–35 the replacement set is not specified. Give answers in radians.

31. $\cos x \tan x + \sin 2x = 0$ **32.** $\sin^2 x - 3 \cos^2 x = 0$

33. $\cos^2 x + \cos^2 x \tan^2 x - 1 = 0$ **34.** $2 \sin^2 x - \sin 2x \tan x = 0$

35. $\cos x \tan x - \sin x + 1 = 0$

36. Find all solutions to the equation

$$\sqrt{2} \, (\sin x + \cos x) = \tan x + \cot x.$$

(Hint: A standard technique for solving an equation is to reduce it to a form in which the right-hand side is zero, and if the left-hand side can be expressed in factored form, then apply the fact that a product can be zero only if one of the factors is zero. See whether you can apply this technique to the given equation. If you are not successful, then try the following.* Apply techniques of the preceding section to show that the given equation can be expressed as

$$(\sin 2x) \sin\left(x + \frac{\pi}{4}\right) = 1.$$

Here the right-hand side is not zero, but now continue with the following reasoning. For every x, both $\sin 2x$ and $\sin(x + \pi/4)$ are numbers in the interval -1 to 1. How can a product of such numbers be equal to 1? Answer and complete the solution.)

37. Solve the equation $\sqrt{3} \sin x + \cos x = \tan x + \cot x$, where the replacement set is given by $0 \le x \le 2\pi$. See the hint for Problem 36.

38. Show that $\sin x + \cos x = \tan x + \cot x$ has no solutions. See the hint given for Problem 36.

39. Solve the equation $2(\sin x + \cos x) = \sec x + \csc x$, where the replacement set is given by $0 \le x \le 2\pi$.

40. Solve the equation $\tan x + \cot x = \sec x + \csc x$, where the replacement set is given by $0° \le x \le 360°$.

5.6 Equations Involving Trigonometric and Algebraic Functions (Optional)

In the preceding sections of this chapter, all the equations considered involved only trigonometric functions. Similarly, in algebra courses, all the equations studied involve only algebraic expressions (such as $x^2 - 2x + 1 = 0$ or $x + \sqrt{3x - 1} = 5$). In this section we consider equations involving both algebraic and trigonometric functions. These are somewhat more difficult to solve, but we shall see that the calculator will help considerably.

EXAMPLE 1   Solve the equation $\sin x + x = 0$.

Solution   In problems of this type we shall rely on graphs to give us some insight into possible solutions. We first write the given equation as $\sin x = -x$ and draw graphs of $y = \sin x$ and $y = -x$ on the same system of coordinates. Solutions to our problem will be given by the x-coordinates of the points of intersection of the two curves. We see from the diagram in Fig. 5.18 that there is only one point of

---

*It is important to realize that the standard textbook methods for solving equations cannot be applied in every situation. The crucial fact about problem solving in mathematics is that by a sequence of logically valid steps the given problem is reduced to one that can be solved. Sometimes it is necessary to introduce your own steps and reasoning.

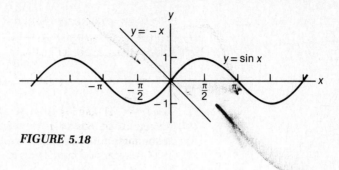

FIGURE 5.18

intersection—the origin—and so the solution set for $\sin x + x = 0$ is $\{0\}$. ∎

EXAMPLE 2 Find the solution set for the equation $\cos x - x = 0$.

Solution   We first write the given equation in the form $\cos x = x$ and then draw the graphs of $y = \cos x$ and $y = x$ on the same system of coordinates, as shown in Fig. 5.19. We see that there is only one point of intersection, so our problem is to find the x-coordinate of that point; we shall denote it by x_0.

There are systematic techniques for finding x_0 to any desired number of decimal places, but these require the study of calculus. The present approach makes use of the calculator and common sense.

Set the calculator in radian mode and then make a reasonable estimate of the value of x_0 from the diagram; call it x_1 and then evaluate $\cos x_1 - x_1$. If this number is positive, then x_1 is to the left of x_0, that is $x_1 < x_0$ (look at the graph); if it is negative, $x_1 > x_0$. Of course, our goal is to find x_0 such that $\cos x_0 - x_0 = 0$. It so happens that there is no finite decimal that has this property, and so we shall be satisfied with an approximate answer, say, correct to three decimal places. We compile a table containing our estimated values of x and the corresponding values of $\cos x - x$, and at each step our estimated x will be based on the previous values of x and $\cos x - x$.

FIGURE 5.19

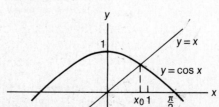

From the diagram a reasonable first guess at x_0 is 0.7.

Estimated x	0.7	0.72	0.74	0.735	0.736	0.739	0.7395
cos x − x	0.065	0.032	−0.0015	0.0068	0.0052	0.00014	−0.0007

We see that 0.7395 is to the right of x_0 and 0.739 is slightly to the left of x_0, and so x = 0.739 is an approximation of x_0 that is correct to three decimal places. (Note: An interesting approach to solving this problem is discussed in Problem 12 of Exercise 5.6.) ■

EXERCISE 5.6

In Problems 1–10, solve the given equation and give the answers correct to two decimal places. If a problem has more than one solution, find the non-zero solution nearest to x = 0 (if there are two such solutions, find the positive one). (Note: In each of the problems, x is necessarily a real number. For example, it does not make sense to solve sin x = x with x in degrees, since sin x is always a real number and cannot be equal to x-degrees. Therefore be certain that your calculator is in radian mode when you solve these problems.)

1. $\sin x + 2x = 0$ 2. $\sin x - \frac{x}{2} = 0$ 3. $\cos x + x = 0$

4. $\cos x = x^2$ 5. $\cos x = \frac{x}{3}$ 6. $\tan x = x$

7. $\tan x + 3x = 0$ 8. $\sin x + x^2 = 0$ 9. $\cos x + 1 = x^2$

10. $\sin x - 3x^2 = 0$

11. Find the smallest positive solution of $x \sin x - 1 = 0$. (Hint: You may wish to write this as $\sin x = 1/x$.)

12. In Example 2 of this section we used a guess approach to find the solution of $\cos x - x = 0$ to three decimal places. Now consider the same problem but try the following approach: Set your calculator in radian mode and start with *any number* in the display (this is the feature that makes this approach interesting), then press (cos). A new number appears in the display; press (cos) again, and again a new number appears in the display. Continue doing this (that is, press the (cos) key repeatedly) and watch the display to see what happens. If you eventually get a number in the display (call it x_0) that is not changing, then the calculator is telling you that $\cos x_0 = x_0$. This is precisely the solution of $\cos x - x = 0$ to the digit capacity of your calculator. Draw graphs of $y = \cos x$ and $y = x$ on the same set of coordinates and see if you can analyze why this technique works.

The student is urged to try the technique described in Problem 12 on other problems. The idea is to write your problem in the form $f(x) = x$ and then start with a guess (say, x_1), evaluate $f(x_1)$, then evaluate f of this number (that is $f(f(x_1))$), and continue this. If your calculator display eventually does not change (call the number in the display x_0), then you have $f(x_0) = x_0$, which is the solution of the given equation. As an example, try this approach to find the solution of $\sin x - x^2 = 0$ by considering $\sqrt{\sin x} = x$. Take $0 < x_1 < \pi$, since we want $\sin x > 0$ for $\sqrt{\sin x}$.

5.7 Equations and Inequalities Involving Inverse Functions

In this section we are interested in two problems: solving equations and solving inequalities involving inverse trigonometric functions. We shall see that being familiar with domains, ranges, and graphs of these functions will be helpful in dealing with such problems. The ideas and techniques are best illustrated by a variety of examples.

EXAMPLE 1   Solve the equation $3 \, \mathrm{Sin}^{-1} x = \pi$.

Solution   The given equation can be written as $\mathrm{Sin}^{-1} x = \pi/3$. Recall that $-\pi/2 \le \mathrm{Sin}^{-1} x \le \pi/2$. Since $\pi/3$ is between $-\pi/2$ and $\pi/2$, we will have a solution. Take the sine of both sides and apply the identity $\sin(\mathrm{Sin}^{-1} x) = x$ for $-1 \le x \le 1$; see Eq. (3.10) in Section 3.4.

$$\sin(\mathrm{Sin}^{-1} x) = \sin\frac{\pi}{3};$$

$$x = \sin\frac{\pi}{3}, \qquad x = \frac{\sqrt{3}}{2}.$$

Thus the given equation has one solution, namely $\sqrt{3}/2$.   ∎

EXAMPLE 2   Solve the equation $\mathrm{Cos}^{-1}(\cos x) = -\pi/3$.

Solution   Since Cos^{-1} evaluated at any number must be in the interval 0 to π and $-\pi/3$ is not such a number, we get no solution.   ∎

EXAMPLE 3   Solve the equations

a] $\cos(\mathrm{Sin}^{-1} x) = \dfrac{1}{2}$ **b]** $\cos(\mathrm{Sin}^{-1} x) = -\dfrac{1}{2}$

Solution   Let $\theta = \mathrm{Sin}^{-1} x$. Then

$$\sin\theta = x \qquad \text{and} \qquad -\frac{\pi}{2} \le \theta \le \frac{\pi}{2}.$$

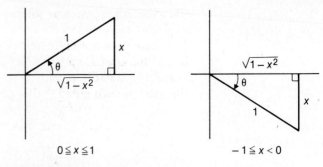

$0 \leq x \leq 1$ $-1 \leq x < 0$

FIGURE 5.20

Thus θ is an angle in the first or fourth quadrants with a reference triangle as shown in Fig. 5.20 for the two cases $0 \leq x \leq 1$, $-1 \leq x \leq 0$. We are interested in $\cos \theta$, and we see from the diagrams that for either case, $\cos \theta = \sqrt{1 - x^2}$. Therefore we have the following identity:

$$\cos (\mathrm{Sin}^{-1} x) = \sqrt{1 - x^2} \qquad \text{for every } x \text{ in } -1 \leq x \leq 1. \quad \textbf{[5.11]}$$

a] Applying the identity given in Eq. (5.11), we see that the given equation can be written as $\sqrt{1 - x^2} = 1/2$. Square both sides and solve for x.

$$1 - x^2 = \frac{1}{4}, \qquad x^2 = \frac{3}{4}, \qquad x = \frac{\sqrt{3}}{2} \quad \text{or} \quad x = -\frac{\sqrt{3}}{2}.$$

The given equation has two solutions, namely, $\sqrt{3}/2$ and $-\sqrt{3}/2$.

b] By using the identity stated in Eq. (5.11) the given equation is equivalent to $\sqrt{1 - x^2} = -1/2$. Since $\sqrt{1 - x^2}$ is nonnegative for every x in $-1 \leq x \leq 1$, it cannot equal $-1/2$. Thus the given equation has no solutions.

    We can also think of the given equation as follows: $\mathrm{Sin}^{-1} x$ is an angle in the first or fourth quadrant, and the cosine of such an angle is positive. Thus $\cos (\mathrm{Sin}^{-1} x)$ cannot equal $-1/2$ for any x. ∎

FIGURE 5.21

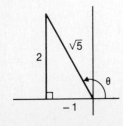

EXAMPLE 4   Solve the equation $\tan (\mathrm{Cos}^{-1} x) = -2$.

Solution   Let $\theta = \mathrm{Cos}^{-1} x$, and so $\cos \theta = x$ and $0 \leq \theta \leq \pi$. In terms of θ, the given equation is $\tan \theta = -2$. Thus θ is an angle determined by $\tan \theta = -2$ and $0 \leq \theta \leq \pi$. The reference triangle for θ is shown in Fig. 5.21. We want x where $x = \cos \theta$. From the diagram we see that $\cos \theta = -1/\sqrt{5}$, and so $x = -1/\sqrt{5}$. ∎

EXAMPLE 5 Find the solution set for $\text{Sin}^{-1}(\sin x) = -\pi/6$.

Solution Let $\mu = \sin x$, and so in terms of μ the given equation can be written as $\text{Sin}^{-1}\mu = -\pi/6$. Since $\text{Sin}^{-1}\mu$ is an angle in the interval $-\pi/2$ to $\pi/2$ and $-\pi/6$ is in this interval, we can solve for μ by taking the sine of both sides and applying the identity $\sin(\text{Sin}^{-1}\mu) = \mu$.

$$\sin(\text{Sin}^{-1}\mu) = \sin\left(-\frac{\pi}{6}\right)$$

$$\mu = \sin\left(-\frac{\pi}{6}\right).$$

Since $\sin(-\pi/6) = -1/2$, $\mu = -1/2$. Now returning to x, we have the familiar problem of solving the equation $\sin x = -1/2$. All solutions are given by $x = (7\pi/6) + k \cdot 2\pi$ or $x = (11\pi/6) + k \cdot 2\pi$, where k is any integer.

Therefore the solution set for the given equation is

$$\left\{x \,\Big|\, x = \frac{7\pi}{6} + k \cdot 2\pi \quad \text{or} \quad x = \frac{11\pi}{6} + k \cdot 2\pi\right\}. \quad \blacksquare$$

EXAMPLE 6 Find the solution set for the inequality $\text{Cos}^{-1}x \le 2$.

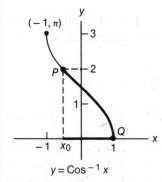

$y = \text{Cos}^{-1}x$

FIGURE 5.22

Solution Let us first draw a graph of $y = \text{Cos}^{-1}x$ and then ask, "For what values of x on the graph is $y \le 2$?" The graph of $y = \text{Cos}^{-1}x$ was drawn in Section 3.4 (see Fig. 3.21), and we reproduce it in Fig. 5.22. For points on the graph between P and Q, $y \le 2$. The values of x that give such points are in $x_0 \le x \le 1$, where $\text{Cos}^{-1}x_0 = 2$, or $x_0 = \cos 2 \approx -0.416$ (to three decimal places). The solution set for the given inequality is

$$\{x \,|\, -0.416 \le x \le 1\}.$$

Warning: It would have been a mistake to take the cosine of both sides of the inequality and get $x \le \cos 2$, or $x \le -0.416$, as the solution. $\blacksquare$

FIGURE 5.23

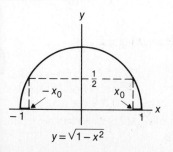

$y = \sqrt{1-x^2}$

EXAMPLE 7 Find the solution set for the inequality $\cos(\text{Sin}^{-1}x) \le 1/2$.

Solution We can apply the identity $\cos(\text{Sin}^{-1}x) = \sqrt{1-x^2}$ given in Eq. (5.11) and write the given inequality as $\sqrt{1-x^2} \le 1/2$. We now have a familiar problem from algebra. The graph of $y = \sqrt{1-x^2}$ is a semicircle as shown in Fig. 5.23. From the graph we see that $y \le 1/2$ for x in $-1 \le x \le -x_0$ or $x_0 \le x \le 1$, where x_0 is given by $1/2 = \sqrt{1-x_0^2}$. Solving for x_0, we get $x_0 = \sqrt{3}/2$. Therefore the solution set for the given inequality is $\{x \,|\, -1 \le x \le -\sqrt{3}/2$ or $\sqrt{3}/2 \le x \le 1\}$. This can be written as $\{x \,|\, \sqrt{3}/2 \le |x| \le 1\}$. $\blacksquare$

> *Note:* In problem solving it is impossible to overemphasize the importance of drawing graphs whenever possible. In the solutions to problems posed in Examples 6 and 7, observe how simple it is to "see" the solutions from the graphs.

EXERCISE 5.7

In Problems 1–24, solve each of the given equations. Use exact numbers in your answers.

1. $6 \, \text{Sin}^{-1} x = \pi$

2. $3 \, \text{Cos}^{-1} x = \pi$

3. $3 \, \text{Cos}^{-1} x - 2\pi = 0$

4. $4 \, \text{Sin}^{-1} x + \pi = 0$

5. $3 \, \text{Tan}^{-1} x + \pi = 0$

6. $3 \, \text{Cos}^{-1} x + 2\pi = 0$

7. $4 \, \text{Tan}^{-1} x - \pi = 0$

8. $2 \, \text{Sin}^{-1} x + \pi = 0$

9. $2 \, \text{Tan}^{-1} x - \pi = 0$

10. $\text{Cos}^{-1} x - \pi = 0$

11. $2 \, \text{Sin}^{-1} x = \dfrac{\pi}{3}$

12. $4 \, \text{Cos}^{-1} x = 0$

13. $\sin(\text{Sin}^{-1} x) = 0.5$

14. $\cos(\text{Cos}^{-1} x) = -0.5$

15. $\text{Sin}^{-1}(\sin x) = \dfrac{\pi}{6}$

16. $\text{Sin}^{-1}(\cos x) = \dfrac{3\pi}{4}$

17. $\tan(\text{Cos}^{-1} x) - 1 = 0$

18. $\text{Sin}^{-1}(\cos x) = \dfrac{\pi}{3}$

19. $\text{Cos}^{-1}(\sin x) = 0$

20. $2 \sin(\text{Cos}^{-1} x) + 1 = 0$

21. $\sin(\text{Tan}^{-1} x) = -0.5$

22. $\text{Tan}^{-1}(\tan x) = \dfrac{\pi}{3}$

23. $\sin(\text{Sin}^{-1} x) - 0.3 = 0$

24. $\cos(\text{Cos}^{-1} x) + 0.6 = 0$

In Problems 25–36, find the *solution sets* for the given equations or inequalities. Use numbers rounded off to two decimal places in your answers.

25. $3 \, \text{Sin}^{-1} x - 4 = 0$

26. $2 \, \text{Cos}^{-1} x - 5 = 0$

27. $\cos(\text{Sin}^{-1} x) - 0.4 = 0$

28. $\sin(\text{Cos}^{-1} x) - 0.8 = 0$

29. $\text{Sin}^{-1} x \geq 0.4$

30. $\text{Cos}^{-1} x \leq 1.2$

31. $\sin(\text{Cos}^{-1} x) \leq 0.6$

32. $\cos(\text{Sin}^{-1} x) \geq 0.3$

33. a] $\text{Sin}^{-1} x = 1.2$
    b] $\text{Sin}^{-1} x \leq 1.2$

34. a] $\text{Cos}^{-1} x = 2$
    b] $\text{Cos}^{-1} x \geq 2$

35. a] $\sin(\text{Sin}^{-1} x) = 0.4$
    b] $\text{Sin}^{-1}(\sin x) = 0.4$

36. a] $\cos(\text{Cos}^{-1} x) = 0.8$
    b] $\text{Cos}^{-1}(\cos x) = 0.8$

Summary

Linear equations

Most of the problems in this chapter ultimately reduce to solving equations of the type

$$af(x) + b = 0,$$

where $f(x)$ is one of the six trigonometric functions and a and b are known numbers. It is a simple matter to solve equations of this type. $f(x) = -b/a$ and then use the inverse of f to get desired solutions.

Quadratic equations

To solve $a[f(x)]^2 + bf(x) = c = 0$, where $f(x)$ is one of the trigonometric functions, factor if possible; otherwise use the quadratic formula to get

$$f(x) = \frac{-b \pm \sqrt{b^2 - 4ac}}{2a}.$$

Equations of the form $a \sin x + b \cos x = c$

If $c = 0$, then the problem reduces to $\tan x = -b/a$.

If $c \neq 0$, then divide both sides by $\sqrt{a^2 + b^2}$ and take

$$\cos \alpha = \frac{a}{\sqrt{a^2 + b^2}}, \qquad \sin \alpha = \frac{b}{\sqrt{a^2 + b^2}}.$$

Then the given equation reduces to

$$\sin(x + \alpha) = \frac{c}{\sqrt{a^2 + b^2}}.$$

Using identities

You should always look for the possibility of simplifying the given equation by applying identities from Chapter 4. Becoming familiar with trigonometric identities is an important part of any course in trigonometry.

Computer Problems (Optional)

Section 5.1

1. Write a program that will give solutions to any quadratic equation, $ax^2 + bx + c = 0$, where $a \neq 0$. Your program should allow the user to enter the three numbers a, b, and c, and then it will be necessary to check $b^2 - 4ac$ to see if it may be negative; if it is, the response should be "no real number solutions." Have your program round off answers to k decimal places, where k can be 0, 1, 2, 3, or 4.

Section 5.2

2. Write a program that will give solutions to the equation $b \sin x + c = 0$, where b and c are any real numbers that can be entered, $b \neq 0$. Have your program take care of any potential error messages (cases in which there are no solutions), and give answers rounded off to two decimal places. Assume that the replacement set is given by

   a] $-\dfrac{\pi}{2} \leq x \leq \dfrac{\pi}{2}$ b] $0 \leq x \leq \pi$ c] $0 \leq x \leq 2\pi$

3. Follow the instructions given in Problem 2 for the equation $b \cos x + c = 0$.

Section 5.3

4. Write a program that will give solutions (to three decimal places) to the equation $a \sin^2 x + b \sin x + c = 0$, $a \neq 0$, where the replacement set is given by

   a] $-\dfrac{\pi}{2} \leq x \leq \dfrac{\pi}{2}$ b] $0 \leq x \leq 2\pi$

   Have your program take care of any potential error messages (cases in which there will be no solutions).

   c] Use your program to get solutions to Problems 9–24 in Exercise 5.3.

5. Follow the instructions given in Problem 4 for the equation $a \cos^2 x + b \cos x + c = 0$.

Section 5.4

6. Write a program that will give the roots of

   $$a \sin x + b \cos x = c,$$

   where $0 \leq x \leq 2\pi$. Your program should allow the user to enter any numbers a, b, and c, and you should have it take care of any potential error messages. For instance, if $c/\sqrt{a^2 + b^2} > 1$ or $c/\sqrt{a^2 + b^2} < -1$, then there are no solutions. Use your program to get solutions to Problems 11–17 in Exercise 5.4.

Section 5.5

7. In Example 8 of Section 5.5 we solved the equation

   $$\sin 3x + \sin x - \cos x = 0,$$

   where $0 \leq x \leq \pi$. As a check on the solutions, write a program that will evaluate $f(x) = \sin 3x + \sin x - \cos x$ at the following values of x: $0, 0.1, 0.2, 0.3, 0.4, \ldots, 3.1, 3.2$. Have your program tell you the consecutive values of x when there is a change of sign for $f(x)$. This will locate the roots of $f(x) = 0$ between consecutive tenths. Check to see whether the results agree with the solutions found in Example 8. Modify your program to locate the roots between consecutive hundredths.

8. Follow a procedure similar to that in Problem 7 to locate the roots of $\sin 3x + \sin 2x - \cos x = 0$ between consecutive tenths, where $0 \leq x \leq \pi$.

Section 5.6

9. In Problem 12, and more generally in the final paragraph of Exercise 5.6, we described a "fixed point" technique for solving certain equations. The idea is to write your equation in the form $f(x) = x$ and then for a reasonable initial value of x, call it x_1, evaluate the following sequence of numbers:

$$x_2 = f(x_1), \quad x_3 = f(x_2), \quad x_4 = f(x_3), \quad \dots$$

In many cases (depending upon the function f and the selection of x_1) the sequence will approach a fixed number, call it x_0. That is, the computations will eventually come to a point at which $f(x_0) = x_0$, and so x_0 is a solution to $f(x) = x$. It is beyond the scope of this book to discuss details of this technique. Here we merely want you to experiment with it by using a computer in a variety of problems. The interested reader will have to wait until after calculus to explore this technique further.

In each of the following problems, write a program that will evaluate the numbers x_2, x_3, $x_4, \dots$. Have your program print out several of these, say the first 20 or 50 or even 100, and see whether these numbers are settling on a fixed number.

a] $\cos x - x = 0$. Use $f(x) = \cos x$ and x_1 any number.

b] $\sqrt{\sin x} - x = 0$. Use $f(x) = \sqrt{\sin x}$ and $x_1 = 1$.

c] $x^3 + 3x + 2 = 0$. Use $f(x) = -2/(x^2 + 3)$ and $x_1 = 1$. Try $f(x) = -\sqrt[3]{3x + 2}$ and $x_1 = 0$.

d] $x^3 + 3x + 5 = y$ where you input a value of y, such as $y = 3$, to get the equation in part (c). Use $f(x) = (y - 5)/(x^2 + 3)$. Here you would be evaluating the inverse function for $g(x) = x^3 + 3x + 5$.

Review Exercises

In each of Problems 1–24, an equation is given. Find the *solution set* where the replacement set is specified for groups of problems.

In Problems 1–8, the replacement set is $\{x \mid 0° \le x \le 360°\}$.

1. $2 \cos x - \sqrt{2} = 0$ **2.** $\sqrt{3} \sin x + \cos x = 0$

3. $\sin x + \sqrt{3} \cos x = 2$ **4.** $\tan x - 3 \cot x = 0$

5. $2 \sin x - \sqrt{3} = 0$ **6.** $2 \sin^2 x - 7 \sin x - 4 = 0$

7. $2 \cos^2 x - \cos x - 1 = 0$ **8.** $2 \sin^2 x - 1 = 0$

In Problems 9–16 the replacement set is $\{x \mid 0 \le x \le 2\pi\}$. Give answers in exact form.

9. $\sin^2 x + \tan^2 x = 4 - \cos^2 x$ **10.** $\sin x \cos x + \cos^2 x = 0$

11. $4 \sin^2 x - 3 = 0$ **12.** $2 \sin x + \sqrt{3} = 0$

13. $\sin x - \cos x = \sqrt{2}$ **14.** $2 \sin x - \cos x = 3$

15. $\sin^2 x + 2 \cos^2 x - 2 \cos x = 0$ **16.** $4 \operatorname{Sin}^{-1} x + \pi = 0$

In Problems 17–24 the replacement set is given by $0 \le x \le \pi$. Give answers with numbers rounded off to two decimal places.

17. $2 \sin x - \cos x = 0$ **18.** $4 \sin x + 3 = 0$

19. $2 \sin^2 x + 2 \sin x - 3 = 0$ **20.** $3 \sin x + 4 \cos x = 5$

21. $3 \sin x \cos x - \cos x = 0$ **22.** $5 \cos x + 12 \sin x = 13$

23. $\sin (\operatorname{Cos}^{-1} x) = 0.6$ **24.** $3 \operatorname{Cos}^{-1} x = 7$

In Problems 25–30, find the solution set for each of the given inequalities. In each case, draw a graph to help you "see" the solution. Give answers with numbers in exact form when reasonable; otherwise round off to two decimal places.

25. $2 \sin x \le 1$ where $0 \le x \le 2\pi$

26. $2 \cos x \le -\sqrt{3}$ where $0 \le x \le 2\pi$

27. $3 \operatorname{Cos}^{-1} x - 4 < 0$ **28.** $\operatorname{Sin}^{-1} x + 0.5 < 0$

29. $\sin (\operatorname{Cos}^{-1} x) \le 0.4$ **30.** $\cos (\operatorname{Sin}^{-1} x) > 0.5$

Graphs of Trigonometric Functions

6

In Sections 3.1 and 3.2 we discussed graphs of the six trigonometric functions and recognized that of these, the sine and cosine are the most basic. Both the sine and cosine are periodic, with period 2π, and vary between -1 and 1. Once we have the shape of the graph for one period, the remainder of the graph is a cyclic repetition of that portion.

In applications, one frequently encounters the problem of graphing more general functions, such as

$$y = 3 \sin\left(2x - \frac{\pi}{4}\right) \qquad \text{or} \qquad y = -2 \cos\left(\pi x + \frac{\pi}{3}\right).$$

These are particular examples of a general class of functions described by the equations:

$$\boxed{\begin{aligned} y &= A \sin(Bx + C), \\ y &= A \cos(Bx + C). \end{aligned}} \qquad \text{[6.1]}$$

where A, B, and C are called *parameters*, given real numbers. For these functions we make the obvious requirement that A and B must be nonzero.

In this section we are interested in exploring the graphs of the functions of the form of Eqs. (6.1). We shall do this by considering a sequence of special cases to determine the role played by each of the parameters in the graphs of such functions. Because each parameter has a different effect and because these functions are important in several different areas of applied work such as analysis of sound waves, electromagnetic radiation, vibrations, and electrical current, there are names attached to each. Unfortunately, the names and an unordered historical development have had the effect of complicating some very simple ideas. If we know *the shape of the sine and cosine curves* and *that each goes through one complete period while the argument increases from 0 to 2π,* then we essentially know all that is necessary to graph *any* function of the type given in Eqs. (6.1).

The parameter A (amplitude)

Once we know the graph of a function $y = f(x)$, we can use the information to sketch the graph of any constant multiple of the function, $y = Af(x)$.

We are working with trigonometric functions in this chapter, but the principles are perfectly general. We would recommend that the student look at familiar graphs such as parabolas, comparing the graphs of the functions of $y = x^2$, $y = 2x^2$, $y = (-1/2)x^2$, and $y = -3x^2$.

To see how changes in the constant A affect the graph of the function $y = A \sin x$, we have listed some values in Table 6.1 and then put them into graphical form in Fig. 6.1.

Suppose we refer to the graph of $y = \sin x$ as the *basic sine graph*. From a comparison of the graphs in Fig. 6.1 we observe that the parameter A changes only the y-coordinates of the basic graph. If A is greater than 1, the effect is to "stretch" the basic graph in the y-direction. Multiplying by a number between -1 and 1 reduces each y-coordinate, thus "compressing" the basic graph in the y-direction. Multiplication by a negative number "tips the basic graph upside-down" in addition to the stretching or compression. Finally, since $A \cdot 0 = 0$, the x-intercepts of the graphs of $y = f(x)$ and $y = Af(x)$ remain the same.

These observations apply to any function and may be summed up by saying that the *general shape of the graph of $y = Af(x)$ is the*

Table 6.1

Function \ x	0	$\dfrac{\pi}{6}$	$\dfrac{\pi}{4}$	$\dfrac{\pi}{3}$	$\dfrac{\pi}{2}$	$\dfrac{2\pi}{3}$	$\dfrac{3\pi}{4}$	$\dfrac{5\pi}{6}$	π
$y = \sin x$	0	$\dfrac{1}{2}$	$\dfrac{\sqrt{2}}{2}$	$\dfrac{\sqrt{3}}{2}$	1	$\dfrac{\sqrt{3}}{2}$	$\dfrac{\sqrt{2}}{2}$	$\dfrac{1}{2}$	0
$y = 3 \sin x$	0	$\dfrac{3}{2}$	$\dfrac{3\sqrt{2}}{2}$	$\dfrac{3\sqrt{3}}{2}$	3	$\dfrac{3\sqrt{3}}{2}$	$\dfrac{3\sqrt{2}}{2}$	$\dfrac{3}{2}$	0
$y = (-1/2) \sin x$	0	$\dfrac{-1}{4}$	$\dfrac{-\sqrt{2}}{4}$	$\dfrac{-\sqrt{3}}{4}$	$\dfrac{-1}{2}$	$\dfrac{-\sqrt{3}}{4}$	$\dfrac{-\sqrt{2}}{4}$	$\dfrac{-1}{4}$	0
$y = -2 \sin x$	0	-1	$-\sqrt{2}$	$-\sqrt{3}$	-2	$-\sqrt{3}$	$-\sqrt{2}$	-1	0

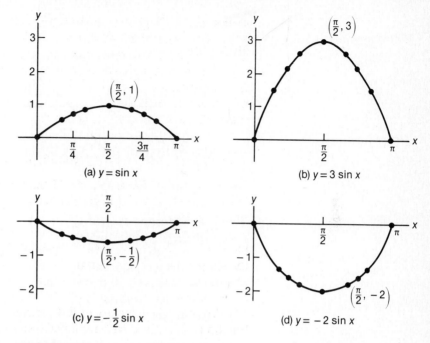

FIGURE 6.1

same as the graph of $y = f(x)$ except for change in the vertical direction.

We are interested in the sine and cosine functions in particular. Curves that oscillate periodically as do the sine and cosine curves are called "sinusoidal." We observe the following properties about the graph of $y = A \sin x$:

1. The graph is sinusoidal (it still looks like a sine curve).
2. The x-intercepts are the same as for $y = \sin x$ (all integer multiples of π).
3. The graph oscillates between the horizontal lines $y = A$ and $y = -A$.

The last property is given a name. The distance from the center of oscillation ($y = 0$) to the extreme values ($y = A$) is called the *amplitude* of a sinusoidal curve. Thus the *parameter A determines the amplitude*. The amplitude of $y = A \sin x$ is the same as the amplitude of $y = -A \sin x$ (both oscillate between $y = -A$ and $y = A$).

Since the graph of the cosine function is just the graph of the sine curve shifted to the left (as was noted in Example 2 of Section 3.1), we would expect graphs of multiples of the cosine to have the same shape as the graph of $y = \cos x$, stretched or compressed. The graph of $y = A \cos x$ is still sinusoidal and oscillates between the horizontal lines $y = -A$ and $y = A$ (whether A is positive or negative).

The *amplitude* of $y = A \sin x$ and of $y = A \cos x$ is $|A|$.

In Fig. 6.2 we show on the same set of axes the graphs of all the functions we examined in Fig. 6.1. Note that all have the same period 2π and the same x-intercepts. Figure 6.3 shows the corresponding multiples of $y = \cos x$.

The parameters *B* and *C* (period and phase shift)

The functions $y = \sin x$ and $y = \cos x$ are periodic of period 2π. That means that the graph of the entire function simply repeats the portion of the graph on the interval $[0, 2\pi]$. We shall refer to that portion of the graph as the *fundamental cycle* of the graph (or function) (see Fig. 6.4).

All functions of the form of either of Eqs. (6.1) are sinusoidal and hence periodic. The distinguishing features of any such graph are its amplitude (determined by A), its frequency of oscillation, and the location of each cycle. The frequency and location are determined by B and C.

FIGURE 6.2

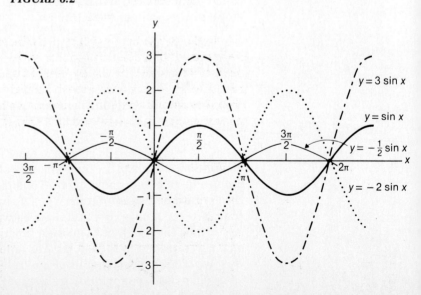

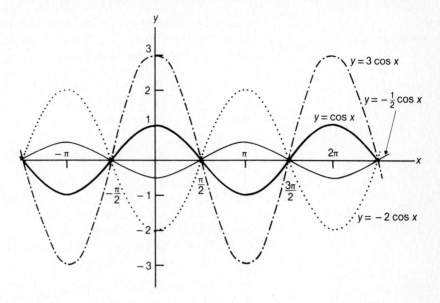

FIGURE 6.3

In discussing the graphs of $y = A\sin(Bx + C)$ and $y = A\cos(Bx + C)$ it is convenient to be able to assume that B is positive. By proper use of the odd–even identities we may always make that assumption because

$$\sin(-Bx + C) = \sin[-(Bx - C)]$$
$$= -\sin(Bx - C) \qquad \text{by identity (I-6),}$$
$$\cos(-Bx + C) = \cos[-(Bx - C)]$$
$$= \cos(Bx - C) \qquad \text{by identity (I-7).}$$

In the equation $y = \sin(Bx + C)$ the expression $Bx + C$ is called the *argument* of the function. Since $\sin 0 = 0 = \sin 2\pi$, the behavior of the function is determined by what happens as the argument varies from 0 to 2π. For convenience of reference we shall call the *fundamental cycle* of a sinusoidal curve the portion of the curve for which the argument is between 0 and 2π (compare Fig. 6.4). Thus the funda-

FIGURE 6.4

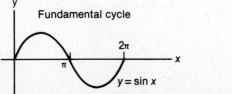

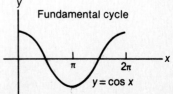

mental cycle of any sinusoidal curve corresponds to the fundamental cycle of $y = \sin x$ or $y = \cos x$; and most important, *the fundamental cycle contains all essential information about the graph* because the graph continues to repeat the fundamental cycle in both directions. The portion of the x-axis corresponding to the fundamental cycle is called the *fundamental interval.* By definition, the fundamental interval is the set $\{x \mid 0 \leq Bx + C \leq 2\pi\}$ or the equivalent

$$\left\{ x \mid -\frac{C}{B} \leq x \leq \frac{2\pi - C}{B} \right\} \qquad \text{(assuming } B > 0\text{)}.$$

We shall look at the graph of $y = \sin(Bx + C)$ first. The fundamental interval is $[-C/B, (2\pi - C)/B]$. The fundamental cycle on the fundamental interval looks just like the fundamental cycle of the sine curve in Fig. 6.4. The x-intercepts are at both ends and the midpoint; the maximum value occurs one quarter of the way across the fundamental interval, and the minimum value is three quarters of the way. A similar analysis applies to $y = \cos(Bx + C)$ in that the fundamental cycle looks like the fundamental cycle of the cosine curve in Fig. 6.4.

Before looking at specific examples we make another suggestion. It may not be difficult to remember the endpoints of the fundamental interval, but we believe that it is of greater value to understand that we are letting the *argument increase from 0 to 2π*. That idea is much more likely to be remembered than a formula for the endpoints of the interval.

EXAMPLE 1   Sketch the graph of $y = 2 \sin x$ and $y = \sin 2x$.

Solution   From our previous discussion, $y = 2 \sin x$ has the same graph as $y = \sin x$ except for the "stretching" factor of 2. The amplitude is 2, and the fundamental cycle is the solid curve in Fig. 6.5(a).

The amplitude of $y = \sin 2x$ is 1, but the parameter $B = 2$ affects the fundamental interval. Setting the argument, $2x$, successively

FIGURE 6.5

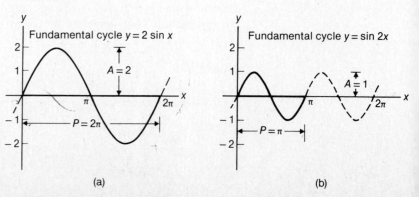

(a)

(b)

equal to 0 and 2π, we find that the fundamental interval is $[0, \pi]$. Repetition of the fundamental cycle (the solid portion) gives the entire graph as in Fig. 6.5(b). ■

A comparison of the two graphs in Fig. 6.5 makes several things clear. The graph of $y = \sin 2x$ is still a sine curve, but it oscillates "faster" (completes a period in a shorter interval). The *period of $y =$ $\sin 2x$ is the *length of the fundamental interval.* Since sinusoidal curves have the shape of waves (and are used to describe wavelike phenomena), the period is also called the *wavelength.* It is the length between successive wave crests. The *frequency* of a sinusoidal curve is the reciprocal of the period. For $y = \sin 2x$ the frequency is $1/\pi$. In general, for the function $y = \sin Bx$, where $B > 0$, we have the following important relations:

$$\text{Frequency} = \frac{1}{\text{Wavelength}} = \frac{1}{\text{Period}} = \frac{B}{2\pi},$$

measured in cycles per unit length.

EXAMPLE 2   Sketch the graphs of $y = \cos(-3x)$ and $y = 2\sin(-2x)$ on the same set of coordinates.

Solution   As suggested above, these functions may be simplified by using the odd–even identities: $\cos(-3x) = \cos 3x$ and $2\sin(-2x) = -2\sin 2x$.

The graph of $y = \cos 3x$ has amplitude 1. We find the fundamental interval by setting $3x = 0$ and $3x = 2\pi$ successively, from which the fundamental interval is $[0, 2\pi/3]$. Now it is simply a matter of sketching a fundamental cycle of a cosine curve on the interval $[0, 2\pi/3]$. In Fig. 6.6 we show the fundamental cycles as solid curves.

FIGURE 6.6

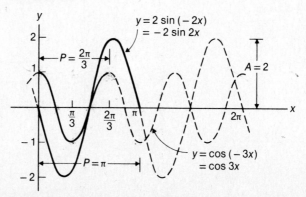

The graph of $y = 2 \sin(-2x) = -2 \sin 2x$ is graphed by noting that the amplitude is $|-2| = 2$, and the fundamental interval is $[0, \pi]$. Since A is negative, the fundamental cycle is a sine cycle "tipped upside-down" as shown in Fig. 6.6. The rest of the graph is a repeat of the fundamental cycle. ∎

EXAMPLE 3   Sketch the graph of $y = -3 \sin(\pi - 2x)$ and find the domain and range of the function.

Solution   For convenience we write $y = -3 \sin(\pi - 2x) = -3 \sin[-(2x - \pi)] = 3 \sin(2x - \pi)$. To get the fundamental interval, we set the argument $(2x - \pi)$ equal to 0 and then to 2π and obtain $[\pi/2, 3\pi/2]$. The amplitude is 3. Thus we have a fundamental cycle, which can be repeated to get the entire graph (see Fig. 6.7). Here the fundamental cycle is "shifted" from the basic sine graph $\pi/2$ units to the right (since the fundamental interval begins at $\pi/2$ rather than 0). This shift is often called the *phase shift* and is affected by both the parameters B and C, since the left-hand end of the fundamental interval for $y = A \sin(Bx + C)$ is given by $Bx + C = 0$, or $x = -C/B$.
From the graph it is easy to see that the domain is the set of all real numbers and the range is the set of numbers $\{y \mid -3 \le y \le 3\}$.

Alternate Solution   We can change the form of the function by the use of the reduction formula $\sin(\pi - \theta) = \sin\theta$. This gives $y = -3 \sin(\pi - 2x) = -3 \sin 2x$.

FIGURE 6.7

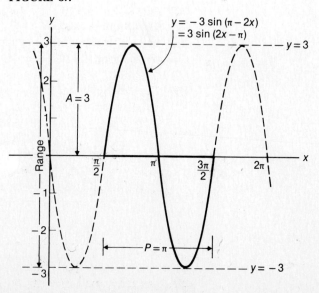

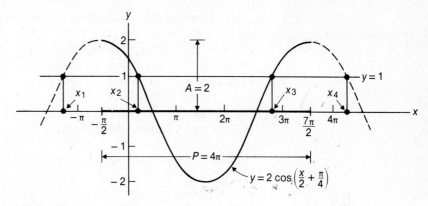

FIGURE 6.8

For $y = -3 \sin 2x$ the fundamental interval is $[0, \pi]$, and the fundamental cycle is an upside-down fundamental sine cycle with amplitude 3. The graph, of course, is the same as that shown in Fig. 6.7. ∎

EXAMPLE 4

a] Sketch the graph of $y = 2 \cos [(x/2) + (\pi/4)]$, and

b] on the graph show the x-values for which $2\cos [(x/2) + (\pi/4)] = 1$.

Solution

a] The amplitude is 2. Solving $(x/2) + (\pi/4) = 0$ and $(x/2) + (\pi/4) = 2\pi$, we get $x = -\pi/2$ and $x = 7\pi/2$, respectively, indicating a phase shift of $\pi/2$ units to the left (from 0 to $-\pi/2$). On the fundamental interval $[-\pi/2, 7\pi/2]$ we sketch a fundamental cosine cycle and repeat, as shown in Fig. 6.8.

b] The points where the graph crosses the line $y = 1$ give the x-values where $2 \cos [(x/2) + (\pi/4)] = 1$. Four such x-values are labelled x_1, x_2, x_3, and x_4 in Fig. 6.8. From the graph it appears that x_2 is near $\pi/6$. It is easily checked that $x = \pi/6$ does satisfy the equation $2 \cos [(x/2) + (\pi/4)] = 1$. A sufficiently accurate graph would enable us to approximate the x-values corresponding to any given y-value in the range of the function. ∎

Summary

Let us summarize the techniques of drawing graphs in this section. To graph any function of the form in Eqs. (6.1):

1. Use odd–even identities if necessary to get

$$y = A \sin (Bx + C) \quad \text{or}$$
$$y = A \cos (Bx + C) \quad \text{with } B \text{ positive.}$$

2. Find the fundamental interval by setting the argument equal to 0 and 2π, successively. This identifies the

$$\text{Period (length of fundamental interval)} = \frac{2\pi}{B},$$

$$\text{Frequency (reciprocal of the period)} = \frac{B}{2\pi},$$

Phase shift (C/B units to the left if $C > 0$ and to the right if $C < 0$).

3. Sketch a fundamental sine or cosine cycle having amplitude $|A|$ on the fundamental interval.

EXERCISE 6.1

In Problems 1–30, determine the fundamental interval, period, frequency, and amplitude for each function. Then sketch a graph, showing at least the fundamental cycle. Appropriate reduction formulas may be helpful in some instances (see the Alternative Solution for Example 3).

1. $y = 2 \sin x$ 2. $y = -3 \sin x$ 3. $y = -4 \cos x$

4. $y = 2 \cos x$ 5. $y = \frac{1}{2} \sin x$ 6. $y = \frac{2}{3} \cos x$

7. $y = \sin 3x$ 8. $y = \sin \frac{x}{2}$ 9. $y = -2 \cos 3x$

10. $y = 3 \sin (-2x)$ 11. $y = -4 \sin \left(\frac{\pi}{4} - x \right)$

12. $y = 3 \sin (\pi x)$ 13. $y = -3 \cos \left(\frac{\pi}{2} x \right)$

14. $y = \sin (-3\pi x)$ 15. $y = -2 \cos (-\pi x)$

16. $y = -2 \sin \left(2x + \frac{\pi}{3} \right)$ 17. $y = \sin \left(x + \frac{\pi}{2} \right)$

18. $y = \sin (2x + \pi)$ 19. $y = 2 \cos \left(3x + \frac{3\pi}{4} \right)$

20. $y = -\cos (\pi - 2x)$ 21. $y = 4 \sin (3x - \pi)$

22. $y = -3 \sin (\pi - 2x)$ 23. $y = 2 \sin \left(2x + \frac{\pi}{2} \right)$

24. $y = \frac{3}{2} \cos \left(-3x + \frac{3\pi}{4} \right)$ 25. $y = -3 \sin \left(2\pi x - \frac{2\pi}{3} \right)$

26. $y = \frac{1}{2} \sin \left(\frac{\pi}{2} - \pi x \right)$ 27. $y = -3 \cos \pi(2x + 1)$

28. $y = \dfrac{1}{2} \sin \pi\!\left(3 - \dfrac{x}{2}\right)$

29. $y = -3 \cos\!\left(2\pi x - \dfrac{\pi}{2}\right)$

30. $y = \sqrt{2} \sin\!\left(\dfrac{x}{2} + \dfrac{\pi}{4}\right)$

In Problems 31–36, sketch a graph for each function; on your graph, show at least two x-values satisfying the accompanying equation (see Example 4). Read these x-values as accurately as you can from your graph and check to see whether they actually are solutions.

31. $y = \sin\!\left(x + \dfrac{\pi}{2}\right)$; $\sin\!\left(x + \dfrac{\pi}{2}\right) = \dfrac{1}{2}$

32. $y = \cos(2x - \pi)$; $\cos(2x - \pi) = \dfrac{\sqrt{2}}{2}$

33. $y = 4 \cos(\pi - x)$; $4 \cos(\pi - x) = -2$

34. $y = \sin \pi(x + 1)$; $\sin \pi(x + 1) = \dfrac{1}{2}$

35. $y = 2 \sin \pi\!\left(\dfrac{1}{2} - x\right)$; $2 \sin \pi\!\left(\dfrac{1}{2} - x\right) = -2$

36. $y = \sqrt{2} \sin\!\left(\dfrac{x}{2} + \dfrac{\pi}{4}\right)$; $\sqrt{2} \sin\!\left(\dfrac{x}{2} + \dfrac{\pi}{4}\right) = -1$

6.2 Graphs of the Remaining Trigonometric Functions

With a solid understanding of the material in the preceding section it is fairly straightforward to graph the other trigonometric functions. We depend, of course, on the fact that the other four functions are definable in terms of the sine and cosine. In this section we consider functions of the following forms:

$$\text{(i)} \quad y = A \tan(Bx + C), \qquad \text{(ii)} \quad y = A \cot(Bx + C),$$
$$\text{(iii)} \quad y = A \sec(Bx + C), \qquad \text{(iv)} \quad y = A \csc(Bx + C). \quad \textbf{[6.2]}$$

In Section 3.2 we saw the graphs of the *basic functions* $y = \tan x$, $y = \cot x$, $y = \sec x$, and $y = \csc x$.

As we would expect on the basis of our knowledge of the general sine and cosine functions, each of the parameters A, B, and C affects the basic graphs differently. The constant multiple A stretches or compresses the graphs, B determines the period and frequency of repetition, and B and C together control the location of the fundamental interval.

There are obvious differences, however. One of the most crucial is that all four of the functions in Eqs. (6.2) are unbounded, so the *idea of amplitude is meaningless.* Closely related is the fact that graphs of all these functions have *vertical asymptotes.* As we mentioned in Section 3.2, a vertical asymptote is a vertical line that is approached by a graph. For the functions in Eqs. (6.2)—for instance $y = A \tan(Bx + C) = A \sin(Bx + C)/\cos(Bx + C)$—there is a denominator, namely, $\cos(Bx + C)$, that becomes zero for infinitely many values of x. At such values of x the function is undefined, but for neighboring values the function becomes *very large* (positive or negative). This behavior characterizes the graphs of the tangent, cotangent, secant, and cosecant functions.

Graphs of the basic functions

The ideas of fundamental interval and fundamental cycle are as useful here as they were in the preceding section, but the pictures of the fundamental cycles for functions given in Eqs. (6.2) are best related to the sine and cosine. Of critical concern in each instance is the location of the vertical asymptotes; these are determined completely by the zeros of the denominator, in each case either sine or cosine. The fundamental cycles for these functions are shown in Fig. 6.9 along with the key facts that we find useful for recalling their properties. Each of the cycles shown is typical of all *positive* multiples of the function. Negative multiples yield graphs that are tipped upside down from the ones in the figure. There is some arbitrary choice in selecting which interval should be considered fundamental, especially for the secant and cosecant. We have made the choice so that the fundamental cycle has only two pieces, each being the reciprocal of one arch of the sine or cosine curve. Since each cycle is repeated over and over for the entire graph, the choice is arbitrary and immaterial. We could describe the general situation for the domain and range of the functions in this section, but we believe that inspection of particular graphs is more instructive.

The *period of the tangent and cotangent is* π, rather than 2π as it is for the other trigonometric functions. Each cycle is symmetric about the midpoint of the fundamental interval. The fundamental cycle of any positive multiple of the tangent function is increasing; that of the cotangent function is decreasing. In Fig. 6.9 the vertical scale in (a) and (b) is set by the fact that $\tan(\pi/4) = \cot(\pi/4) = 1$.

As may be seen from Figs. 6.9(c) and 6.9(d), the fundamental cycles for secant and cosecant are identical except for location. The fundamental interval for the cosecant coincides with that for the sine function, $[0, 2\pi]$. We choose $[-\pi/2, 3\pi/2]$ for the secant so that we are taking the reciprocal of a period of the cosine consisting of two

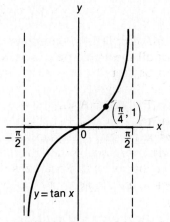

(a) Fundamental cycle of $y = \tan x$

Fundamental interval $= \left(-\dfrac{\pi}{2}, \dfrac{\pi}{2}\right)$

Period $= \pi$

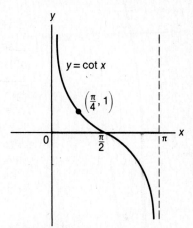

(b) Fundamental cycle for $y = \cot x$

Fundamental interval $= (0, \pi)$

Period $= \pi$

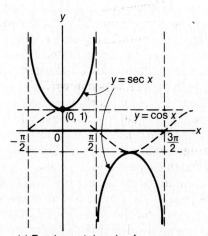

(c) Fundamental cycle of $y = \sec x$

Fundamental interval $= \left(-\dfrac{\pi}{2}, \dfrac{3\pi}{2}\right)$

Period $= 2\pi$

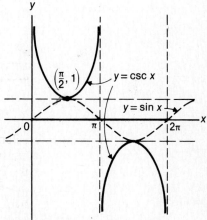

(d) Fundamental cycle of $y = \csc x$

Fundamental interval $= (0, 2\pi)$

Period $= 2\pi$

FIGURE 6.9

complete arches. Since the period of both the secant and cosecant is 2π, the length of each fundamental interval is also 2π. The minimum value (for vertical scale) on the positive piece of the fundamental cycle occurs one quarter of the way across the fundamental interval, at the point corresponding to $x = 0$ for the secant and to $x = \pi/2$ for the cosecant.

Graphs of the general functions _____

Having the fundamental cycles for the basic functions available for reference, we may look at some examples of graphing.

EXAMPLE 1 Sketch the graph of $y = 3 \tan [(\pi - x)/2]$.

Solution Since B is negative ($B = -1/2$), we first change the form of the function, using the identity $\tan(-t) = -\tan t$:

$$y = 3 \tan \left(\frac{\pi - x}{2} \right) = 3 \tan [-(x - \pi)/2]$$
$$= -3 \tan [(x - \pi)/2] = -3 \tan (x/2 - \pi/2).$$

Thus we wish to graph $y = -3 \tan (x/2 - \pi/2)$.

To identify the fundamental interval and cycle, we look at Fig. 6.9(a) for the tangent curve. The fundamental interval corresponds to $[-\pi/2, \pi/2]$. Setting the argument $(x/2 - \pi/2)$ successively equal to $-\pi/2$ and $\pi/2$, we find $x = 0$ and $x = 2\pi$.

Thus we will have an upside-down fundamental tangent cycle (because of the factor -3) in the interval $[0, 2\pi]$. This means that the function is decreasing. There are vertical asymptotes at each end of the fundamental cycle, and $(\pi, 0)$ is the point of symmetry for the cycle. We can draw a fairly accurate graph when we locate a point to fix the vertical scale. Since $\tan \pi/4 = 1$, we go to the corresponding point three quarters of the way across the fundamental interval (in this case at $x = 3\pi/2$) and evaluate $y = 3 \tan [(\pi - 3\pi/2)/2] = 3 \tan (-\pi/4) = -3$. This gives us the points $(3\pi/2, -3)$ and $(\pi/2, 3)$ (by symmetry). Repeating the fundamental cycle gives us the entire graph (see Fig. 6.10).

Alternate Solution The *fundamental cycle* in Fig. 6.10 appears to be a cotangent cycle. As a matter of fact, we may use a reduction identity, $\tan(\pi/2 - \theta) = \cot \theta$, to change our approach to this problem, setting

$$y = 3 \tan (\pi/2 - x/2) = 3 \cot (x/2).$$

This means that we want a (positive) cotangent cycle on the interval corresponding to the interval $[0, \pi]$ in Fig. 6.9. Setting the argument $x/2$ successively equal to 0 and π gives us a *fundamental interval* of $[0, 2\pi]$ for $\cot (x/2)$, vertical asymptotes at each end, and when $x = \pi/2$ (one quarter of the *fundamental interval*), $y = 3 \cot (\pi/4) = 3$, giving us the same graph we obtained in Fig. 6.10. ■

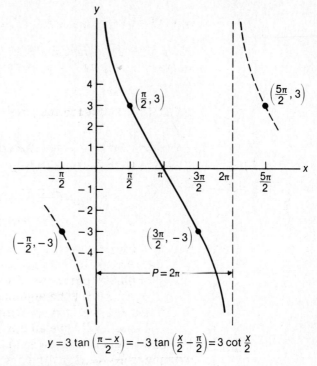

$$y = 3 \tan\left(\frac{\pi - x}{2}\right) = -3 \tan\left(\frac{x}{2} - \frac{\pi}{2}\right) = 3 \cot \frac{x}{2}$$

FIGURE 6.10

EXAMPLE 2   If $f(x) = 2 \sec(\pi x - \pi/2)$, sketch the graph of $y = f(x)$ and find the domain and range of f.

Solution   From Fig. 6.9(c) the fundamental interval for the secant corresponds to $[-\pi/2, 3\pi/2]$. Setting the argument $(\pi x - \pi/2)$ equal successively to $-\pi/2$ and $3\pi/2$ yields $x = 0$ and $x = 2$. Thus the fundamental cycle occurs in the interval $[0, 2]$, and we will have vertical asymptotes at $x = 0$, $x = 1$, and $x = 2$. Since this cycle is repeated, the graph will have vertical asymptotes at every integer value of x. The low point of the first (left) piece will be halfway between the first two asymptotes, at $x = 1/2$, where the value of y is $2 \sec(0) = 2$. See Fig. 6.11, where the solid portion of the curve shows a fundamental cycle.

From the graph we may easily read the domain and range. The function is clearly defined everywhere except at the integers (where the graph has vertical asymptotes). The range consists of all numbers except those between -2 and 2:

$D(f) = \{x \mid x \text{ is a real number and } x \text{ is not an integer}\}$,
$R(f) = \{y \mid y \leq -2 \quad \text{or} \quad y \geq 2\}$.

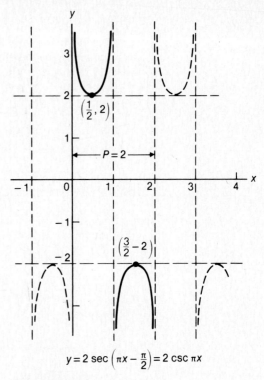

$$y = 2 \sec\left(\pi x - \frac{\pi}{2}\right) = 2 \csc \pi x$$

FIGURE 6.11

Alternate Solution   We can use the reduction formula $\sec(\theta - \pi/2) = \csc\theta$ to express the given equation as

$$y = 2 \sec\left(\pi x - \frac{\pi}{2}\right) = 2 \csc \pi x.$$

The graph of $y = 2 \csc \pi x$ is the same as that shown in Fig. 6.11.   ∎

Summary

To graph any of the functions of the form described in this section (Eqs. 6.2):

1. If B is negative, then first use an appropriate odd–even reduction identity to get an equivalent form with a positive B.

2. Find the *fundamental interval* by setting the argument $Bx + C$ equal to each endpoint of the fundamental interval of the appropriate function as shown in Fig. 6.9:

$$\left(-\frac{\pi}{2}, \frac{\pi}{2}\right) \quad \text{for } \tan x; \qquad \left(-\frac{\pi}{2}, \frac{3\pi}{2}\right) \quad \text{for } \sec x,$$

$$(0, \pi) \quad \text{for } \cot x; \qquad (0, 2\pi) \quad \text{for } \csc x.$$

The *length* of the fundamental interval is the *period* of the function.

3. Determine the vertical asymptotes in relation to the fundamental interval. Fix the vertical scale by locating a point corresponding to a known point on the basic graph such as

$$\tan \pi/4 = 1 = \cot \pi/4, \qquad \sec 0 = 1, \qquad \csc \pi/2 = 1.$$

4. Sketch a fundamental cycle for the appropriate function (Fig. 6.9) on the fundamental interval, and repeat.

EXERCISE 6.2

For Problems 1–22, sketch a graph of the given function by identifying (a) the fundamental interval (corresponding to the appropriate interval in Fig. 6.9) and vertical asymptotes and (b) the vertical scale. Then sketch the fundamental cycle as a solid portion and repeat with a broken portion of the curve.

1. $y = 3 \tan x$
2. $y = -3 \tan 2x$
3. $y = -2 \cot x$

4. $y = 3 \cot \dfrac{x}{2}$
5. $y = 2 \tan \left(-\dfrac{x}{2}\right)$

6. $y = -3 \tan \left(\dfrac{\pi}{2}x\right)$
7. $y = -4 \tan \left(2x + \dfrac{\pi}{2}\right)$

8. $y = -3 \cot \left(2x - \dfrac{\pi}{4}\right)$
9. $y = 2 \tan \left(\pi x - \dfrac{\pi}{2}\right)$

10. $y = \dfrac{1}{2} \cot \left(\dfrac{\pi}{2}x - \dfrac{\pi}{3}\right)$
11. $y = \sqrt{3} \tan \pi \left(x + \dfrac{3}{2}\right)$

12. $y = 3 \tan (2x + 1)$
13. $y = 3 \sec x$

14. $y = -2 \sec x$
15. $y = -3 \sec 2x$

16. $y = 4 \csc 3x$
17. $y = \csc (3\pi x)$

18. $y = 2 \sec (2\pi x)$
19. $y = 3 \csc \left(2\pi x + \dfrac{\pi}{2}\right)$

20. $y = 4 \sec \left(\pi x + \dfrac{\pi}{2}\right)$
21. $y = -2 \sec \left(2x - \dfrac{\pi}{2}\right)$

22. $y = 2 \sec (3x - \pi)$

6.3 More Graphs: Adding and Multiplying Ordinates

In the preceding two sections we considered graphs of trigonometric functions individually. In this section we also discuss the problem of drawing graphs of equations that involve algebraic functions as well as trigonometric functions. The techniques employed are best illustrated by examples.

Adding ordinates

We illustrate the technique of drawing graphs by addition of ordinates in the next two examples.

EXAMPLE 1   Draw a graph of $y = 2 + \sin x$.

Solution   We first draw a graph of $y = \sin x$ (call it C). This is shown by the broken curve in Fig. 6.12. Since the y-values for the curve we want are greater by 2 units than the corresponding y-value on C, it should be clear that our curve can be obtained by adding 2 to each of the y-values of C, that is, by moving C vertically upward two units. This gives the solid curve shown in Fig. 6.12.   ■

EXAMPLE 2   Draw a graph of $y = x/3 + \sin x$.

Solution   A direct approach would be to make a table of corresponding values of x and y that satisfy the given equation, plot these points, and then draw the graph. This is essentially what we are going to do, except that we shall draw two auxiliary curves and use them to draw the graph of the given equation.

Let C_1 denote the graph of $y_1 = x/3$ and C_2 the graph of $y_2 = \sin x$, as shown by the dashed curves in Fig. 6.13.

It is clear that for each x the corresponding value of y is the sum of y_1 and y_2 for that value of x. And so we geometrically add the corresponding ordinates of C_1 and C_2 to get the ordinates of the graph

FIGURE 6.12

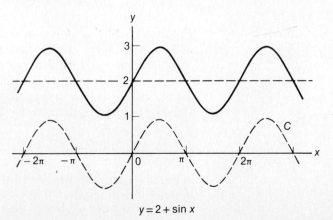

$y = 2 + \sin x$

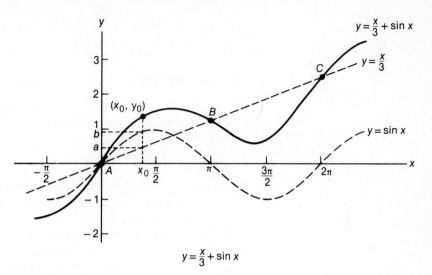

FIGURE 6.13

for the given equation. This is illustrated in Fig. 6.13 for $x = x_0$; the corresponding value of $y = y_0$ is obtained by adding a and b. We also make an observation concerning the key points labeled A, B, and C in Fig. 6.13. If we take the values $x = 0, \pi, 2\pi$, then in each case the corresponding value of $y_2 = \sin x$ is zero, so the value of y is $y_1 = x/3$. Thus the curve passes through points on the line $y_1 = x/3$ given by $(k\pi, k\pi/3)$, where k is any integer. That is, the graph of $y = x/3 + \sin x$ is a curve winding around the line $y = x/3$, as shown in Fig. 6.13. ∎

Multiplying ordinates

EXAMPLE 3 Draw the graph of $y = x(\cos 2x)$.

Solution Multiplication of ordinates is generally so complicated that point plotting may be necessary; but when one of the factors is sine or cosine, multiplication can be a reasonable approach. We know that the cosine varies between -1 and 1. Consequently, the function $x(\cos 2x)$ will vary between $x(-1) = -x$ and $x(1) = x$, and the graph of $y = x(\cos 2x)$ will oscillate between the lines $y = -x$ and $y = x$.

In a manner similar to that used in Example 2 we first draw the graphs of $y = x$, $y = -x$, and $y = \cos 2x$ (which has the fundamental interval $[0, \pi]$) with dashed lines all on the same set of coordinates. The graph of $y = x(\cos 2x)$ will be tangent to ("touch") the line $y = x$ whenever $\cos 2x = 1$ (hence the importance of the techniques of Sec-

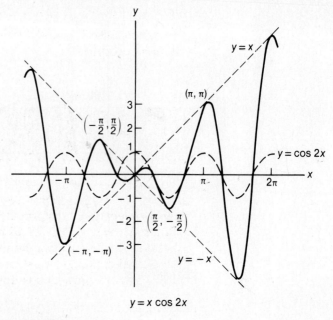

$$y = x \cos 2x$$

FIGURE 6.14

tion 6.1 to have a good picture of the graph of $y = \cos 2x$). The graph will touch the line $y = -x$ when $\cos 2x = -1$, and the graph will cross the x-axis whenever $\cos 2x = 0$. It also crosses the axes at the origin where the other factor x is zero. The graph is shown in Fig. 6.14. We should also note that when $x = t$ (any number), $y = t \cos 2t$; and for $x = -t$, $y = -t \cos(-2t) = -t \cos 2t$. This tells us that the graph is symmetric with respect to the origin. For instance, pairs of points such as (π, π), $(-\pi, -\pi)$ and $(\pi/2, -\pi/2)$, $(-\pi/2, \pi/2)$ are on the graph. ■

The graph in Fig. 6.14 is typical of a number of graphs of functions having a sine or cosine factor. All such graphs oscillate between the graph of the other factor (in this example the other factor is x) and its negative, so that the effect is a sinusoidal type of curve squeezed between two other curves.

EXERCISE 6.3

In Problems 1–16, draw a graph of the given equation.

1. $y = 1 + \sin x$ **2.** $y = 2 - \cos x$ **3.** $y = \dfrac{x}{2} + 2 \sin x$

4. $y = 2x + \sin x$ **5.** $y = 2x + \cos x$ **6.** $y = x - 2 \cos x$

7. $y = \sin x - 1$ **8.** $y = 2 \cos x - 3$ **9.** $y = x \sin x$

10. $y = 2x \cos x$ **11.** $y = -x \cos 2x$ **12.** $y = x \sin(-2x)$

13. $y = \sqrt{x} \sin x$ **14.** $y = \sqrt{x} + \sin x$ **15.** $y = \sqrt{x} + \sin(-x)$

16. $y = \sqrt{x} \cos 2x$

6.4 The Use of Identities in Graphing (Optional)

There have been several instances in this textbook when our approach to solving problems involved a sequence of steps in which the given problem was transformed into an equivalent one with a known solution. In this section we discuss problems of drawing graphs of equations in which trigonometric identities are used to transform the given equation to an equivalent one whose graph may be familiar to us. We illustrate how this is done with the following examples.

EXAMPLE 1   Draw the graph of $y = (\sin x + \cos x)^2$.

Solution   We first write the given equation in the following equivalent forms:

$$y = \sin^2 x + 2 \sin x \cos x + \cos^2 x \quad \text{(by algebra)}$$
$$= 1 + 2 \sin x \cos x \quad \text{(by identity (I-5))}$$
$$= 1 + \sin 2x \quad \text{(by identity (I-25))}.$$

We recognize the final form, $y = 1 + \sin 2x$, as an equation that is easily graphed by the methods of the preceding section. The function $y = \sin 2x$ was graphed in Example 1 of Section 6.1. The graph of the function $y = (\sin x + \cos x)^2 = 1 + \sin 2x$ is the same as the graph of $y = \sin 2x$ translated vertically upward 1 unit so that we get a sine curve of period π that winds about the line $y = 1$, as shown in Fig. 6.15.   ■

FIGURE 6.15

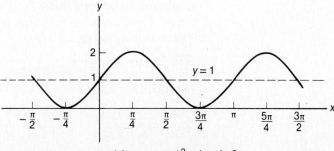

$$y = (\sin x + \cos x)^2 = 1 + \sin 2x$$

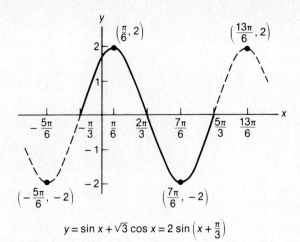

$$y = \sin x + \sqrt{3} \cos x = 2 \sin\left(x + \frac{\pi}{3}\right)$$

FIGURE 6.16

EXAMPLE 2   Sketch the graph of $y = \sin x + \sqrt{3} \cos x$.

Solution   The RHS of this equation is an example of an expression of the form $a \sin x + b \cos x$. In the preceding chapter (see Section 5.4) we discovered that by factoring out (or multiplying and dividing by) the quantity $\sqrt{a^2 + b^2}$ we may always write such an expression in the form $\sqrt{a^2 + b^2} \sin(x + \alpha)$, where $\cos\alpha = a/\sqrt{a^2 + b^2}$ and $\sin\alpha = b/\sqrt{a^2 + b^2}$.

Using that procedure here, we have $\sqrt{1^2 + (\sqrt{3})^2} = \sqrt{4} = 2$, and so

$$y = 2\left(\frac{1}{2}\sin x + \frac{\sqrt{3}}{2}\cos x\right) = 2\left(\sin x \cos\frac{\pi}{3} + \cos x \sin\frac{\pi}{3}\right)$$

$$= 2\sin\left(x + \frac{\pi}{3}\right) \qquad \text{(by identity (I-10)).}$$

In this form we may use the techniques of Section 6.1. The graph is a sine curve with amplitude 2 and a fundamental interval of $[-\pi/3, 5\pi/3]$, as shown in Fig. 6.16, where the solid portion of the curve corresponds to the fundamental cycle.   ■

EXAMPLE 3   Use graphical methods to find the solution set for the following equations:

a] $\sqrt{2}(\sin x + \cos x) = \tan x + \cot x$

b] $\sin x + \cos x = \sqrt{2}(\tan x + \cot x)$

Solution   These equations are of the type discussed in Section 5.5; in fact (a) appeared as Problem 36 in Exercise 5.5. If each of the functions defined by the left-hand side (LHS) and right-hand side (RHS) can be readily graphed, the solutions of the equation are given by the intersections of the graphs. This can be very helpful in identifying the nature of the solutions, if there are any, and can provide a very useful check on other methods used to solve the equations.

a] If we apply the techniques from Section 5.4 (see the preceding example), we may factor out $\sqrt{1^2 + 1^2} = \sqrt{2}$ from the LHS:

$$\text{LHS:} \quad \sqrt{2}\,(\sin x + \cos x) = \sqrt{2} \cdot \sqrt{2}\left(\frac{1}{\sqrt{2}}\sin x + \frac{1}{\sqrt{2}}\cos x\right)$$

$$= 2\left(\sin x\,\cos\frac{\pi}{4} + \cos x\,\sin\frac{\pi}{4}\right)$$

$$= 2\,\sin\left(x + \frac{\pi}{4}\right).$$

The RHS may also be expressed in terms of sine and cosine:

$$\text{RHS:} \quad \tan x + \cot x = \frac{\sin x}{\cos x} + \frac{\cos x}{\sin x} = \frac{\sin^2 x + \cos^2 x}{\cos x \,\sin x}$$

$$\text{(by algebra)}$$

$$= \frac{1}{\sin x\,\cos x} = \frac{2}{2\,\sin x\,\cos x} = \frac{2}{\sin 2x}$$

$$\text{(by identities (I-5) and (I-25))}$$

$$= 2\,\csc 2x.$$

Equating the two sides and dividing by 2, we have the equivalent equation

$$\sin\left(x + \frac{\pi}{4}\right) = \csc 2x.$$

The graphs of $y = \sin(x + \pi/4)$ and $y = \csc 2x$ are readily drawn by the methods of this chapter. The first is a sine curve with fundamental interval $[-\pi/4, 7\pi/4]$ and amplitude 1 and is shown in Fig. 6.17. The second has a fundamental interval $[0, \pi]$ and vertical asymptotes at all integer multiples of $\pi/2$. The graph is shown in Fig. 6.17(a). It is clear that the graphs intersect only where $x = \pi/4 \pm 2k\pi$. A check in the original equation verifies the correctness of the solution. Also check this solution with that obtained analytically in Problem 36 of Exercise 5.5.

b] The only difference between equation (a) and equation (b) is the location of the factor $\sqrt{2}$, so exactly the same algebraic and trigonometric identities apply. We keep the LHS and RHS in parallel.

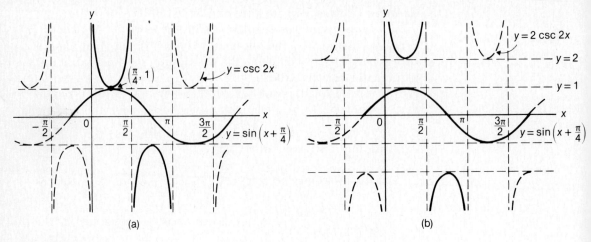

FIGURE 6.17

$$\sin x + \cos x = \sqrt{2}\,(\tan x + \cot x),$$

$$\sqrt{2}\left(\frac{1}{\sqrt{2}}\sin x + \frac{1}{\sqrt{2}}\cos x\right) = \sqrt{2}\left(\frac{\sin x}{\cos x} + \frac{\cos x}{\sin x}\right),$$

$$\sqrt{2}\,\sin\left(x + \frac{\pi}{4}\right) = \sqrt{2}\,(2\,\csc 2x),$$

and dividing by $\sqrt{2}$, we want the solution set for

$$\sin\left(x + \frac{\pi}{4}\right) = 2\,\csc 2x.$$

We now draw the graphs of $y = \sin(x + \pi/4)$ and $y = 2\,\csc 2x$ as shown in Fig. 6.17(b). From the graphs it is apparent that the graph of $y = 2\,\csc 2x$ lies entirely outside the strip where $-2 < y < 2$. The two graphs never intersect, and the solution set is the empty set. ∎

EXAMPLE 4 Draw the graph of $y = \cot\left(\text{Cos}^{-1}\dfrac{x}{\sqrt{1+x^2}}\right)$.

Solution Let

$$\theta = \text{Cos}^{-1}\frac{x}{\sqrt{1+x^2}}, \qquad \text{then} \qquad \cos\theta = \frac{x}{\sqrt{1+x^2}}.$$

Since Cos^{-1} is the inverse cosine function, the angle θ must lie within $0 \le \theta \le \pi$; so we draw θ in the first quadrant (if $x > 0$) or in the second quadrant (if $x < 0$) (see Fig. 6.18a). In either case, $\cot\theta = x$;

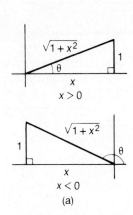

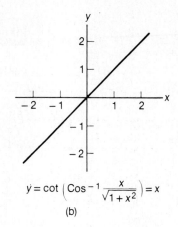

$$y = \cot\left(\text{Cos}^{-1}\frac{x}{\sqrt{1+x^2}}\right) = x$$

(a)   (b)

FIGURE 6.18

therefore the given equation is equivalent to $y = x$. Thus the graph is the straight line shown in Fig. 6.18(b). Note that for any real number x, $-1 < \dfrac{x}{\sqrt{1+x^2}} < 1$.

EXERCISE 6.4

In Problems 1–20, (a) determine the domain of the given function; (b) draw a graph; (c) use the graph to help determine the range of the function.

1. $y = (\cos x - \sin x)^2$
2. $y = \cos^4 x - \sin^4 x$

3. $y = 2\cos x \tan x$
4. $y = \sin x \cos x$

5. $y = 2\sin^2 x \cot x$
6. $y = \sqrt{3}\sin x - \cos x$

7. $y = \sin x + \cos x$
8. $y = \sin x - \sqrt{3}\cos x$

9. $y = \cos x - \sin x$
10. $y = (1 - \tan^2 x)\tan 2x$

11. $y = (1 - 2\sin^2 x)\sec 2x$
12. $y = 3^{\cos^2 x} \cdot 3^{\sin^2 x}$

13. $y = \cos 2x - \sqrt{3}\sin 2x$
14. $y = \cos^2 x + \cos^2 x \tan^2 x$

15. $y = \cos(\text{Cos}^{-1} x)$
16. $y = \sin(\text{Sin}^{-1} x)$

17. $y = \cot x - \tan x$
18. $y = 2\cos^2 \dfrac{x}{2}$

19. $y = \tan\left(\text{Sin}^{-1}\dfrac{x}{\sqrt{1+x^2}}\right)$
20. $y = \cos\left(\text{Sin}^{-1}\dfrac{\sqrt{x^2 - 1}}{x}\right)$

In Problems 21 and 22, use graphical techniques as illustrated in Example 3 to solve the given equations. See Problems 37 and 38 of Exercise 5.5.

21. $\sin x + \cos x = \tan x + \cot x$
22. $\sqrt{3}\sin x + \cos x = \tan x + \cot x$

Summary

In each graph below, a fundamental cycle (for $A > 0$) is shown as a solid curve. The continuation of the graph is suggested by the broken curve. To calculate the fundamental interval, we assume that $B > 0$.

(a) $y = A \sin (Bx + C)$

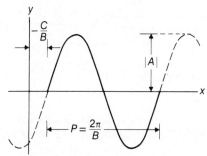

Period $= \dfrac{2\pi}{B}$

Frequency $= \dfrac{B}{2\pi}$

Amplitude $= |A|$

Fundamental interval $= \left[-\dfrac{C}{B}, \dfrac{2\pi - C}{B} \right]$

(b) $y = A \cos (Bx + C)$

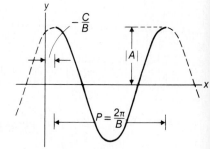

Period $= \dfrac{2\pi}{B}$

Frequency $= \dfrac{B}{2\pi}$

Amplitude $= |A|$

Fundamental interval $= \left[-\dfrac{C}{B}, \dfrac{2\pi - C}{B} \right]$

(c) $y = A \tan (Bx + C)$

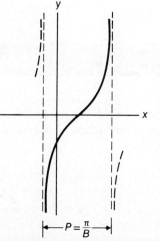

Period $= \dfrac{\pi}{B}$

Fundamental interval $= \left(\dfrac{-\dfrac{\pi}{2} - C}{B}, \dfrac{\dfrac{\pi}{2} - C}{B} \right)$

(d) $y = A \cot (Bx + C)$

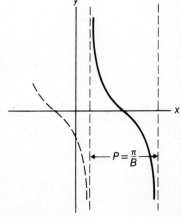

Period $= \dfrac{\pi}{B}$

Fundamental interval $= \left(-\dfrac{C}{B}, \dfrac{\pi - C}{B} \right)$

(e) $y = A \sec (Bx + C)$

(f) $y = A \csc (Bx + C)$

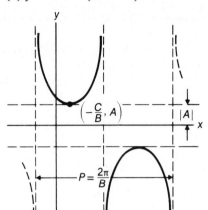

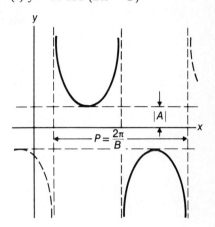

Period $= \dfrac{2\pi}{B}$

Fundamental interval $= \left(\dfrac{-\frac{\pi}{2} - C}{B}, \dfrac{\frac{3\pi}{2} - C}{B} \right)$

Period $= \dfrac{2\pi}{B}$

Fundamental interval $= \left(-\dfrac{C}{B}, \dfrac{2\pi - C}{B} \right)$

Graph of $y = a \sin x + b \cos x$

By using appropriate identities this can be written as $y = \sqrt{a^2 + b^2} \sin (x + \alpha)$, where α is determined by

$$\cos \alpha = \frac{a}{\sqrt{a^2 + b^2}} \quad \text{and} \quad \sin \alpha = \frac{b}{\sqrt{a^2 + b^2}}.$$

Apply (a) above to draw a graph of $y = \sqrt{a^2 + b^2} \sin (x + \alpha)$.

Review Exercises

In Problems 1–30, each equation defines a function. If the function is periodic, (a) find the period and the fundamental interval and (b) then sketch the graph, showing the fundamental cycle if the function is a general trigonometric function.

1. $y = 2 \cos x$

2. $y = 3 \sin (-x)$

3. $y = \dfrac{1}{2} \sin 3x$

4. $y = 3 \sin \dfrac{1}{2}x$

5. $y = -2 \cos x$

6. $y = 2 \cos \pi x$

7. $y = \tan 2x$

8. $y = 2 \cot \dfrac{1}{2}x$

9. $y = 1 + \tan x$

10. $y = 1 - \sec \pi x$

11. $y = \dfrac{1}{2} \cot 2x$

12. $y = -4 \sin x \cos x$

13. $y = x + \cos x$ **14.** $y = x - \sin 2x$ **15.** $y = \sin x - x$

16. $y = \sin 2\left(x - \dfrac{\pi}{4}\right)$ **17.** $y = -\tan\left(2x - \dfrac{\pi}{2}\right)$

18. $y = 2 \csc\left(x + \dfrac{\pi}{4}\right)$ **19.** $y = \sin \pi\left(x - \dfrac{1}{3}\right)$

20. $y = 2 \cos\left(\dfrac{\pi}{4} - \dfrac{\pi x}{2}\right)$ **21.** $y = \sin x + \cos x$

22. $y = \sqrt{3}\, \sin x + \cos x$ **23.** $y = \sin\left(x - \dfrac{\pi}{4}\right) + \cos\left(x - \dfrac{\pi}{4}\right)$

24. $y = 2 \sin \dfrac{x}{2} \cos \dfrac{x}{2}$ **25.** $y = \sin\left(\text{Sin}^{-1} x\right)$

26. $y = \text{Sin}^{-1}\left(\sin x\right)$ **27.** $y = \dfrac{1}{2} x \sin x$

28. $y = -x \cos\left(-x\right)$ **29.** $y = \cos^2 2x - \sin^2 2x$

30. $y = \dfrac{\tan x + 1}{1 - \tan x}$ (*Hint*: $\tan \dfrac{\pi}{4} = 1$, and use an appropriate identity.)

Solving Triangles

7

A triangle has six parts—three sides and three angles. Suppose we are given a sufficient number of these to determine a unique triangle. We are interested in developing techniques that will allow us to determine the remaining parts.

In Section 2.4 we were able to solve a variety of problems by formulating each in terms of right triangles and then applying definitions from right triangle trigonometry. In this chapter we are interested in the more general situation in which problems involve oblique (non-right) triangles. It is true that we can deal with problems of oblique triangles by expressing them in terms of appropriate right triangles, but it will be simpler to derive some general formulas that can be applied directly. The most important formulas are given special names: *Law of Sines* and *Law of Cosines.* We develop these in the first two sections of this chapter and see how they can be applied to solving triangles.

In elementary geometry we learn that if corresponding lengths of the three sides of two triangles are equal, then the triangles are congruent. This suggests that if the given information consists of the lengths of the three sides, then we have a unique triangle, and we can proceed to find the three angles. This is sometimes referred to as the SSS case. By contrast, if the three angles are given, then such information merely describes similar triangles, and we cannot determine the lengths of the sides uniquely.

Similarly, with notation from geometry, the two cases SAA (or ASA) and SAS determine unique triangles. There is one additional case, SSA, in which the given information determines either no triangle, one triangle, or two triangles.

We can classify problems of solving triangles into the following four cases depending upon the three given parts:

1. One side and two angles (SAA or ASA)
2. Two sides and the included angle (SAS)

3. Three sides (SSS)

4. Two sides and an angle opposite one of them (SSA).

The Law of Sines will be applied to solve triangles of the type described in Case 1, while the Law of Cosines is suitable for the types of Case 2 and 3. Case 4 is referred to as the *ambiguous case*, since the given information will determine either no triangle, one triangle, or two triangles, depending upon the relative sizes of the two given sides and the given angle. In Section 7.3 we shall see that either the Law of Sines or the Law of Cosines can be applied to problems of this type.

Labeling parts of a triangle

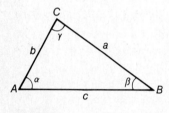

FIGURE 7.1

It is conventional to use the following scheme for labeling the parts of a triangle. Suppose A, B, and C are vertices of a triangle as shown in Fig. 7.1, where the Greek letters α, β, and γ denote the three angles and a, b, and c represent the lengths of the three sides. Note that angle α has vertex at A and that side a is opposite angle α; similarly for B, β, b and C, γ, c. *Throughout this chapter it is understood that we are employing this system of labeling.*

It is customary to use *degree measure* when referring to angles α, β, and γ of a triangle.

Rounding off numbers

Included in Appendix B are some guidelines in working with approximate numbers. In this chapter we apply "rules" given there for determining the level of accuracy of answers in examples and exercises. Also see Section 1.2 for additional comments concerning significant digits and rounding off numbers.

7.1 Law of Sines

FIGURE 7.2

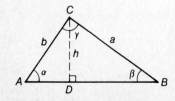

In triangle ABC shown in Fig. 7.2, let $\overline{CD}$ be the altitude from vertex C and h its length. From the two right triangles shown, $\triangle ACD$ and $\triangle BCD$,

$$\sin \alpha = \frac{h}{b} \quad \text{and} \quad \sin \beta = \frac{h}{a}.$$

Eliminating h between these two equations gives $b \sin \alpha = a \sin \beta$. This can be written as

$$\frac{a}{\sin \alpha} = \frac{b}{\sin \beta}.$$

In a similar manner we can show that

$$\frac{a}{\sin \alpha} = \frac{c}{\sin \gamma} \quad \text{and} \quad \frac{b}{\sin \beta} = \frac{c}{\sin \gamma}.$$

Note that if the altitude falls outside the triangle (as would be the case for the altitude from vertices A or B for the triangle shown in Fig. 7.2 where γ is an obtuse angle), then the derivation will require application of a reduction formula, such as $\sin(180° - \gamma) = \sin \gamma$, but the final results are the same. We leave it to the reader to fill in the details. See Problem 24 of Exercise 7.1.

The three equations given above are referred to as the *Law of Sines* and can be written in compact form as

$$\boxed{\frac{a}{\sin \alpha} = \frac{b}{\sin \beta} = \frac{c}{\sin \gamma}.} \qquad \textbf{[7.1]}$$

Note that the Law of Sines can be used to solve Case 1 triangles discussed above where two angles and a side are given. For instance, if α, γ, and b are given, then we can determine by $\beta = 180° - (\alpha + \gamma)$. Thus α, β, and γ are known. Looking at Eq. (7.1), we have

$$\frac{a}{\sin \textcircled{\alpha}} = \frac{\textcircled{b}}{\sin \textcircled{\beta}} = \frac{c}{\sin \textcircled{\gamma}}$$

where the circled parts are known. We can use the first equality to get

$$a = \frac{\textcircled{b} \, \sin \textcircled{\alpha}}{\sin \textcircled{\beta}}$$

and the second to get

$$c = \frac{\textcircled{b} \, \sin \textcircled{\gamma}}{\sin \textcircled{\beta}}.$$

Also, note that the Law of Sines is not suitable for solving Case 2 triangles. For instance, if α, b, and c are given, then when we circle these in Eq. (7.1), we get

$$\frac{a}{\sin \textcircled{\alpha}} = \frac{\textcircled{b}}{\sin \beta} = \frac{\textcircled{c}}{\sin \gamma}.$$

No matter which of the three equations we use in this case, we get an equation with two "unknowns."

In a similar manner we can argue that the Law of Sines is not suitable for solving triangles of the Case 3 type (given three sides). In the following section we shall derive other formulas (Law of Cosines) that can be applied in solving triangles of the types given in Cases 2 and 3.

EXAMPLE 1 (Given two angles and a side) Suppose $b = 5.8$, $\alpha = 64°$, and $\gamma = 47°$. Find a, c, and β.

Solution   We can find β by using $\beta = 180° - (\alpha + \gamma)$ as follows:

$$\alpha + \gamma = 64° + 47° = 111°,$$
$$\beta = 180° - 111° = 69°.$$

To find a, from Eq. (7.1) we get $a = (b \sin \alpha)/(\sin \beta)$:

$$a = \frac{5.8 \sin 64°}{\sin 69°} \approx 5.5839.$$

Similarly, for c, $c = (b \sin \gamma)/(\sin \beta)$:

$$c = \frac{5.8 \sin 47°}{\sin 69°} \approx 4.5436.$$

Therefore the solution is given by $\beta = 69°$, $a \approx 5.6$, $c \approx 4.5$, where a and c are rounded off to two significant digits.   ∎

EXAMPLE 2 (Area of a triangle)   Given $\alpha = 42°$, $a = 74$, and $b = 61$, find the area of triangle ABC.

FIGURE 7.3

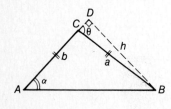

Solution   Triangle ABC is shown in Fig. 7.3 with the given parts identified for easy reference. Let h be the altitude from vertex B. We shall get the area by using

$$\text{Area} = \tfrac{1}{2}(\text{base}) \times (\text{altitude}),$$

where $|\overline{AC}| = b$ is the base. That is, area $= \tfrac{1}{2}bh = \tfrac{1}{2}(61)h$, and so we need first to find h. From the right triangle BCD we get $h = a \sin \theta = 74 \sin \theta$. We now have

$$\text{Area} = \tfrac{1}{2}(61)(74 \sin \theta),$$

and so now we must find θ. But $\theta = \alpha + \beta$ (the exterior angle of a triangle is the sum of the opposite interior angles). Since α is known, we need to determine β. We can do this by applying the Law of Sines:

$$\frac{a}{\sin \alpha} = \frac{b}{\sin \beta} \qquad \text{or} \qquad \sin \beta = \frac{b \sin \alpha}{a}.$$

Therefore

$$\sin \beta = \frac{61 \sin 42°}{74} \quad \text{and} \quad \beta = \text{Sin}^{-1}\left(\frac{61 \sin 42°}{74}\right) \approx 33.48°.$$

This gives $\theta = \alpha + \beta = 42° + \text{Sin}^{-1}[(61 \ \sin 42°)/74] \approx 42° + 33.48° = 75.48°$. Finally, area $\approx \frac{1}{2}(61)(74)\sin 75.48° \approx 2184.912$.* Rounded off to two significant digits, area ≈ 2200. ∎

Suggestion concerning computations _____

In the solution to the problem in Example 2 we recorded some intermediate numbers (rounded off). This was done primarily for purposes of clarity. We recommend that you get a final exact form result (such as that in the footnote) and then do all of your calculator computations at one time. This will minimize roundoff error and will result in greater accuracy generally. It also provides excellent practice with algebraic manipulations of trigonometric quantities. In addition, you will be using the full digit capacity of your calculator in arriving at the final answer. It must always be remembered, however, that in general, the digits in the final display have no more significance than the accuracy of the initial information justifies.

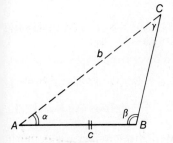

FIGURE 7.4

EXAMPLE 3 A surveyor wishes to determine the distance from a point A to an inaccessible point C. He locates a point B to which the distance from A can be measured. Also the angles at A and B are measured. In Fig. 7.4 the measured quantities are given by $c = 164$ m, $\alpha = 38.1°$, and $\beta = 108.4°$. Find the distance from A to C.

Solution From the Law of Sines we get

$$\frac{b}{\sin \beta} = \frac{c}{\sin \gamma} \quad \text{or} \quad b = \frac{c \sin \beta}{\sin \gamma}.$$

We first determine γ:

$$\gamma = 180° - (38.1° + 108.4°) = 180° - 146.5° = 33.5°.$$

Therefore

$$b = \frac{164 \sin 108.4°}{\sin 33.5°} \approx 281.94.$$

The distance from A to C is 282 meters (three significant digits). ∎

---

*In our computations we used rounded-off numbers in the final calculation. However, we could have expressed the area in exact form as

$$\text{Area} = \tfrac{1}{2}(61)(74) \sin\left[42° + \text{Sin}^{-1}\left(\frac{61 \sin 42°}{74}\right)\right].$$

Evaluating this, we get area $\cong 2184.868$. The two answers to three decimal places disagree in the decimal part. However, they agree when rounded off to the appropriate accuracy.

Identities for checking computations (Optional) ⎯⎯⎯⎯

In solving triangles there are often so many calculations that it is helpful to have an independent check on the results of computations. There are some identities that relate all six parts of a triangle and thus allow us to check a complete solution at once. Their derivation also uses the Law of Sines in an interesting and essential way.

We state four identities in terms of the standard labeling of parts of a triangle. It should be obvious that we can use any permutation of sides a, b, and c in the left-hand side of any of these identities if we make the corresponding changes in the right-hand side. The second and third identities (Eqs. 7.3 and 7.4) are also known as *Mollweide's equations*:

$$\frac{a \pm b}{c} = \frac{\sin\alpha \pm \sin\beta}{\sin\gamma}, \qquad [7.2]$$

$$\frac{a - b}{c} = \frac{\sin\left(\dfrac{\alpha-\beta}{2}\right)}{\cos(\gamma/2)}, \qquad [7.3]$$

$$\frac{a + b}{c} = \frac{\cos\left(\dfrac{\alpha-\beta}{2}\right)}{\sin(\gamma/2)}, \qquad [7.4]$$

$$\frac{a^2 - b^2}{c^2} = \frac{\sin(\alpha-\beta)}{\sin\gamma}. \qquad [7.5]$$

Before proving Eq. (7.2) it is handy to have some facts about the angles in a triangle (whose sum is 180°):

$$\sin\frac{\gamma}{2} = \sin\left(\frac{180° - (\alpha+\beta)}{2}\right) = \sin\left(90° - \frac{\alpha+\beta}{2}\right) = \cos\frac{\alpha+\beta}{2}, \quad [7.6]$$

$$\cos\frac{\gamma}{2} = \cos\left(\frac{180° - (\alpha+\beta)}{2}\right) = \cos\left(90° - \frac{\alpha+\beta}{2}\right) = \sin\frac{\alpha+\beta}{2}. \quad [7.7]$$

We will establish one of the equations in Eq. (7.2) and use it to prove the first of Mollweide's equations. The proof of the other Mollweide equation and Eq. (7.5) are left as exercises (see Problems 31–35 of Exercise 7.1).

Proof of equations (7.2) and (7.3) ⎯⎯⎯⎯⎯⎯

From the Law of Sines, $a/\sin\alpha = b/\sin\beta = c/\sin\gamma$, we may rewrite the first equality in the form $a/b = \sin\alpha/\sin\beta$. We shall use this in the following:

$$\frac{a-b}{c} = \frac{a-b}{c} \cdot \frac{b}{b} = \frac{a-b}{b} \cdot \frac{b}{c} = \left(\frac{a}{b} - 1\right) \cdot \frac{b}{c}$$

$$= \left(\frac{\sin \alpha}{\sin \beta} - 1\right)\frac{b}{c} = \frac{\sin \alpha - \sin \beta}{\sin \beta} \cdot \frac{b}{c}$$

$$= \frac{\sin \alpha - \sin \beta}{c} \cdot \frac{b}{\sin \beta} = \frac{\sin \alpha - \sin \beta}{c} \cdot \frac{c}{\sin \gamma}$$

$$= \frac{\sin \alpha - \sin \beta}{\sin \gamma}.$$

This is Eq. (7.2) with the negative sign. It should be clear that the same steps are valid if the $-$ is changed to $+$.

To prove Eq. (7.3), we simply continue from Eq. (7.2) by the use of a sum to product identity and double-angle formula:

$$\frac{a-b}{c} = \frac{\sin \alpha - \sin \beta}{\sin \gamma} = \frac{2 \cos\left(\frac{\alpha + \beta}{2}\right) \sin\left(\frac{\alpha - \beta}{2}\right)}{2 \sin(\gamma/2) \cos(\gamma/2)}$$

(by identities (I-22) and (I-25))

$$= \frac{\cos\left(\frac{\alpha + \beta}{2}\right) \sin\left(\frac{\alpha - \beta}{2}\right)}{\cos\left(\frac{\alpha + \beta}{2}\right) \cos(\gamma/2)} = \frac{\sin\left(\frac{\alpha - \beta}{2}\right)}{\cos(\gamma/2)}$$

(by Eq. (7.6) and simplifying).

EXAMPLE 4 Use identity (7.2) to check the triangle solution obtained in Example 1.

Solution In Example 1 we obtained the data $a = 5.6$, $b = 5.8$, $c = 4.5$, (rounded off) $\alpha = 64°$, $\beta = 69°$, $\gamma = 47°$. Substitution of these numbers into Eq. (7.2) gives

LHS: $\dfrac{a-b}{c} \approx \dfrac{5.6 - 5.8}{4.5} \approx -0.044,$ RHS: $\dfrac{\sin \alpha - \sin \beta}{\sin \gamma} \approx -0.048.$

Comparison of these results shows that we do not get precise agreement. Had we used the closer approximations $a \approx 5.5839$ and $c \approx 4.5436$, the agreement would be closer. However, we have no justification for claiming more than two significant digits for the given data. More pertinent for our purposes is the fact that *we want a quick rough check* on all of our answers. It is easier and quicker to use the rounded answers, and -0.044 and -0.048 are certainly close enough to indicate that our answers are essentially correct. ∎

EXERCISE 7.1

In Problems 1–8, use the given data to find the remaining three parts of the triangle. You may use any of the triangle identities (Eqs. 7.2–7.5) as a check on your solutions. Answers should be rounded off to be consistent with given data.

1. $\alpha = 27°$, $\beta = 73°$, $a = 16$ 2. $\beta = 67.0°$, $\gamma = 26.0°$, $a = 463$
3. $\alpha = 47°$, $\gamma = 112°$, $c = 81$ 4. $\alpha = 51.2°$, $\beta = 70.6°$, $c = 133$
5. $\alpha = 32°30'$, $\beta = 55°15'$, $a = 32.5$ 6. $\beta = 61°45'$, $\gamma = 82°15'$, $b = 63.5$
7. $\alpha = 73.46°$, $\beta = 25.75°$, $c = 4.875$ 8. $\alpha = 35.4°$, $\gamma = 73.5°$, $b = 3.75$

9. Triangle ABC has measurements $a = 41.3$ cm, $\alpha = 43.5°$, and $\beta = 73.4°$. Find the length of the longest side.

10. In triangle ABC, $a = 4666$, $c = 2730$, and $\alpha = 35.82°$. Find angle γ.

In Problems 11–14, three parts of a triangle are given. Find the area of the triangle.

11. $\alpha = 47°$, $\beta = 67°$, and $a = 16$ 12. $\beta = 32°$, $\gamma = 73°$, and $a = 25$
13. $\gamma = 47°$, $a = 8.3$, and $b = 6.7$ 14. $\beta = 36°$, $a = 3.7$, and $b = 4.6$

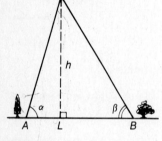

FIGURE 7.5

15. In order to measure the height of clouds at night, two observers are located 126 meters apart at points A and B; the spotlight is at point L in line with A and B. A vertical beam of light from L is reflected from the bottom of the clouds at point C, and the angles of elevation are measured from A and B. These are $\alpha = 74.3°$ and $\beta = 58.5°$ as shown in Fig. 7.5. How far above the ground is the bottom of the clouds?

16. From point A on top of a building the angle of depression of point C on the ground is observed to be $\alpha = 54°$, while from a window at point B (15 meters directly below A) the angle of depression to the same point is $\beta = 42°$. Find the height of the building.

17. A surveyor wishes to find the distance from point A to a point C on the opposite side of the river. He locates a point B on his side of the river and measures the distance $\overline{AB}$ and the two angles α and β, as shown in Fig. 7.6. The measurements are $\overline{AB} = 132$ m, $\alpha = 78.3°$ and $\beta = 53.2°$. Find the distance $\overline{AC}$.

18. Assume that satellite S shown in Fig. 7.7 is traveling around the earth in a circular orbit with center E (the earth's center). Points A and B are

FIGURE 7.6

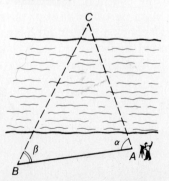

FIGURE 7.7

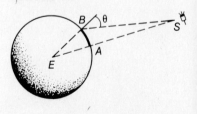

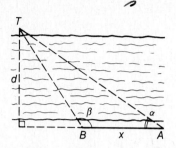

FIGURE 7.8

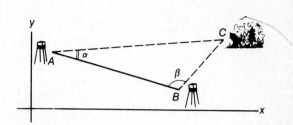

FIGURE 7.9

on the surface of the earth, and B is 524 miles directly north of A (arc $\widehat{AB}$ is 524 miles). At the instant when the satellite is directly over A, angle θ between the vertical and line of sight of S from B is found to be 20.5°. Find the height of the satellite above the earth's surface (distance $|\overline{AS}|$). Assume that the radius of the earth is 3960 miles. The diagram is not drawn to scale.

19. A surveyor wishes to find the width of a river. He notices a tree T on the opposite bank, so he takes two points A and B along the bank on his side of the river. He measures the distance x between A and B, and the two angles α and β, as shown in Fig. 7.8 and finds $x = 19$ meters, $\alpha = 33°$, $\beta = 124°$. From these measurements, calculate the width d of the river.

20. On a rectangular set of coordinates the locations of two forest-ranger stations are given as $A(15, 32)$, $B(84, 15)$. A fire is spotted at point C, and angles $\alpha = 20°$, $\beta = 117°$ are measured, as shown in Fig. 7.9. Locate the fire by finding the coordinates of C.

21. Suppose a triangle ABC is inscribed in a circle, as shown in Fig. 7.10(a). Show that the ratio appearing in the Law of Sines,

$$\frac{a}{\sin \alpha} = \frac{b}{\sin \beta} = \frac{c}{\sin \gamma},$$

is equal to the diameter of the circle, that is,

$$\text{Diameter} = \frac{a}{\sin \alpha}.$$

FIGURE 7.10

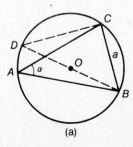

(a)　　　　　　(b)

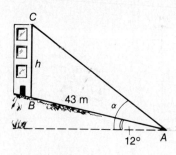

FIGURE 7.11   **FIGURE 7.12**

(*Hint*: Point D is selected so that side $\overline{DB}$ passes through the center O of the circle. Recall from geometry that angle CDB is equal to angle CAB (angle α). Also, angle DCB is a right angle and $\overline{DB}$ is a diameter.)

22. Triangle ABC is inscribed in a circle, as shown in Fig. 7.10(b), where Q is the center of the circle, α is one angle, and a is the opposite side. Prove that the diameter d of the circle is given by $d = a/\sin\alpha$.
    (*Hint*: Triangle QBC is an isosceles triangle with angle 2α opposite side a and two sides each of length $d/2$. This problem also appears as Problem 21. However, the solution suggested there is quite different.)

23. A vertical tower BC is located on a hill inclined at $12°$ (Fig. 7.11). From point A (43 meters down the hill from base B of the tower) the angle of elevation of point C at the top of the tower is $\alpha = 37°$. Find the height of the tower.

24. In the derivation of the Law of Sines, Fig. 7.2 was used, in which both angles α and β are acute. Derive the same law using a diagram in which angle α is obtuse (Fig. 7.12). Use the fact that $\sin(180° - \alpha) = \sin\alpha$.

25. A railroad crosses the highway at point C at an angle of $40°$, as shown in Fig. 7.13. An observer at point A on the highway (1.5 km from C) notices that it takes a train 20 seconds to travel from P to Q and that the angles α and β are $\alpha = 45°$, $\beta = 75°$. How fast is the train traveling? Give your answer in kilometers per hour.

26. In Fig. 7.14, $|\overline{AB}| = 205$, $\alpha = 31.1°$, $\beta = 24.2°$, $\gamma = 23.5°$, and $\delta = 71.4°$. Find the length of $\overline{AD}$.

FIGURE 7.13   **FIGURE 7.14**

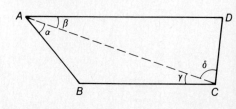

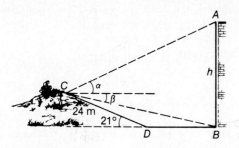

FIGURE 7.15

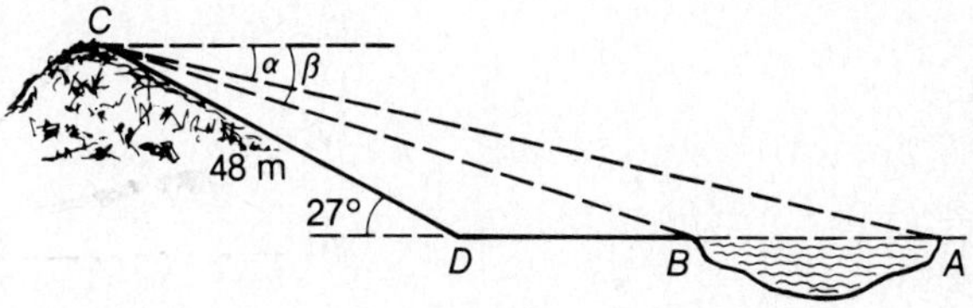

FIGURE 7.16

27. From point C located on a hill $21°$ steep the elevation angle of the top A of a nearby building is observed to be $\alpha = 25°$, and the angle of depression of the base B is $\beta = 12°$. If the distance between C and the bottom of the hill D is 24 meters, find the height of the building (Fig. 7.15).

28. Points A and B are located on opposite sides of a lake (Fig. 7.16). From point C on a nearby hill the angles of depression of A and B are observed to be $\alpha = 12°$ and $\beta = 17°$, respectively. If the hill is $27°$ steep and point D at the base of the hill is 48 meters from C, find the width of the lake.

29. A technique for determining the height of an inaccessible point is the following: A surveyor locates two points A and B and measures the distance between them. Then the angles α, β, and θ are measured. This is illustrated by Fig. 7.17, in which points A, B, and C are in the plane of the ground; D is directly above C; angle θ is the angle of elevation of point D from B; and α and β are angles of triangle ABC. Show that

$$a = \frac{c \sin \alpha}{\sin (\alpha + \beta)} \quad \text{and} \quad h = \frac{c \sin \alpha \tan \theta}{\sin (\alpha + \beta)}.$$

FIGURE 7.17

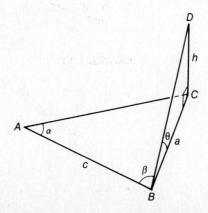

30. In Problem 29, suppose that we wish to determine the height h of a mountain peak, and points A and B are such that $c = 463$ meters, $\beta = 63°10'$, $\alpha = 46°40'$, and $\theta = 47°20'$. Find h.

31. Prove the second Mollweide equation, Eq. (7.4):

$$\frac{a+b}{c} = \frac{\cos\left(\frac{\alpha-\beta}{2}\right)}{\sin(\gamma/2)}.$$

Use Eqs. (7.2) and (7.7) and the appropriate sum to product identity from Section 4.5.

32. Prove identity (7.5):

$$\frac{a^2-b^2}{c^2} = \frac{\sin(\alpha-\beta)}{\sin\gamma}$$

from Eqs. (7.3) and (7.4).

Hint: $\dfrac{a^2-b^2}{c^2} = \left(\dfrac{a+b}{c}\right)\left(\dfrac{a-b}{c}\right)$

33. Prove triangle identity (7.2):

$$\frac{a+b}{c} = \frac{\sin\alpha+\sin\beta}{\sin\gamma}$$

directly from the Law of Sines as was done in the text for the identity $(a-b)/c = (\sin\alpha - \sin\beta)/\sin\gamma$.

34. Is the equation obtained by changing the signs in Eq. (7.5) an identity, that is, $(a^2+b^2)/c^2 = \sin(\alpha+\beta)/\sin\gamma$?

35. Is $(a+b)/c = (\cos\alpha + \cos\beta)/\cos\gamma$ an identity?

7.2 Law of Cosines

In the preceding section we observed that the Law of Sines is not suitable for solving triangles in which the given information consists of two sides and the included angle or the three sides. In this section we develop formulas that can be applied in these situations.

Suppose angle γ and sides a and b are given. We wish to derive a formula to determine side c. Figure 7.18 shows a triangle in which D is the base of the altitude from vertex A. Let $h = |\overline{AD}|$ and $x = |\overline{CD}|$. From right triangle ADC we get

$$x = b\cos\gamma \quad\text{and}\quad h = b\sin\gamma.$$

Applying the Pythagorean Theorem to right triangle ADB, we have

$$c^2 = h^2 + (a-x)^2 = h^2 + a^2 - 2ax + x^2.$$

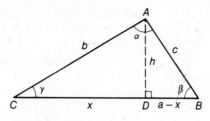

FIGURE 7.18

Substituting $x = b \cos \gamma$ and $h = b \sin \gamma$ gives

$$c^2 = (b \sin \gamma)^2 + a^2 - 2a(b \cos \gamma) + (b \cos \gamma)^2$$
$$= a^2 + b^2 [(\sin \gamma)^2 + (\cos \gamma)^2] - 2ab \cos \gamma$$
$$= a^2 + b^2 - 2ab \cos \gamma,$$

where in the last step we replaced $(\sin \gamma)^2 + (\cos \gamma)^2$ by 1. Thus we have

$$c^2 = a^2 + b^2 - 2ab \cos \gamma.*$$

Note: For $\gamma = 90°$ this becomes the equality in the Pythagorean Theorem.

In a similar manner we can develop analogous formulas for a^2 and b^2. The three equations are listed in Eqs. (7.8), and these are called the *Law of Cosines* for triangle ABC:

$$\boxed{\begin{aligned} a^2 &= b^2 + c^2 - 2bc \cos \alpha, \\ b^2 &= a^2 + c^2 - 2ac \cos \beta, \\ c^2 &= a^2 + b^2 - 2ab \cos \gamma. \end{aligned}}$$ **[7.8]**

We now consider examples in which two sides and the included angle are known and we wish to determine the remaining three parts of the triangle. In Example 1 we illustrate a technique of solution using only the Law of Cosines. In Example 2 we consider the same problem but apply the Law of Sines to complete the solution. We include a "faulty solution," which illustrates the need to be cautious when using the Law of Sines.

EXAMPLE 1 (*Given two sides and the included angle*)   Suppose $a = 82.0$, $b = 65.0$, and $\gamma = 36.5°$. Find c, α, and β.

---

* In the derivation of this formula we used a figure in which the altitude is inside the triangle (the case in which γ is an acute angle). We get the same result from the case in which γ is an obtuse angle and the altitude is outside the triangle. The reader is encouraged to supply the details of the derivation in this case. See Problem 23 of Exercise 7.2.

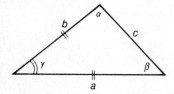

FIGURE 7.19

Solution We first draw a triangle showing the given parts (Fig 7.19). To find c, apply the Law of Cosines (third equation of Eqs (7.8)).

$$c = \sqrt{a^2 + b^2 - 2ab\,\cos\gamma} = \sqrt{82^2 + 65^2 - 2(82)(65)\cos 36.5°}.$$

Evaluate by calculator and store the result so that we can use c to maximum decimal accuracy in further computations. Here we merely recognize that to three significant digits, $c \approx 48.8$, but the value of c that has been stored is 48.7840739.

To find α and β, we use the first two formulas of Eqs. (7.8) as follows:

$$\cos\alpha = \frac{b^2 + c^2 - a^2}{2bc} = \frac{65^2 + c^2 - 82^2}{2(65)c}.$$

Evaluating the right-hand side, we see -0.018782 in the calculator display; then pressing (INV), (Cos) (or (Cos^{-1})) gives $\alpha \approx 91.1°$.

Similarly for β,

$$\cos\beta = \frac{82^2 + c^2 - 65^2}{2(82)c},$$

and so $\beta \approx 52.4°$.

Note that we could have found β by using $\beta = 180° - (\alpha + \gamma)$, but we prefer to use this as an independent check on our computations. That is, using the computed values of α and β and the given value of γ, we see that $\alpha + \beta + \gamma \approx 91.1° + 52.4° + 36.5° = 180°$. We could also use any of the identities (7.2)–(7.5) given in Section 7.1.

Thus our solution is given by $c \approx 48.8$, $\alpha \approx 91.1°$, and $\beta \approx 52.4°$. ∎

EXAMPLE 2 (Same as problem stated in Example 1 but a different method of solution) Suppose $a = 82.0$, $b = 65.0$, and $\gamma = 36.5°$. Find c, α, and β.

Solution To find c, use the Law of Cosines as we did in Example 1 to get $c \approx 48.8$, with the full decimal value of 48.7840739 stored in the calculator. We can now determine α by applying the Law of Sines:

$$\sin\alpha = \frac{a\,\sin\gamma}{c} = \frac{82\,\sin 36.5°}{c} \approx 0.9998236.$$

If we take $\alpha = \text{Sin}^{-1}(0.9998236)$, we get $\alpha \approx 88.9°$. In a similar fashion we can get β:

$$\beta = \text{Sin}^{-1}\left(\frac{65\,\sin 36.5°}{c}\right) \approx 52.4°.$$

Now as a check we evaluate $\alpha + \beta + \gamma$ for the above values and get $\alpha + \beta + \gamma \approx 88.9° + 52.4° + 36.5° = 177.8°$. Since we do not get a sum of 180°, we suspect that there is an error in our computations. Checking the above calculations, we find no error. The discrepancy between 177.8° and 180° is great enough that calculator roundoff is not the explanation.* Finally, we observe that there are two angles α for which $\sin \alpha = 0.9998236$, namely, 88.92° and $180° - 88.92° = 91.08°$. The second is the one that fits our problem, and so we conclude that $\alpha \approx 91.1°$ is the desired solution. With this value of α our check gives $\alpha + \beta + \gamma \approx 91.1° + 52.4° + 36.5° = 180°$. ■

> *Warning:* The solution demonstrated in Example 2 shows that care must be exercised when applying the Law of Sines in determining an angle.

We note that if the smaller of the two unknown angles (the one opposite the shorter side) had been found first (angle β in the above example) and then $\alpha = 180° - (\beta + \gamma)$ was used to find α, we would not have had the difficulty encountered in Example 2. However, we would have had no independent check ($\alpha + \beta + \gamma = 180°$) in our calculations. In that case we should use an identity such as Eq. (7.2) to provide us with a check.

When we compare solution techniques shown in Examples 1 and 2, we see that the Law of Sines is conceptually simpler to apply than the Law of Cosines, but we also note that it is necessary to check whether the angle in question is acute or obtuse. In most situations it is sufficient to draw a reasonably accurate diagram of the triangle (such as Fig. 7.19) and use it to decide whether an angle is greater than 90°. However, if the angle is near 90° (as is the case for α in the above example), then we may not be able to rely completely on the diagram.

EXAMPLE 3 (*Given three sides*) Suppose $a = 56.84$, $b = 83.45$, and $c = 51.63$. Find angles α, β, and γ.

Solution

$$\cos \alpha = \frac{b^2 + c^2 - a^2}{2bc} = \frac{(83.45)^2 + (51.63)^2 - (56.84)^2}{2(83.45)(51.63)}.$$

This gives $\alpha \approx 42.0491° \approx 42°03'$.

$$\cos \beta = \frac{a^2 + c^2 - b^2}{2ac} = \frac{(56.84)^2 + (51.63)^2 - (83.45)^2}{2(56.84)(51.63)}.$$

---

* Depending upon the number of decimal places retained in the final results, we should not expect always to get a sum of precisely 180° as the sum of the three angles because of roundoff.

Thus $\beta \approx 100.4788° \approx 100°29'$.

$$\cos\gamma = \frac{a^2 + b^2 - c^2}{2ab} = \frac{(56.84)^2 + (83.45)^2 - (51.63)^2}{2(56.84)(83.45)}.$$

We get $\gamma \approx 37.4721° \approx 37°28'$.

As a check, we add the computed values of α, β, γ and get

$$\alpha + \beta + \gamma \approx 42°03' + 100°29' + 37°28' = 180°. \qquad \blacksquare$$

EXAMPLE 4   A surveyor wishes to determine the distance between two points A and C that cannot be measured directly. However, he can locate a point B and measure the distances from A to B and B to C, as well as $\angle ABC$. The measurements are $|\overline{AB}| = 32$ m, $|\overline{BC}| = 47$ m and $\angle ABC = 112°$. Find the distance from A to C.

Solution   The given information is shown in Fig. 7.20, where $a = 47$, $c = 32$, and $\beta = 112°$. We wish to determine b. This can be done by applying the Law of Cosines, $b^2 = a^2 + c^2 - 2ac \cos\beta$.

$$b = \sqrt{47^2 + 32^2 - 2(47)(32)\cos 112°} \approx 66.029.$$

Thus the distance from A to C is 66 meters (to two significant digits).

EXAMPLE 5   Determine the area of triangle ABC where $b = 5.3$, $c = 8.1$, and $\alpha = 125°$.

Solution   The area of a triangle is given by area $= \frac{1}{2}$ (base) × (altitude). Let us take the altitude from vertex C to base c (extended) as shown in Fig. 7.21. From right triangle ADC we get $h = b \sin(180° - \alpha) = b \sin\alpha$. Therefore

$$\text{Area} = \tfrac{1}{2}ch = \tfrac{1}{2}c(b \sin\alpha) = \tfrac{1}{2}(8.1)(5.3)\sin 125° \approx 17.583.$$

Rounding off to two significant digits, we have area ≈ 18. $\qquad \blacksquare$

FIGURE 7.20

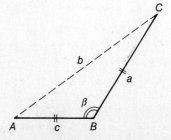

FIGURE 7.21

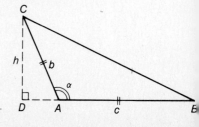

EXERCISE 7.2

In Problems 1–12, three parts of a triangle are given. Find the remaining parts. Answers should be rounded off to be consistent with the given data.

1. $a = 36$, $b = 67$, $\gamma = 43°$
2. $b = 24$, $c = 73$, $\alpha = 130°$
3. $a = 85$, $c = 42$, $\beta = 83°$
4. $a = 41.32$, $b = 57.56$, $\gamma = 61°15'$
5. $b = 33$, $c = 65$, $\alpha = 48°$
6. $a = 17$, $b = 12$, $c = 23$
7. $a = 288$, $b = 175$, $c = 337$
8. $a = 31.5$, $b = 63.4$, $c = 42.6$
9. $a = 0.23$, $b = 0.16$, $c = 0.31$
10. $a = 73.4$, $b = 45.6$, $c = 68.7$
11. $b = 34.3$, $c = 28.5$, $\alpha = 32.8°$
12. $a = 43.2$, $c = 28.6$, $\beta = 48.5°$

13. If $a = 32.6$, $b = 56.3$, and $c = 36.8$, find the measure of the smallest angle of the triangle correct to the nearest ten minutes.

14. If $a = 391$, $b = 172$, and $c = 427$, find the measure of the largest angle of the triangle correct to the nearest ten minutes.

15. If $a = 3.76$, $b = 5.34$, and $\gamma = 48°50'$, find the altitude to side b and then determine the area of the triangle.

16. If $b = 34.52$, $c = 76.81$, and $\alpha = 121°35'$, find the altitude to c and then find the area of the triangle.

17. An equilateral triangle is inscribed in a circle of radius 4.56. Find the perimeter of the triangle.

18. A square is inscribed in a circle of radius 4.56. Find the area of the square.

19. A ship sails due east from point A for a distance of 48.6 km; then it changes direction southward by an angle of $16°40'$. After sailing 37.8 km in the new direction, how far is the ship from point A?

20. A triangular slab of marble has sides of length 120 cm, 156 cm, and 173 cm. If it is placed vertically, so that the longest edge is on the ground, how high from the ground will it reach?

21. The coordinates of the vertices of triangle ABC are $A(3,2)$, $B(7,5)$, $C(5,8)$. Find the angle at A in degrees rounded off to one decimal place.

22. In the preceding problem, suppose M is the midpoint of $\overline{BC}$. Determine angle BAM in degrees rounded off to one decimal place.

23. In this section the Law of Cosines was derived by using Fig. 7.18 where angle γ is acute. Suppose γ is an obtuse angle. Derive the Law of Cosines for this case by showing that
$$c^2 = a^2 + b^2 - 2ab \cos \delta.$$

24. If triangle ABC is a right triangle with $\gamma = 90°$, show that the third expression of Eqs. (7.8) reduces to $c^2 = a^2 + b^2$ (Pythagorean Theorem).

25. Use the Law of Cosine equations given in Eqs. (7.8) as follows: Replace term b^2 in the first equation by that given by the right side of the second equation, and then simplify to get
$$a \cos \beta + b \cos \alpha = c.$$

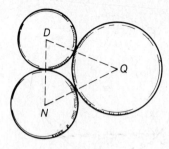

FIGURE 7.22

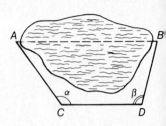

FIGURE 7.23

In a similar manner, if we use the first and third and then the second and third equations of Eqs. (7.8), we can get

$$c \cos \alpha + a \cos \gamma = b,$$
$$b \cos \gamma + c \cos \beta = a.$$

26. A dime, a nickel, and a quarter are placed on a table so that they just touch each other, as shown in Fig. 7.22. The diameters of the dime, nickel, and quarter are 1.75 cm, 2.25 cm, and 2.50 cm, respectively. Find the length of the smaller part of the circumference of the quarter between the two points where it touches the dime and the nickel. (In the diagram, D, N, and Q are respective centers.)

27. A surveyor wishes to determine the distance between points A and B on opposite sides of a lake. He does this by taking points C and D (Fig. 7.23) and gets the following measurements: $|\overline{AC}| = 205$ m, $|\overline{CD}| = 263$ m, $|\overline{DB}| = 185$ m, $\alpha = 126°$, and $\beta = 104°$. Using this information, find the distance across the lake to three significant digits.

28. Triangle ABC is inscribed in a circle of radius 20 cm as shown in Fig. 7.24 where Q is the center of the circle. If the length of side $\overline{BC}$ is 24 cm, find angle α to the nearest degree. (*Hint:* What is the measure of angle BQC?) It appears that this problem does not have a unique solution, since vertex A can be moved on the circle to get triangles other than the one shown in the diagram. Convince yourself that the measure of angle α is the same regardless of where point A is taken on the larger arc from B to C.

29. The lengths of two of the sides of a parallelogram are 4.3 cm and 8.2 cm, and one angle is 67°. Find the lengths of the two diagonals.

30. Two ships leave the same port at the same time with an angle of 25° between their courses. The first travels at a speed of 15 knots (nautical miles per hour) and the second at a speed of 20 knots. How far apart are they at the end of 5 hours? Give answer in nautical miles.

31. a] Find the central angle of a circular sector (see Section 1.2) where the radius is 25 and the chord is 16. Give the answer in radians.
    b] Determine the area of the circular sector.

32. Consider a regular pentagon $ABCDE$ with sides of unit length, as shown in Fig. 7.25. Let r be the length of a diagonal (such as CE),

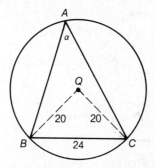

FIGURE 7.24   FIGURE 7.25

a] Show that each of two angles α is equal to 36°, and each of the angles β is 72°. Thus triangles $\underline{ACE}$ and BCF are similar.

b] Show that $|CF| = 1$ and $|BF| = r - 1$; then, using the corresponding-ratios property of similar triangles, prove that r satisfies the equation $r^2 - r - 1 = 0$. Solve this equation and get

$$r = \frac{1 + \sqrt{5}}{2}.$$

This is a well-known number called the *golden ratio*.

c] Apply the Law of Cosines to triangle ACE to find $\cos 72°$ and show that

$$\cos 72° = \frac{1}{2r} = \frac{1}{1 + \sqrt{5}}.$$

Thus we have expressed $\cos 72°$ in exact form (in fact, in simple terms involving the golden ratio). As a check, evaluate $\cos 72°$ directly with your calculator and then evaluate $1/(1 + \sqrt{5})$, and see whether the two numbers are equal (at least to the digit capacity of your calculator).

33. In each of the following, find angles α and γ. Verify that in each case $\gamma = 2\alpha$.

a] $a = 4, b = 5, c = 6$ **b]** $a = 16, b = 33, c = 28$

c] $a = 9, b = 16, c = 15$ **d]** $a = 16, b = 9, c = 20$

34. The triangles in Problem 33 all have the property that one angle is twice another angle of the triangle. Show that, in general,

a] if $\gamma = 2\alpha$, then $c^2 = a^2 + ab$, and

b] if $c^2 = a^2 + ab$, then $\gamma = 2\alpha$.

(*Hint*: If $\gamma = 2\alpha$, then $\sin \gamma = \sin 2\alpha = 2 \sin \alpha \cos \alpha$. Apply the Law of Sines to express $\sin \gamma$ in terms of $\sin \alpha$, and use the Law of Cosines to express $\cos \alpha$ in terms of a, b, c.)

35. *Right-angle (Pythagorean) triples and double-angle triples.* For triangles whose sides have integer lengths a, b, c, we say that

(a, b, c) is a *right-angle triple* if and only if $c^2 = a^2 + b^2$
(see Eq. 1.3)

(a, b, c) is a *double-angle triple** if and only if $c^2 = a^2 + ab$
(see Problem 34)

_____

* Double-angle triples are discussed in an article, "Integer-Sided Triangles with One Angle Twice Another," by R. S. Luthar in the January 1984 issue of *The College Mathematics Journal*, pp. 55–56.

Formulas that generate right-angle triples have long been known, and there are similar formulas that generate double-angle triples. Verify (by algebra) that if m and n are any positive integers with $m > n$, then
a] $a = m^2 - n^2$, $b = 2mn$, $c = m^2 + n^2$ is a right-angle triple, and
b] $a = m^2$, $b = 2mn + n^2$, $c = m(m + n)$ is a double-angle triple.

36. Use the formulas in Problem 35 to complete the following table.

		Right-angle triples			Double-angle triples		
m	n	a	b	c	a	b	c
2	1						
3	1						
3	2						
4	3						
4	2						
5	2						
5	3						
15	8						

37. Based on the numbers in Problem 36,
a] make a guess as to what conditions on m and n will give a right-angle triple in which a, b, and c have no common factor, and
b] make a guess as to what conditions on m and n will give a double-angle triple in which a, b, and c have no common factor.
You may wish to try additional examples to check your guesses.

38. It is obvious that the formulas for right-angle triples in Problem 35 require that $m > n$ to make a positive. It is less apparent that $m > n$ is also necessary for double-angle triples. Show that unless $m > n$, the formulas for double-angle triples do not determine a triangle.
(*Hint:* Show that $a + c \leq b$ if $m \leq n$.)

39. Show that there is no right-angle triple that is also a double-angle triple.

7.3 The Ambiguous Case

In the introductory remarks of this chapter we classified triangle-solving problems into four groups depending upon the given information. In the preceding two sections we saw that the Law of Sines is suitable for solving the first type (SAA), while the Law of Cosines can be applied to the second and third types (SAS and SSS). The fourth type, given two sides and an angle opposite one of them, can be solved by either the Law of Sines or the Law of Cosines. It is called the *ambiguous case* because there may be either no solution, one solution, or two solutions, depending upon the relative sizes of the given parts of the triangle. We shall illustrate this and the methods of solution in the following examples. First let us discuss in general the technique of using the Law of Sines and then the method in which the Law of Cosines is applied.

Method using the Law of Sines

Suppose a, b, and α are given and we wish to determine the other three parts, namely, β, γ, and c.

1. Find β by using $\sin \beta = (b \sin \alpha)/a$.

2. Now that we have β, find γ by the formula $\gamma = 180° - (\alpha + \beta)$.

3. After γ is determined, find c by $c = (a \sin \gamma)/\sin \alpha$.

In step 1 we must consider the possibility of getting two angles β. Since $\sin(180° - \beta) = \sin \beta$, β or $180° - \beta$ might be angles of a triangle. This will be illustrated in Example 1.

Method using the Law of Cosines

Suppose a, b, and α are given and we wish to determine c. The formula in Eqs. (7.8) that involves these four parts is

$$a^2 = b^2 + c^2 - 2bc \cos \alpha.$$

The "unknown" in this equation is c, and so we need to solve a *quadratic equation* in c that can be written in standard form as

$$c^2 - (2b \cos \alpha)c + (b^2 - a^2) = 0.$$

Applying the quadratic formula and simplifying, we get

$$c = \frac{2b \cos \alpha \pm \sqrt{(2b \cos \alpha)^2 - 4(b^2 - a^2)}}{2}$$
$$= b \cos \alpha \pm \sqrt{a^2 - (b \sin \alpha)^2},$$

where under the radical we used the identity $1 - \cos^2 \alpha = \sin^2 \alpha$.

Therefore if a, b, and α are known, a formula that can be used to determine c is*

$$\boxed{c = b \cos \alpha \pm \sqrt{a^2 - (b \sin \alpha)^2}.}$$
[7.9]

---

* See Fig. 7.28 for a geometrical interpretation of the result in Eq. (7.9).

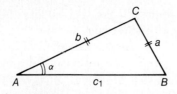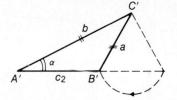

FIGURE 7.26

We can get a better understanding of the result in Eq. (7.9) by looking at the geometry of the problem. First it appears that we have two values for c, which is indeed the case if we have a situation such as that shown in Fig. 7.26, where we have two triangles, both of which have the same given values of a, b, and α.

In triangle ABC, $c_1 = b \cos\alpha + \sqrt{a^2 - (b \sin\alpha)^2}$, while in triangle $A'B'C'$, $c_2 = b \cos\alpha - \sqrt{a^2 - (b \sin\alpha)^2}$. Note that we can get $\Delta A'B'C'$ from ΔABC by rotating $\overline{CB}$ clockwise about point C as indicated in the diagram.

Number of solutions

Depending upon the relative values of the given parts (a, b, and α), we get the following possibilities.

Two solutions

If when we apply Eq. (7.9) we get two positive values for c, then we get two solutions (as shown in Fig. 7.26).

One solution

If we get only one positive answer in applying Eq. (7.9), then we get one solution. This occurs in one of two ways: The quantity under the radical, $a^2 - (b \sin\alpha)^2$, is zero (in which case we have a right triangle), or $b \cos\alpha - \sqrt{a^2 - (b \sin\alpha)^2}$ is a negative number, and so we do not get a second solution (since the length of each side must be a positive number).

No solution

If the quantity under the radical in Eq. (7.9) is negative (that is, $a^2 - (b \sin\alpha)^2 < 0$), then we get no solution. A geometric interpretation in this case is illustrated in Fig. 7.27, where side $\overline{BC}$ is "too short" to reach side $\overline{AB}$. In Fig. 7.27 we have also drawn right triangle $A'B'C'$ where $|\overline{B'C'}| = b \sin\alpha$ and, as illustrated, $a < b \sin\alpha$. But this is precisely the situation for which the expression under the radical, $a^2 - (b \sin\alpha)^2$, is a negative number.

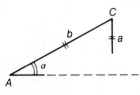 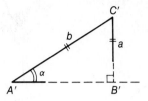

FIGURE 7.27

Summarizing the ambiguous case

Suppose we apply Eq. (7.9) to find c (or its counterpart, in case two other sides and an angle opposite one of them are given).

1. If we get two positive values for c, then we have *two solutions.*

2. If we get only one positive value of c, then there is only *one solution.*

3. If we get complex number answers for c that are not real numbers $(a^2 - (b \sin \alpha)^2 < 0)$, then we get *no solution.*

Geometric interpretation of Equation (7.9)

It may be helpful to interpret the formula in Eq. (7.9) geometrically. For the case shown in Fig. 7.28 we see that from right triangle ADC, $h = b \sin \alpha$ and $|\overline{AD}| = b \cos \alpha$, while from right triangle BDC, $|\overline{DB}| = \sqrt{a^2 - h^2} = \sqrt{a^2 - (b \sin \alpha)^2}$. Thus $c_1 = |\overline{AB}| = |\overline{AD}| + |\overline{DB}| = b \cos \alpha + \sqrt{a^2 - (b \sin \alpha)^2}$. Similarly, from $\triangle A'B'C'$ we see that

$$c_2 = |\overline{A'B'}| = |\overline{A'D'}| - |\overline{B'D'}| = b \cos \alpha - \sqrt{a^2 - (b \sin \alpha)^2}.$$

EXAMPLE 1 Suppose $a = 17$, $b = 26$, and $\alpha = 37°$. Find c, β, and γ.

Solution

   Method 1: Apply the Law of Cosines

   We first find c by applying Eq. (7.9):

$$c = 26 \cos 37° \pm \sqrt{17^2 - (26 \sin 37°)^2}.$$

FIGURE 7.28

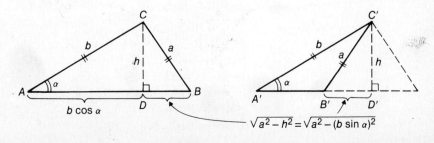

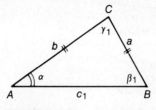
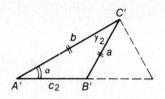

FIGURE 7.29

Evaluating, we get

$$c_1 \approx 27.4102 \quad \text{and} \quad c_2 \approx 14.1188.^*$$

Thus we get two solutions as shown in Fig. 7.29. We can now find β
and γ by applying the Law of Cosines as we did in the preceding
section.

For $\triangle ABC$ (where $c_1 \cong 27.4102$) we have

$$\cos\beta_1 = \frac{17^2 + (27.4102)^2 - 26^2}{2(17)(27.4102)} \quad \text{and} \quad \cos\gamma_1 = \frac{17^2 + 26^2 - (27.4102)^2}{2(17)(26)}.$$

This gives $\beta_1 \approx 66.988°$ and $\gamma_1 \approx 76.012°$. Thus the rounded-off answers are given by $c_1 \approx 27$, $\beta_1 \approx 67°$, and $\gamma_1 \approx 76°$. In a similar manner
the second solution is given by

$$c_2 \approx 14, \qquad \beta_2 \approx 113°, \qquad \gamma_2 \approx 30°.$$

Method 2: *Apply the Law of Sines*

First find β:

$$\sin\beta = \frac{b\,\sin\alpha}{a} = \frac{26\,\sin 37°}{17}.$$

So $\beta \approx 66.988°$. Then $\gamma = 180° - (\alpha + \beta) \approx 76.012°$. Now find c by
using

$$c = \frac{a\,\sin\gamma}{\sin\alpha} = \frac{17\,\sin 76.012°}{\sin 37°} \approx 27.4102.$$

This completes one solution, $c_1 \approx 27$, $\beta_1 \approx 67°$, $\gamma_1 \approx 76°$.

To get the second solution, we note that there are two possible
values of β from $\sin\beta = (26 \sin 37°)/17$. The first is $66.988°$, and the
second is $\beta_2 \approx 180° - 66.988° = 113.012°$. We can apply the Law of
Sines, using $\beta_2 = 113.012°$ to get γ_2 and c_2. This gives the same answers as above under Method 1, namely, $\beta_2 \approx 113°$, $\gamma_2 \approx 30°$, and
$c_2 \approx 14$. ∎

---

* *Note on efficient use of calculators:* We recorded intermediate values for c_1 and c_2. This was
done only to make it easier to follow through the solution. We suggest storing the computed value
of c_1 to full capacity in your calculator and then recalling it as needed in subsequent computations.
Do the same for c_2.

EXAMPLE 2 Suppose $b = 3.2$, $c = 2.5$, and $\beta = 43°$. Determine side a and angles α and γ.

Solution

Method 1

We first write a formula corresponding to that in Eq. (7.9) where we replace α by β, a by b, b by c, and c by a to get

$$a = c \cos \beta \pm \sqrt{b^2 - (c \sin \beta)^2}.$$

We suggest that you draw a diagram similar to that in Fig. 7.28 and get this equation directly from the diagram. Substituting the given quantities, we have

$$a = 2.5 \cos 43° \pm \sqrt{(3.2)^2 - (2.5 \sin 43°)^2}.$$

Evaluating, we get $a_1 \approx 4.5363$ (store to full decimal capacity in your calculator) and $a_2 \approx -0.8796$. Since the length of side a cannot be negative, there is only one solution, namely, $a \approx 4.5363$.

To find α and γ, we can use the Law of Cosines formulas,

$$\cos \alpha = \frac{b^2 + c^2 - a^2}{2bc} \quad \text{and} \quad \cos \gamma = \frac{a^2 + b^2 - c^2}{2ab}.$$

This gives $\alpha \approx 104.80°$ and $\gamma \approx 32.20°$. (At this point, check to see whether the sum $\alpha + \beta + \gamma$ equals $180°$.) Alternative procedure: Calculate γ using $\gamma = 180° - (\alpha + \beta)$ and then apply Eq. (7.2) as a check.

Rounding off to the appropriate number of digits gives the answer:

$$a \approx 4.5, \quad \alpha \approx 105°, \quad \text{and} \quad \gamma \approx 32°.$$

Method 2

First find γ by using the Law of Sines:

$$\sin \gamma = \frac{c \sin \beta}{b} = \frac{2.5 \sin 43°}{3.2}.$$

This gives $\gamma \approx 32.1956°$. For α we have $\alpha = 180° - (\beta + \gamma) \approx 104.8044°$. To find side a, use the Law of Sines:

$$a = \frac{b \sin \alpha}{\sin \beta} \approx 4.5363.$$

Rounding off, we have $a \approx 4.5$, $\alpha \approx 105°$, and $\gamma \approx 32°$.

Note that if we looked for a second solution, we would have $\gamma_1 \approx 180° - 32.1956° = 147.8044°$ and $\alpha_1 = 180° - (\beta + \gamma_1) \approx -10.8044°$. Since an angle of a triangle cannot be negative, we do not have a second solution. ∎

EXAMPLE 3   Given $a = 27$, $b = 64$, and $\alpha = 68°$, find c.

Solution   Substitute into Eq. (7.9) to get

$$c = 64 \cos 68° \pm \sqrt{(27)^2 - (64 \sin 68°)^2},$$
$$c \approx 23.97 \pm \sqrt{-2792.21}.$$

Since the values for c are not real numbers, we get no solution. Geometrically, side a is not "long enough" to reach side c.

Suggestion: Try solving the problem in Example 3 by using the Law of Sines as we did in the preceding example (Method 2).   ∎

EXERCISE 7.3

In Problems 1–10, three parts of a triangle are given. Solve for the remaining parts. In each case begin by drawing a diagram showing the given information.

1. $a = 86$, $b = 63$, $\alpha = 53°$
2. $a = 48$, $b = 24$, $\alpha = 115°$
3. $a = 3.6$, $b = 1.5$, $\alpha = 124°$
4. $a = 5.7$, $b = 4.6$, $\alpha = 34°$
5. $a = 31$, $b = 42$, $\alpha = 25°$
6. $b = 4.5$, $c = 3.6$, $\beta = 32°$
7. $b = 54$, $c = 23$, $\beta = 64°$
8. $a = 5.6$, $c = 3.1$, $\gamma = 55°$
9. $a = 56$, $b = 83$, $\beta = 117°$
10. $a = 6.3$, $b = 5.3$, $\alpha = 73°$

In Problems 11–16, two sides and an angle opposite one of them are given. In each case, find *the number of possible triangles* (0, 1, or 2) that are determined by the given data.

11. $a = 17.4$, $b = 27.3$, $\alpha = 65°40'$
12. $a = 19.3$, $b = 26.5$, $\beta = 63°30'$
13. $b = 4.7$, $c = 8.7$, $\gamma = 106°$
14. $b = 3.7$, $c = 6.5$, $\gamma = 114°$
15. $a = 1.3$, $c = 2.4$, $\alpha = 25°$
16. $a = 0.15$, $b = 0.33$, $\alpha = 53°$

17. In Fig. 7.30, ABC is a right triangle, $\alpha = 35°$, $|\overline{BD}| = 6.4$, and $|\overline{AD}| = 4.5$. Find the lengths of $\overline{BC}$ and $\overline{CD}$.

18. Assume that the orbits of the earth E and the planet Venus V are circular with the sun S at the center and radii 9.3×10^7 miles and 6.8×10^7 miles, respectively. Let θ be the angle that an observer on earth measures between the lines of sight E to S and E to V.
    a] When $\theta = 18°$, how far is Venus from the earth? There are two possible answers.
    b] What is the largest possible value of θ? For this value of θ, how many solutions are there to the problem stated in (a)?

19. A vertical tower $\overline{BC}$ is 30 meters tall and is located on top of a hill as shown in Fig. 7.31. At a point A 200 meters down the hill from B, the angle between the line of sight to B and that to C is 8°.
    a] Find the distance from A to C.
    b] Find the angle of inclination of the hill (angle θ).

FIGURE 7.30

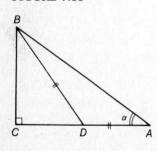

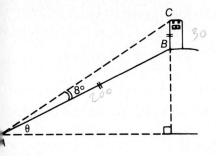

FIGURE 7.31

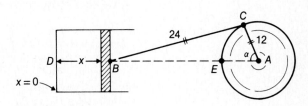

FIGURE 7.32

20. A vertical antenna on top of a building is 18.4 meters tall. The distance from a point A on the ground to the bottom of the antenna is 124 meters, and the angle at A between the lines of sight to the top and to the bottom of the antenna is 5.2°. How tall is the building?

21. A piston is driven by a rotating wheel of radius 12 cm as shown in Fig. 7.32, where the driving arm $\overline{BC}$ has length 24 cm and is attached to the wheel by a pivot at C. Suppose the wheel is rotating clockwise at the rate of 10 deg/sec and it starts with C at E. Let x denote the displacement of the piston as shown in the diagram. Find the displacement x at the end of 2 seconds ($\alpha = 20°$) and at the end of 12 seconds ($\alpha = 120°$). Find a formula that gives x as a function of time t.
(*Hint:* Let $a = |\overline{BC}| = 24$, $b = |\overline{AC}| = 12$, and $c = |\overline{AB}|$. Apply Eq. (7.9) to get c in terms of α.* Then $x = |\overline{AD}| - |\overline{AB}| = 36 - |\overline{AB}| = 36 - c$.)

7.4 Area of a Triangular Region

We have already encountered the problem of finding areas of given specific triangles[†] in Sections 7.1 and 7.2. In each case we applied the fundamental formula

$$\text{Area} = \tfrac{1}{2}(\text{base}) \times (\text{altitude}). \qquad \textbf{[7.10]}$$

In this section we are interested in deriving general formulas that can be applied to finding such areas.

Although the formulas developed here are useful in many situations, we strongly recommend that for most of the problems of this

---

* Since Eq. (7.9) applies to triangles, $0° < \alpha < 180°$. Show that your answer holds for $\alpha = 0°$ and $\alpha = 180°$ (by inspection) and then that it is also valid for $\alpha > 180°$.

[†] We use the expression "area of a triangle" to mean "area of the region enclosed by the triangle."

section you draw a diagram of the triangle described by the given data and select one of the given sides as the base, find the corresponding altitude and then apply the formula given in Eq. (7.10).

Given two angles and a side

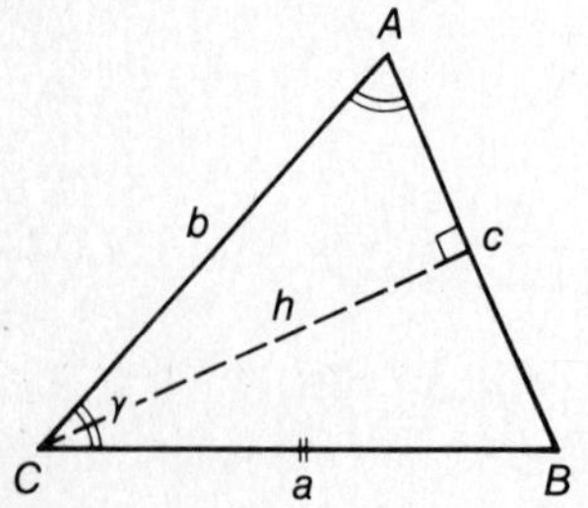

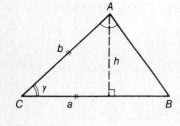

FIGURE 7.33

Suppose α, γ, and a are given as shown in Fig. 7.33. Using the Law of Sines, we obtain

$$c = \frac{a \, \sin \gamma}{\sin \alpha}.$$

The altitude h can be determined from the right triangle involving β:

$$h = a \, \sin \beta.$$

Therefore

$$\text{Area} = \tfrac{1}{2}\left(\frac{a \, \sin \gamma}{\sin \alpha}\right)(a \, \sin \beta).$$

$$\boxed{\text{Area} = \frac{a^2 \, \sin \gamma \, \sin \beta}{2 \, \sin \alpha} \, * \qquad \text{where } \beta = 180° - (\alpha + \gamma).} \quad \textbf{[7.11]}$$

Given two sides and the included angle

FIGURE 7.34

Suppose a, b, and γ are given. Let h be the altitude to $\overline{CB}$, as shown in Fig. 7.34. From the right triangle containing angle γ, we have $h = b \sin \gamma$. Therefore

$$\boxed{\text{Area} = \tfrac{1}{2}ab \, \sin \gamma.*} \qquad \textbf{[7.12]}$$

If γ is an obtuse angle, the diagram shown in Fig. 7.34 is different, but the formula still holds (see Problem 15 of Exercise 7.4).

Given three sides

For this case we derive the famous Heron's formula, named after the Greek philosopher-mathematician Heron (also known as Hero) of Alexandria (75 B.C.).

Suppose a, b, and c are given. We wish to derive a formula for area in terms of the three sides. We can use Eq. (7.12) given above, provided $\sin \gamma$ can be expressed in terms of a, b, and c. We can use

---

* There are other formulas similar to Eqs. (7.11) and (7.12) depending on the given information. All of the formulas are listed in the chapter summary on p. 291.

the identity $(\sin\gamma)^2 = 1 - (\cos\gamma)^2$ as follows:

$$(\sin\gamma)^2 = 1 - (\cos\gamma)^2 = (1 + \cos\gamma)(1 - \cos\gamma). \qquad \textbf{[7.13]}$$

We now get $\cos\gamma$ in terms of a, b, and c by using the Law of Cosines.

$$\cos\gamma = \frac{a^2 + b^2 - c^2}{2ab}. \qquad \textbf{[7.14]}$$

By substituting Eqs. (7.13) and (7.14) into Eq. (7.12) we get

$$\text{Area} = \tfrac{1}{2}ab\sqrt{\left(1 - \frac{a^2 + b^2 - c^2}{2ab}\right)\left(1 + \frac{a^2 + b^2 - c^2}{2ab}\right)}. \qquad \textbf{[7.15]}$$

This result can be written in compact form by introducing the quantity s, denoting one half of the perimeter of the triangle (called the *semiperimeter*), that is,

$$s = \tfrac{1}{2}(a + b + c).$$

Then, as a consequence of a good exercise in algebra (see Problem 14 of this section), we get *Heron's formula*:

$$\boxed{\text{Area} = \sqrt{s(s - a)(s - b)(s - c)}.} \qquad \textbf{[7.16]}$$

EXAMPLE 1 Find the area of the triangle that has $b = 3.57$, $c = 4.83$, and $\alpha = 49°38'$.

Solution Using an equivalent form of Eq. (7.12), we obtain

$$\text{Area} = \tfrac{1}{2}bc\,\sin\alpha = \tfrac{1}{2}(3.57)(4.83)\sin 49°38' \approx 6.57. \qquad \blacksquare$$

EXAMPLE 2 If $a = 34.75$, $b = 48.38$, and $c = 28.46$, find the area of the triangle.

Solution Use Heron's formula stated in Eq. (7.16):

$$\text{Area} = \sqrt{s(s - a)(s - b)(s - c)},$$

where $s = \tfrac{1}{2}(a + b + c) = \tfrac{1}{2}(34.75 + 48.38 + 28.46) = 55.795$. Put this result in the memory of the calculator. The remaining calculation can be carried out by using the (RCL) key to recall s when needed. Evaluate

$$\text{Area} = \sqrt{s(s - 34.75)(s - 48.38)(s - 28.46)} \approx 487.9$$

(to four significant digits). $\blacksquare$

EXAMPLE 3 The area of triangle ABC is 32.47 m², $\alpha = 23.1°$, and $\gamma = 112.4°$. Find the length of side b to three significant digits.

Solution   The formula given in Eq. (7.11) involves the area, one side, and the three angles. Thus we can use it in the form

$$\text{Area} = \frac{b^2 \sin \alpha \, \sin \gamma}{2 \sin \beta}.$$

We can now solve for b:

$$b = \sqrt{\frac{2(\text{Area}) \sin \beta}{\sin \alpha \, \sin \gamma}}.$$

Substituting $\beta = 180° - (23.1° + 112.4°) = 44.5°$ and the given data, we get

$$b = \sqrt{\frac{2(32.47) \sin 44.5°}{\sin 23.1° \, \sin 112.4°}} \approx 11.2019.$$

Therefore the desired answer is $b \approx 11.2$ m.   ∎

EXERCISE 7.4

In Problems 1–6 find the areas of the given triangles.

1. $\alpha = 37°10'$, $\beta = 65°20'$, $a = 34.6$ **2.** $\alpha = 42°10'$, $\gamma = 96°30'$, $b = 483$

3. $a = 32.7$, $b = 73.2$, $\gamma = 57°30'$ **4.** $b = 73.63$, $c = 87.65$, $\alpha = 124°47'$

5. $a = 73.5$, $b = 84.8$, $c = 58.5$ **6.** $a = 0.433$, $b = 0.632$, $c = 0.543$

7. A farm consists of a triangular plot of land bounded on three sides by roads, in which $\gamma = 47.0°$, $a = 254$ m, and $b = 531$ m (Fig. 7.35). Find the area of the farm and also the amount of fencing required to completely enclose it.

8. A level lot is in the shape of a quadrilateral with dimensions shown in Fig. 7.36. If land sells for $3.50 per square meter, find the cost of the lot. Round off to the nearest hundred dollars.

9. A farm is triangular; the rectangular coordinates of its vertices are $A(247, 123)$, $B(72, 411)$, and $C(328, 438)$; and the unit of measurement is the meter. Find the area rounded off to three significant digits.

10. The area of triangle ABC is 246.3 m², $a = 31.4$ m, and $b = 17.5$ m. Find angle γ to the nearest ten minutes.

FIGURE 7.35

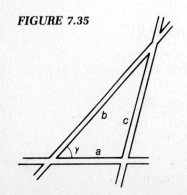

FIGURE 7.36

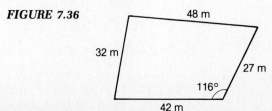

11. If the area of triangle ABC is 25.46 m^2, $\alpha = 46.2°$, and $\beta = 82.4°$, find the lengths of the three sides.

12. Suppose that $\alpha = 53°$, $c = 35$ cm, and the area of triangle ABC is 387 cm^2. Find b and a.

13. The area of triangle ABC is 254.6 cm^2. Find the area of the new triangle if
a] each side of ABC is doubled;
b] each side of ABC is tripled.

14. In the development of Heron's formula we introduced the quantity $s = \frac{1}{2}(a + b + c)$ and indicated that after some algebraic manipulation, Eq. (7.15) can be written in the form given by Eq. (7.16). To do this, we go through the following steps:

$$1 - \cos\gamma = 1 - \frac{a^2 + b^2 - c^2}{2ab} = \frac{c^2 - (a - b)^2}{2ab} = \frac{(c - a + b)(c + a - b)}{2ab}$$

$$= \frac{(a + b + c - 2a)(a + b + c - 2b)}{2ab} = \frac{2(s - a)(s - b)}{ab}.$$

Complete the problem by going through similar steps for $1 + \cos\gamma$, and then obtain the formula given in Eq. (7.16).

15. In this section we derived a formula for the area of a triangle when two sides a and b and the included angle γ are given. In Fig. 7.34, angle γ appears as an acute angle. Derive the same formula using a diagram in which γ is an obtuse angle.

16. Given a circle of radius 8.4 and a central angle $\theta = 52°$, find the area of the shaded region between the chord and arc, as shown in Fig. 7.37.

17. The shaded region shown in Fig. 7.37 is called a *segment* of the circle. Find a general formula for the area of a segment of a circle of radius r and central angle θ where $0 < \theta < \pi$.

18. A dime, a nickel, and a quarter are placed on the table so that they just touch each other as shown in Fig. 7.38. The diameters of the dime, nickel, and quarter are 1.75 cm, 2.25 cm, and 2.50 cm, respectively.
a] Find the area of the triangular region NQD.
b] Find the area of the trianglelike region with curved sides enclosed by the three coins.

FIGURE 7.37   *FIGURE 7.38*

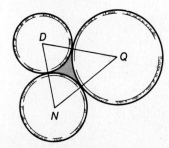

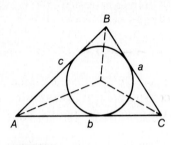

FIGURE 7.39   FIGURE 7.40

19. Suppose a circle is inscribed in a triangle with sides a, b, c (Fig. 7.39). Show that the radius of the circle is given by

$$r = \sqrt{\frac{(s-a)(s-b)(s-c)}{s}}.$$

(*Hint:* From geometry, recall that the bisectors of the three angles of a triangle are concurrent and their point of intersection is the center of the circle. Express the area of triangle ABC by using Eq. (7.16) and also as the sum of the three triangles shown in Fig. 7.39.)

20. Quadrilateral $OABC$ is inscribed in a quarter circle, as shown in Fig. 7.40, where $|\overline{AB}| = 2$ and $|\overline{BC}| = 4$. Find the area of the quadrilateral $OABC$ and express the answer as $a + b\sqrt{c}$, where a, b, and c are *positive integers.*
(*Hint:* You do not need any more information (such as the radius of the circle), and you should first convince yourself that angle BOC is *not* twice angle AOB.)

FIGURE 7.41

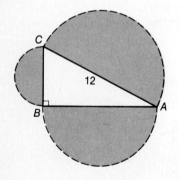

21. Suppose ABC is a right triangle with hypotenuse 12. One such triangle is shown in Fig. 7.41. What is the sum of the areas of the three semicircular regions shown in the diagram?
(*Hint:* The given information is sufficient.)

22. Follow the instructions given in Problem 21 but replace the three semicircular regions with squares.

23. Follow the instructions given in Problem 21 but replace the three semicircular regions with equilateral triangles.

24. Follow the instructions given in Problem 21 but replace the semicircles with half of regular polygons of an even number of sides (say $2n$ sides, $n = 2, 3, 4, \dots$) where each side of the triangle is a diagonal between opposite vertices of the polygon.

25. Evaluate the result obtained in Problem 24 for $n = 25, 50, 100, 500$. Compare the four numbers you get with the answer to Problem 21.

7.5 Vectors: Geometric Approach
(Optional)

In order to introduce the concept of scalar and vector quantities we first consider a simple example. Suppose a particle travels from point A to point B, as shown in Fig. 7.42. We ask two questions:

1. How far did the particle travel?
2. What is its displacement from A to B?

The answer to question 1 depends upon the path taken by the particle in going from A to B. In any case the answer will be given as a distance (that is, a number accompanied by a unit of measure, such as 24 cm).

In question 2 we are actually asking: "How far and in what direction is B from A?" We say that B is displaced from A by 17 cm in the direction of 60° east of north. When we talk about displacement, we ignore the actual path taken by the particle and focus our attention on the change in position.

This example illustrates two types of quantities that occur frequently in applications. The distance actually traveled can be described by giving a number and a unit of measure; such a quantity is called a *scalar*. Displacement requires a *number* (with the unit of measure) *and the direction* for its description; such a quantity is called a *vector*.

In general any quantity that can be described in terms of *magnitude only* (a number with a unit) is called a *scalar* quantity. Examples of scalar quantities are distance, mass, time, temperature, area, and volume. We shall also include real numbers as scalars; for example, 3, π, $\sqrt{2}$, 17,... will be called scalars even though there is no

FIGURE 7.42

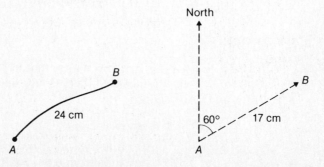

unit of measurement involved. Quantities that can be described by *magnitude and direction* are called *vector* quantities. Examples of vectors are displacement, force, velocity, and acceleration.

Notation

To distinguish a vector quantity from a scalar it is customary to write the symbol for a vector in boldface type or with an arrow. In this book we use the arrow notation such as $\vec{v}$ or $\vec{u}$. In the example above we can use the symbol $\overrightarrow{AB}$ as a vector to represent the displacement of B from A, or A to B.

In most problems it is convenient to draw a diagram in which a vector is represented by a directed line segment whose length is equal to the magnitude of the vector (drawn to scale). The magnitude (or length) of a vector is called the *absolute value* of a vector and is denoted by $|\vec{v}|$.

A vector having magnitude 1 is called a *unit vector*.

Algebra of vectors

We are already familiar with the algebra of scalars, since scalars are essentially real numbers. The algebra of vectors is different; for example, we do not get the sum of two vectors by merely adding their magnitudes, and so it will be necessary to define addition of vectors. However, we first ask the question: "When are two vectors equal?"

Equality of vectors

We can get some insight for defining equality and sum of two vectors by returning to the example given at the beginning of this section, in which a particle travels from A to B (see Fig. 7.42). The displacement from A to B is denoted by $\overrightarrow{AB}$ and described as a vector of magnitude 17 cm in the direction 60° east of north.

Now suppose a second particle travels from C to D, as shown in Fig. 7.43; its displacement is denoted by $\overrightarrow{CD}$—a vector described as having magnitude 17 cm in the direction 60° east of north. We see that the descriptions of both $\overrightarrow{AB}$ and $\overrightarrow{CD}$ are exactly the same. Therefore, we shall say that they are equal and write:

$$\overrightarrow{AB} = \overrightarrow{CD}.$$

In general, we say that *two vectors are equal* (regardless of their location in the plane) if they have the same magnitudes and the same directions.

Sum of vectors

To define the sum of two vectors, we let our particle travel from A to B and then from B to E, as shown in Fig. 7.43. The displacement

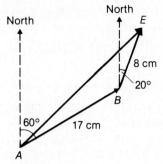
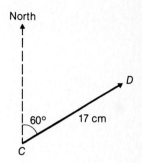

FIGURE 7.43

from A to E can be described in terms of the two displacements: A to B and B to E; we say that the sum of vectors $\overrightarrow{AB}$ and $\overrightarrow{BE}$ is equal to $\overrightarrow{AE}$, and write

$$\overrightarrow{AB} + \overrightarrow{BE} = \overrightarrow{AE}.$$

In general, we define the *sum of two vectors* geometrically as follows: Suppose $\vec{v}$ and $\vec{u}$ are two vectors, as shown in Fig. 7.44; move $\vec{u}$ parallel to itself until its initial point coincides with the terminal point of $\vec{v}$ (Fig. 7.44(a)).* Then the vector drawn from the initial point of $\vec{v}$ to the terminal point of $\vec{u}$ is the sum of $\vec{v}$ and $\vec{u}$, and is represented by $\vec{v} + \vec{u}$. The vector $\vec{v} + \vec{u}$ is called the *resultant vector*.

Suppose we move $\vec{u}$ parallel to itself, so that its initial point coincides with the initial point $\vec{v}$, and then draw the parallelogram (Fig. 7.44(b)). The sum $\vec{v} + \vec{u}$ will be represented by the diagonal, as shown.

This method of adding vectors geometrically is referred to as the *parallelogram law*. It should be clear that vector addition is commutative, that is,

$$\vec{v} + \vec{u} = \vec{u} + \vec{v}.$$

FIGURE 7.44

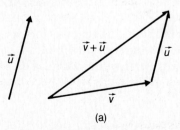
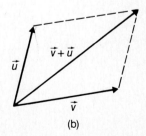

(a) (b)

_____

*If we consider $\vec{u}$ as a displacement vector, then moving $\vec{u}$ parallel to itself gives an equivalent displacement (a vector with the same direction and magnitude).

Also, the associative property holds for addition of vectors (see Problem 17 of this section); that is,

$$(\vec{u} + \vec{v}) + \vec{w} = \vec{u} + (\vec{v} + \vec{w}).$$

Product of a scalar and a vector

We introduce the idea of a vector multiplied by a scalar through the following examples.

$2\vec{v}$ denotes a vector of magnitude $2|\vec{v}|$ in the same direction as $\vec{v}$,

$-2\vec{v}$ denotes a vector of magnitude $2|\vec{v}|$ in the opposite direction of $\vec{v}$,

$(-1)\vec{v}$ will be denoted by $-\vec{v}$,

$0 \cdot \vec{v}$ is a vector of zero magnitude and no specific direction. It is called the *zero* (or *null*) *vector* and is written as $\vec{0}$. Thus $0 \cdot \vec{v} = \vec{0}$.

We define the *subtraction* of vectors in terms of addition as follows:

$$\vec{u} - \vec{v} = \vec{u} + (-\vec{v}).$$

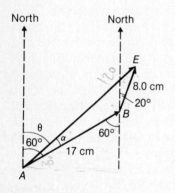

FIGURE 7.45

EXAMPLE 1 In Fig. 7.45, find the sum of $\overrightarrow{AB}$ and $\overrightarrow{BE}$.

Solution We can describe the sum $\overrightarrow{AB} + \overrightarrow{BE} = \overrightarrow{AE}$ by giving the length of line segment $\overline{AE}$ and the angle θ. Thus we need to solve triangle ABE for side $\overline{AE}$ and angle α. Using the Law of Cosines, we get

$$(\overline{AE})^2 = 8^2 + 17^2 - 2 \cdot 8 \cdot 17 \cos 140°,$$
$$|\overline{AE}| \approx 23.69 \text{ cm} \approx 24 \text{ cm}.$$

To find angle α, we use the Law of Sines:

$$\sin \alpha = \frac{8 \sin 140°}{23.69}.$$

This gives $\alpha \approx 12.54°$, and so $\theta = 60° - \alpha \approx 60° - 12.54° = 47.46° \approx 47°$.

Thus the sum of $\overrightarrow{AB}$ and $\overrightarrow{BE}$ can be described as a vector having magnitude 24 cm in the direction of 47° east of north. ∎

EXAMPLE 2 Suppose vectors $\vec{u}$ and $\vec{v}$ are as follows: $\vec{u}$ has magnitude 3.5 units in direction 20° east of south and $\vec{v}$ has magnitude 5.1 units in direction 76° west of north. Find

a] $\vec{u} + \vec{v}$ **b]** $-3\vec{u}$

Solution

a] We first draw a diagram showing $\vec{u}$, $\vec{v}$, and $\vec{u} + \vec{v}$ (Fig. 7.46). We can describe $\vec{u} + \vec{v}$ in terms of the length of line segment $\overline{CB}$ and

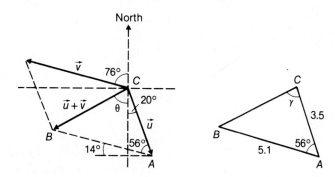

FIGURE 7.46

angle θ. Thus we isolate triangle ABC. First check to see that $\angle A = 56°$. Using the Law of Cosines, we get

$$|\overline{CB}|^2 = (3.5)^2 + (5.1)^2 - 2(3.5)(5.1) \cos 56°,$$
$$|\overline{CB}| \approx 4.2775.$$

Using the Law of Sines, we get

$$\sin \gamma \approx \frac{5.1 \sin 56°}{4.2775}.$$

This gives $\gamma \approx 81.3°$. Therefore $\theta \approx 81.3° - 20° = 61.3°$. Thus $\vec{u} + \vec{v}$ is a vector with magnitude 4.3 units in the direction 61° west of south.

b] $-3\vec{u}$ is a vector with magnitude $3(3.5) = 10.5$ units and the direction opposite to $\vec{u}$, that is, 20° west of north. ∎

EXAMPLE 3 Using the map given in Fig. 7.48 on p. 284, find the displacement from Los Angeles to Reno.

Solution The coordinates of Reno and Los Angeles are $R(-649, -175)$, $L(-618, -828)$. Thus the relative positions of R and L are as shown in Fig. 7.47. We wish to find vector $\overrightarrow{LR}$. In the right triangle we have

$$|\overline{RC}| = |-649 - (-618)| = 31, \qquad |\overline{LC}| = |-175 - (-828)| = 653.$$

Therefore

$$|\overrightarrow{LR}| = \sqrt{31^2 + 653^2} \approx 653.74,$$
$$\tan \theta = \frac{31}{653}, \qquad \theta \approx 2.72°.$$

Thus the displacement of Reno from Los Angeles is 654 km in the direction 2.7° west of north. ∎

FIGURE 7.47

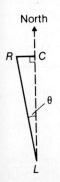

FIGURE 7.48

EXERCISE 7.5

1. A man walks 2.4 km north and then 1.5 km west. Construct a vector diagram and describe his displacement from the starting point.

2. A car travels 60 km east and then 83 km northeast. Draw a vector diagram and describe its displacement from the starting point.

3. Vectors $\vec{u}$ and $\vec{v}$ are as follows: $\vec{u}$ has magnitude 1.5 cm in direction of 60° east of north, and $\vec{v}$ has magnitude 2.0 cm in direction of 75° east

of north. Using a protractor and ruler, determine (by measurements) each of the following vectors:

a] $\vec{u} + \vec{v}$ **b]** $\vec{u} - \vec{v}$

4. Do Problem 3 by computing the vectors, and then compare with the answers obtained in Problem 3.

5. Using the map in Fig. 7.48, find the displacement from Logan to Phoenix.

6. Using the map in Fig. 7.48, find the displacement from Denver to Las Vegas.

7. Point B is displaced north of point A by 24 m, and point C is displaced from B by 15 m in the northeast direction. Find the displacement from A to C; then give the displacement from C to A.

8. A boat travels east 47 km and then turns 25° toward the south and travels 65 km. Find its displacement from the starting point.

9. A golfer takes two putts to get his ball into the hole. The first one rolls the ball 3.4 m in the northeast direction, and the second putt sends the ball north 1.2 m into the hole. How far and in what direction should he have aimed the first putt to get the ball into the hole with one stroke?

10. A girl walks 1 km southeast, then 3 km in the direction 30° west of south, and then 4 km in the direction 50° west of north. Using a protractor and ruler, draw a vector diagram (to scale) and determine (by measuring) the distance and direction in which she should walk to return to the starting point.

11. Suppose A and B are two points in the plane with rectangular coordinates $A(2, 5)$, $B(3, 7)$. If O is the origin and vectors $\vec{u}$ and $\vec{v}$ are given by $\vec{u} = \overrightarrow{OA}$, $\vec{v} = \overrightarrow{OB}$, find

a] $|\vec{u}|$ **b]** $|\vec{v}|$ **c]** $|\vec{u} + \vec{v}|$

12. Points A and B are on the opposite ends of a lake. Starting at A, a man walks to B by taking the route shown in Fig. 7.49: A to C (56 m in a southeast direction), C to D (40 m due east), D to B 85 m due north). If he went by boat directly from A to B, how far and in what direction would he go?

13. Consider two displacements having magnitudes 8 m and 15 m, respectively. Determine directions in which they should be taken so that the magnitude of the resultant displacement is

a] 23 m   **b]** 7 m   **c]** 17 m

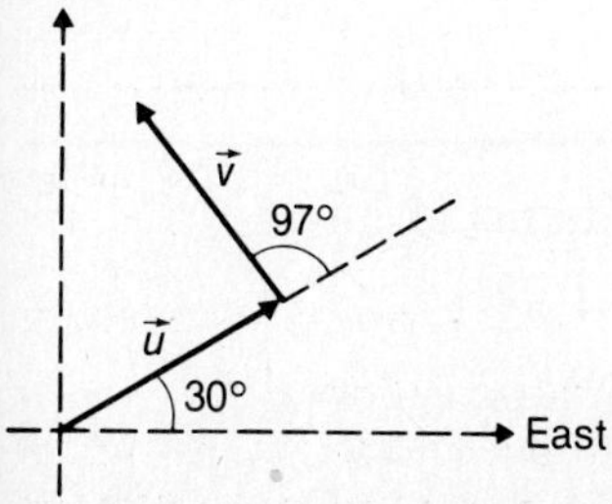

14. Vectors $\vec{u}$ and $\vec{v}$ both have magnitude 40 km. If they are oriented as shown in Fig. 7.50, find the direction and magnitude of $\vec{u} + \vec{v}$.

15. Using the map in Fig. 7.48, find the coordinates of a point that is 200 km southeast of Cheyenne.

16. A plane travels from Seattle to Denver and then continues in the same direction for another 400 km. Using the map in Fig. 7.48, find the coordinates of its position.

17. Using a geometrical argument, prove that addition of vectors is commutative and associative; that is, show that

$$\vec{u} + \vec{v} = \vec{v} + \vec{u} \quad \text{and} \quad (\vec{u} + \vec{v}) + \vec{w} = \vec{u} + (\vec{v} + \vec{w}).$$

FIGURE 7.49

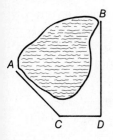

FIGURE 7.50

7.6 Vector Algebra: Analytic Approach (Optional)

In the preceding section we introduced the concept of vector addition as a geometric operation (the parallelogram rule). As may be apparent from the problems in Exercise 7.5, the process of adding vectors geometrically is awkward. In this section we introduce an analytic technique that simplifies addition of vectors.

In all of the examples of Section 7.5 the description of vectors was given in relation to a compass orientation. We now introduce a rectangular coordinate system in which the positive x-axis is in the east direction and the positive y-axis is in the north direction. The direction of any vector $\vec{v}$ can now be described by giving the angle θ (measured counterclockwise) from the positive x-axis to the vector.

Let $\vec{i}$ and $\vec{j}$ be unit vectors (of length one) in the positive x- and y-directions, respectively, as shown in Fig. 7.51. Any vector in the plane can be expressed as a linear combination of these two unit vectors, as shown in the diagram, where (x, y) represents the coordinates of the terminal point of $\vec{v}$.

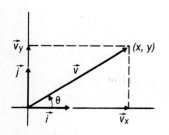

FIGURE 7.51

Thus we have two vectors $\vec{v}_x = x\vec{i}$ and $\vec{v}_y = y\vec{j}$ such that their sum is $\vec{v}$. That is,

$$\vec{v} = \vec{v}_x + \vec{v}_y = x\vec{i} + y\vec{j}.$$

Vectors $\vec{v}_x$ and $\vec{v}_y$ are called the components of $\vec{v}$ in the x- and y-directions, respectively. The process of expressing $\vec{v}$ as the sum of $\vec{v}_x$ and $\vec{v}_y$ is known as resolution of $\vec{v}$ into its x- and y-components (or $\vec{i}$- and $\vec{j}$-directions). The magnitude of $\vec{v}$ is given by $|\vec{v}| = \sqrt{x^2 + y^2}$.

Using $\cos\theta = x/|\vec{v}|$ and $\sin\theta = y/|\vec{v}|$, we see from the right triangle shown in Fig. 7.51 that

$$x = |\vec{v}| \cos\theta \qquad \text{and} \qquad y = |\vec{v}| \sin\theta.$$

Thus any vector $\vec{v}$ can be written in the form

$$\boxed{\vec{v} = (|\vec{v}| \cos\theta)\vec{i} + (|\vec{v}| \sin\theta)\vec{j}.}$$

Addition of vectors

Suppose vectors $\vec{u}$ and $\vec{v}$ are expressed in terms of $\vec{i}, \vec{j}$ as

$$\vec{u} = a\vec{i} + b\vec{j}, \qquad \vec{v} = c\vec{i} + d\vec{j}.$$

Vector addition is associative and commutative (see Problem 17 of Exercise 7.5), and so we have

$$\vec{u} + \vec{v} = (a\vec{i} + b\vec{j}) + (c\vec{i} + d\vec{j}) = (a + c)\vec{i} + (b + d)\vec{j}.$$

Note that we also used the distributive property of scalar multiplication over vector addition.

To add vectors, we merely add their corresponding $\vec{i}$ and $\vec{j}$ components:

$$\vec{u} + \vec{v} = (a + c)\vec{i} + (b + d)\vec{j}.$$

EXAMPLE 1 Suppose $\vec{v}$ is a vector with magnitude 4 and direction $\theta = 120°$. Resolve $\vec{v}$ into its x- and y-components.

Solution

$$a = |\vec{v}| \cos \theta = 4 \cos 120° = -2,$$
$$b = |\vec{v}| \sin \theta = 4 \sin 120° = 2\sqrt{3}.$$

Thus

$$\vec{v} = -2\vec{i} + 2\sqrt{3}\vec{j}. \qquad \blacksquare$$

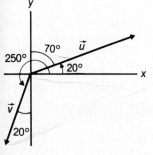

FIGURE 7.52

EXAMPLE 2 Suppose $\vec{u}$ is a vector of length 5 in the direction of 70° east of north and $\vec{v}$ has length 3 in the direction of 20° west of south (Fig. 7.52). Find the sum of $\vec{u}$ and $\vec{v}$; then find $|\vec{u} + \vec{v}|$.

Solution We first express $\vec{u}$ and $\vec{v}$ in $\vec{i}, \vec{j}$ form:

$$\vec{u} = 5 \cos 20°\vec{i} + 5 \sin 20°\vec{j} \approx 4.70\vec{i} + 1.71\vec{j}$$
$$\vec{v} = 3 \cos 250°\vec{i} + 3 \sin 250°\vec{j} \approx -1.03\vec{i} - 2.82\vec{j}.$$

Therefore

$$\vec{u} + \vec{v} \approx 3.67\vec{i} - 1.11\vec{j}.$$

To find $|\vec{u} + \vec{v}|$, we have

$$|\vec{u} + \vec{v}| \approx \sqrt{(3.67)^2 + (-1.11)^2} \approx 3.83. \qquad \blacksquare$$

EXAMPLE 3 Suppose $\vec{u} = 2\vec{i} + 3\vec{j}$ and $\vec{v} = 4\vec{j} - \vec{j}$. Find the vector $3\vec{u} - 5\vec{v}$.

Solution

$$3\vec{u} - 5\vec{v} = 3(2\vec{i} + 3\vec{j}) - 5(4\vec{i} - \vec{j}) = (6\vec{i} + 9\vec{j}) + (-20\vec{i} + 5\vec{j}).^*$$

---

* Note that in replacing $3(2\vec{i} + 3\vec{j})$ by $6\vec{i} + 9\vec{j}$ we used the distributive property

$$3(2\vec{i} + 3\vec{j}) = 3(2\vec{i}) + 3(3\vec{j}),$$

and the associative property

$$3(2\vec{i}) = (3 \cdot 2)\vec{i} \qquad \text{and} \qquad 3(3\vec{j}) = (3 \cdot 3)\vec{j}.$$

These properties hold in general and are basic in the study of *vector spaces*.

Thus

$$3\vec{u} - 5\vec{v} = -14\vec{i} + 14\vec{j}. \quad \blacksquare$$

EXAMPLE 4 If the displacement from Las Vegas to Havre is given by the vector $\overrightarrow{LH} = 541\vec{i} + 1383\vec{j}$, find the coordinates of Havre on the map of Fig. 7.48 (p. 284). The given distances are in kilometers.

Solution We wish to find the coordinates of H as shown in Fig. 7.53. We can do this by finding vector

$$\overrightarrow{OH} = \overrightarrow{OL} + \overrightarrow{LH}.$$

From information given on the map we have

$$\overrightarrow{OL} = -331\vec{i} - 622\vec{j}.$$

Therefore

$$\overrightarrow{OH} = (-331\vec{i} - 622\vec{j}) + (541\vec{i} + 1383\vec{j}) = 210\vec{i} + 761\vec{j}.$$

The coordinates of Havre are $(210, 761)$. $\quad \blacksquare$

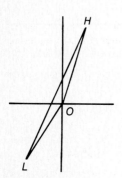

FIGURE 7.53

EXERCISE 7.6

In Problems 1–10, vectors $\vec{u}$, $\vec{v}$, and $\vec{w}$ are given by

$$\vec{u} = \vec{i} + \vec{j}, \qquad \vec{v} = 2\vec{i} - 5\vec{j}, \qquad \vec{w} = -2\vec{i} + \vec{j}.$$

In each case, draw a diagram illustrating the problem geometrically and then determine the given vector in $\vec{i}, \vec{j}$ form:

1. $\vec{u} + \vec{v}$
2. $\vec{u} - \vec{v}$
3. $2\vec{u} + 3\vec{w}$
4. $\vec{u} + \vec{v} + \vec{w}$
5. $3\vec{u} - 2\vec{v} + 4\vec{w}$
6. Find a vector that gives $\vec{w}$ when added to $\vec{u}$.
7. Find a vector that gives $\vec{v}$ when subtracted from $\vec{w}$.
8. Find
    a] $|\vec{u}|$
    b] $|\vec{v}|$
    c] $|\vec{u} + \vec{v}|$
9. Find $|2\vec{u} - 3\vec{v}|$
10. Find $|3\vec{u} + 2\vec{v} - 5\vec{w}|$

In Problems 11–14, suppose the x, y-coordinate system corresponds to compass directions and the direction angle θ is measured as described in this section. Let $\vec{u}$ and $\vec{v}$ be given as follows: $\vec{u}$ has magnitude 1.5 cm in direction 60° east of north, and $\vec{v}$ has magnitude 3.2 cm in direction 20° west of north.

11. Draw a diagram illustrating vectors $\vec{u}$ and $\vec{v}$, and then give the direction of each in terms of the corresponding θ angle.

12. Resolve $\vec{u}$ and $\vec{v}$ into their x, y-components.

13. Find the sum of $\vec{u}$ and $\vec{v}$ and describe the resultant in terms of compass direction.

14. Find $2\vec{u} - \vec{v}$ and give the result in terms of its magnitude and compass direction.

In Problems 15–19, use information from the map given in Fig. 7.48 (p. 284).

15. Find the displacement of Boise from Portland as a vector in $\vec{i}, \vec{j}$-form.

16. Point P is 200 km from Albuquerque in the direction of 54° east of north. Find the coordinates of P.

17. Find the displacement vector of El Paso from Missoula in $\vec{i}, \vec{j}$-form. Get an approximate check on your result by using a ruler and protractor on the map.

18. The displacement of point P from San Diego is given by the vector

$$\overrightarrow{DP} = 321\vec{i} + 175\vec{j}.$$

Find the displacement vector of P from Logan.

19. Determine the direction in which a plane should fly to travel directly from Los Angeles to Salt Lake City (assuming no wind effect).

20. Find the magnitude and direction of a vector whose x-component is 32 units and whose y-component is 24 units.

21. A girl walks 2 km in the southwest direction, then 1.5 km east, and then 3 km in the direction 30° east of north. Find her displacement from the starting point. Give the answer in terms of distance and compass direction.

22. What are the x- and y-components of a vector with magnitude 16 cm and the direction given by $\theta = 210°$?

23. Find a unit vector with the same direction as $\vec{u} = 3\vec{i} + \vec{j}$.

24. Find a unit vector perpendicular to vector $\vec{u} = 3\vec{i} + \vec{j}$.

25. If $\vec{u} = 3\vec{i} - 2\vec{j}$ and $\vec{v} = 2\vec{i} + \vec{j}$, find
a] the angle between $\vec{u}$ and $\vec{v}$
b] the angle between $\vec{u} + \vec{v}$ and $\vec{u} - \vec{v}$

FIGURE 7.54

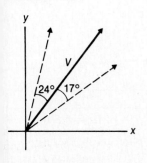

26. Find a unit vector parallel to the line through points $(3, 5)$ and $(2, -1)$.

27. Express vector $\vec{v} = 3i + 4j$ as the sum of two vectors with directions shown by broken lines in Fig. 7.54.

28. Find the coordinates of point P whose displacement from point $(3, 1)$ is of magnitude 4 in the direction of 136° with the positive x-axis.

29. A particle moving in the x, y-plane is photographed each second and its x, y-components for the first five seconds are given by the following table:

t (sec)	0	1	2	3	4	5
x (cm)	10	14	21	27	29	31
y (cm)	0	5	8	12	22	30

a] Draw a diagram that illustrates the displacements for successive seconds.
b] Find the displacement from $t = 0$ to $t = 4$ sec.
c] Find the displacement from $t = 1$ to $t = 5$ sec.

30. Suppose that the coordinates of a particle moving in the x, y-plane are given by

$$x = 3t - 5t^2, \qquad y = -4t^2 + t^3,$$

where t is in seconds and x, y are in centimeters. Find
a] the displacement of the particle from $t = 0$ to $t = 4$ sec,
b] the displacement of the particle from $t = 2$ to $t = 4$ sec.

31. A plane travels from Seattle to Missoula and then 450 km in the southeast direction. Using the map in Fig. 7.48 (p. 284), find how far and in what direction the plane is from Seattle.

32. On a par 4 hole a golfer scores a birdie with the following three strokes

the first travels 84 m at an angle of 54° east of south;
the second goes 21 m in the direction 10° west of south;
the third is a putt of 2.5 m in the northwest direction.

How far and in what direction should he have hit his drive to get a hole in one?

Summary

Important formulas used in solving triangles

$$\textit{Law of Sines:} \qquad \frac{a}{\sin \alpha} = \frac{b}{\sin \beta} = \frac{c}{\sin \gamma}$$

$$\textit{Law of Cosines:} \quad a^2 = b^2 + c^2 - 2bc \cos \alpha$$
$$b^2 = a^2 + c^2 - 2ac \cos \beta$$
$$c^2 = a^2 + b^2 - 2ab \cos \gamma$$

Solving triangles classified into four groups

	Given	Apply
SAA or ASA	One side and two angles	Law of Sines
SAS	Two sides and the included angle	Law of Cosines
SSS	Three sides	Law of Cosines
SSA	Two sides and an angle opposite one of them	Law of Sines or Law of Cosines (Check number of solutions)

Law of Cosine formulas for the ambiguous case

$$c = b\cos\alpha \pm \sqrt{a^2 - (b\,\sin\alpha)^2} = a\cos\beta \pm \sqrt{b^2 - (a\,\sin\beta)^2}$$
$$b = a\cos\gamma \pm \sqrt{c^2 - (a\,\sin\gamma)^2} = c\cos\alpha \pm \sqrt{a^2 - (c\,\sin\alpha)^2}$$
$$a = b\cos\gamma \pm \sqrt{c^2 - (b\,\sin\gamma)^2} = c\cos\beta \pm \sqrt{b^2 - (c\,\sin\beta)^2}$$

Formulas for area of a triangle

$$\text{Area} = \tfrac{1}{2}bc\sin\alpha = \tfrac{1}{2}ac\sin\beta = \tfrac{1}{2}ab\sin\gamma$$
$$\text{Area} = \frac{a^2\sin\beta\sin\gamma}{2\sin\alpha} = \frac{b^2\sin\alpha\sin\gamma}{2\sin\beta} = \frac{c^2\sin\alpha\sin\beta}{2\sin\gamma}$$
$$\text{Area} = \sqrt{s(s-a)(s-b)(s-c)} \qquad \text{where } s = (a+b+c)/2$$

Triangle identity formulas for checking solutions

$$\frac{a \pm b}{c} = \frac{\sin\alpha \pm \sin\beta}{\sin\gamma}$$

$$\frac{a^2 - b^2}{c^2} = \frac{\sin(\alpha - \beta)}{\sin\gamma}$$

$$\frac{a - b}{c} = \frac{\sin\left(\frac{\alpha - \beta}{2}\right)}{\cos(\gamma/2)}$$

$$\frac{a + b}{c} = \frac{\cos\left(\frac{\alpha - \beta}{2}\right)}{\sin(\gamma/2)}$$

Computer Problems (Optional)

Section 7.1

1. *Given two angles and a side.* Write a program that will allow you to enter measures of two angles in degrees and the length of one side of a triangle so that the output will be the third angle and the lengths of the other two sides. Use your program to check your answers for several of the problems you solved in this section using a calculator.

Section 7.2

2. *Given two sides and the included angle.* Write a program that will allow you to enter the lengths of the two sides and the included angle in degrees so that the output will be the third side and the two angles.* If you apply the technique illustrated in Example 2 of Section 7.2, include in your program a check to make certain that it avoids the difficulty encountered there.

---

* Check the computer problems at the end of Chapter 3 for identities giving the inverse sine and inverse cosine functions in terms of the inverse tangent function.

3. *Given three sides.* Follow a procedure similar to that in the preceding two problems except that you enter the lengths of the three sides and the output includes the three angles in degrees. Include a check to see whether the sum of the three computed angles is equal to 180°. Also you should include at the beginning of your program a check to see that the three numbers entered do indeed give a triangle (see Chapter computer problems).

Section 7.3

4. Write a program that will tell you *"how many solutions"* you will get in the ambiguous case. That is, for every a, b, and α that you enter, you should get one of the following responses: "no solution," "one solution," "two solutions."

5. In Problem 21 of Exercise 7.3 we can show that the displacement of the piston is given by

$$x = 36 - 12\left[\cos\frac{\pi t}{18} + \sqrt{4 - \left(\sin\frac{\pi t}{18}\right)^2}\right].$$

Note that we changed from degrees to radians for the purposes of the computer. Write a program that will give values of x corresponding to time t at $0, 1, 2, 3, \ldots 36$ seconds (one complete revolution of the wheel). Have your program round off the x-values to two decimal places. From your table of t, x-values, approximate the times when the displacement is 16 centimeters.

6. Write a program using triangle identity (7.2) to check a solution for any triangle in Exercises 7.1, 7.2, or 7.3.

Section 7.4

7. Write a program that will give the area of a triangle when two angles and a side are given. You may wish to use the formula given by Eq. (7.11).

8. Write a program that will give the area of a triangle when two sides and the included angle are given. Use the formula given by Eq. (7.12).

9. Use Heron's formula to write a program that will give the area of a triangle when three sides are given. If your program allows the user to input any three positive numbers for the lengths of the sides, then have it first check to see if they actually do form a triangle.

Section 7.5

10. Write a program that will give the distance between any two cities shown on the map in Fig. 7.48 (p. 284). That is, the input data will consist of the coordinates of the two cities, and the output will be the distance between them.

Review Exercises

In Problems 1–12, three parts of triangle ABC are given. Determine the remaining three parts.

1. $\alpha = 43°$, $\gamma = 78°$, $b = 8.4$ **2.** $\beta = 83°$, $\gamma = 47°$, $c = 0.43$

3. $a = 37.4$, $c = 64.2$, $\beta = 115.4°$ **4.** $a = 5.3$, $b = 7.6$, $c = 6.3$

5. $a = 35$, $b = 24$, $c = 31$ **6.** $b = 17.3$, $c = 38.4$, $\alpha = 112.3°$

7. $\alpha = 27.4°$, $\beta = 132.4°$, $a = 56.3$ **8.** $b = 68$, $c = 48$, $\beta = 75°$

9. $a = 63$, $b = 34$, $\gamma = 55°$ **10.** $a = 47$, $b = 32$, $\gamma = 90°$

11. $b = 32.3$, $c = 63.2$, $\gamma = 90.0°$ **12.** $a = 32.5$, $b = 43.7$, $c = 37.2$

In Problems 13–18, *how many* triangles are there (if any) with the given data as parts?

13. $a = 3.5$, $b = 7.2$, $\gamma = 73°$ **14.** $a = 5.6$, $b = 4.3$, $\beta = 26°$

15. $\alpha = 40°$, $\beta = 110°$, $\gamma = 30°$ **16.** $a = 7.3$, $b = 2.5$, $c = 4.1$

17. $a = 8.3$, $b = 3.4$, $c = 4.8$ **18.** $\alpha = 23°$, $\beta = 87°$, $\gamma = 50°$

In Problems 19–24, three parts of a triangle ABC are given. Find the area of the triangle.

19. $a = 7.3$, $b = 8.7$, $\gamma = 72°$ **20.** $a = 45$, $b = 41$, $c = 52$

21. $b = 17$, $\alpha = 45°$, $\gamma = 71°$ **22.** $b = 3.5$, $c = 7.3$, $\alpha = 130°$

23. $a = 2.3$, $b = 4.5$, $c = 5.4$ **24.** $c = 28$, $\alpha = 37°$, $\beta = 77°$

25. The coordinates of the vertices of triangle ABC are given by

$$A(3,7), \qquad B(5,-2), \qquad C(8,4).$$

Find angle BAC in degrees rounded off to one decimal place.

26. An equilateral triangle is inscribed in a circle of radius 12.4 cm. What is the length of a side of the triangle?

In Problems 27 and 28, vectors $\vec{v}$ and $\vec{u}$ are given by $\vec{v} = 3\vec{i} - 4\vec{j}$ and $\vec{u} = 2\vec{i} + 5\vec{j}$.

27. Find   **a]** $\vec{v} + \vec{u}$ **b]** $|\vec{v} + \vec{u}|$

28. Find   **a]** $\vec{v} - 3\vec{u}$ **b]** $|\vec{v} - 3\vec{u}|$

29. Suppose $\vec{v}$ is a vector having magnitude 8 units and $\theta = 240°$ (see Fig. 7.51). Express $\vec{v}$ in terms of $\vec{i}$ and $\vec{j}$.

30. B is displaced from A 15 miles in the northeast direction, and C is displaced from B 20 miles due north. What is the displacement of C from A?

31. Suppose $P(x_1, y_1)$ and $Q(x_2, y_2)$ are two points in the plane as shown in Fig. 7.55. Let M be the midpoint of line segment $\overrightarrow{PQ}$. Use vectors to find the coordinates of M. (*Hint:* $\overrightarrow{OM} = \overrightarrow{OP} + \frac{1}{2}\overrightarrow{PQ}$ where $\overrightarrow{PQ} = \overrightarrow{OQ} - \overrightarrow{OP}$.)

32. Follow the instructions of Problem 31 when point M is one third of the way from P to Q.

FIGURE 7.55

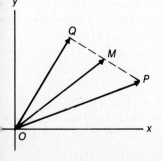

Polar
Coordinates
8

8.1 Introduction

Many problems involve equations relating two variables. We have seen that it is helpful to have geometrical representations of such relationships, since these can frequently provide insights that are not readily apparent from the equation itself. In some problems the situation is reversed in that we have a problem described geometrically and it becomes useful to consider it in an algebraic setting, which usually means an equation relating two variables. The form of the equation we get depends to a large degree on the reference (or coordinate) system we decide to use. So far, all our geometrical representations have been relative to a rectangular (or cartesian) system of coordinates. This has served us well for most problems. However, there are situations in which a given geometrical problem translates into a cumbersome equation when rectangular coordinates are used. A system of coordinates known as polar coordinates can be particularly useful in many situations.

As indicated at the beginning of this book, our geometrical considerations are restricted to a given plane (in future courses the student will encounter problems requiring three-dimensional geometry). A rectangular system of coordinates begins with two perpendicular lines. It is customary to take these lines as horizontal and vertical and call them the x-axis and y-axis, respectively. On each axis we have a one-to-one correspondence between points and real numbers. This provides us with a system that has a one-to-one correspondence between pairs of real numbers (x, y) and points P in the plane.

For the *system of polar coordinates* we begin with a ray (half line), which we call the *polar axis*; its endpoint is called the *polar origin* (point O), as shown in Fig. 8.1

Polar axis

Polar origin

FIGURE 8.1

Let point P be any point (other than O) in the plane. Consider the ray $\overrightarrow{OP}$ (see Fig. 8.2(a)) as the terminal side of the directed angle θ obtained by rotating the polar axis about point O through the angle of measure θ. We call $\overrightarrow{OP}$ the θ ray. If the distance from O to P is denoted by r, where r is a positive number, then polar coordinates of P consist of the ordered pair r and θ, denoted by $[r, \theta]$.* This is shown in Fig. 8.2(b).

In many situations it is convenient to allow the first member of the ordered pair $[r, \theta]$ to be a negative number. Suppose we consider the ordered pair $[-r, \theta + \pi]$, where r is a positive number. The pair $[-r, \theta + \pi]$ represents the point that is a directed distance of $-r$ along the $(\theta + \pi)$ ray; we interpret this as meaning r units in the opposite direction, which is along the θ ray. This puts us at $P[r, \theta]$. Therefore both $[r, \theta]$ and $[-r, \theta + \pi]$ are names in polar coordinates of the same point P, as shown in Fig. 8.2(c).

It is clear that the θ ray and the $(\theta + 2\pi)$ ray are the same; so $[r, \theta]$ and $[r, \theta + 2\pi]$ represent the same point. In fact, the point P shown in Fig. 8.2 can be represented by any of the ordered pairs $[r, \theta + 2k\pi]$ or $[-r, \theta + (2k + 1)\pi]$, where k is any integer.

The above discussion indicates how we name any point P in the plane in terms of polar coordinates. The special case where P is the *polar origin* is denoted by $[0, \theta]$, where θ can have any value.

Note that in polar coordinates we do not have the luxury we have in rectangular coordinates, in which there is a one-to-one correspondence between points in the plane and ordered pairs of real

FIGURE 8.2

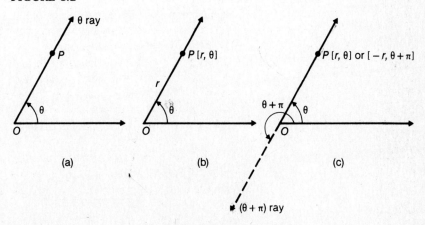

(a) (b) (c)

θ ray

$(\theta + \pi)$ ray

---

* We use the bracket notation $[r, \theta]$ as the name of a point in polar coordinates corresponding to the name (x, y) in rectangular coordinates.

numbers. In polar coordinates, each point P can be represented b
infinitely many ordered pairs; however, a given ordered pair corre
sponds to exactly one point. Although the lack of a one-to-one corre
spondence is an undesirable feature of polar coordinates, it does nc
create a serious problem.

We remind the reader that the *definition of equality of orderer
pairs* is given by

$$(a, b) = (c, d) \qquad \text{if and only if} \qquad a = c \quad \text{and} \quad b = d.$$

We retain this definition for ordered pairs $[r, \theta]$, and *we do not sa
that $[r, \theta]$ equals $[-r, \theta + \pi]$* even though they both represent the sam
point.

In our discussion so far we have used radian measure for θ, anc
this is consistent with the study of polar coordinates in calculus
However, in the following examples and exercises we shall occa
sionally use degree measure for θ.

EXAMPLE 1 For each of the following, draw a diagram to illustrat
the given ray.

a] $30°$ ray   **b]** $480°$ ray   **c]** $-\dfrac{5\pi}{6}$ ray   **d]** $\dfrac{5\pi}{4}$ ray

Solution See Fig. 8.3. ■

EXAMPLE 2 In each of the following, give two other names for the
given ray.

a] $45°$ ray   **b]** π ray   **c]** 2.5 ray   **d]** -2.5 ray

FIGURE 8.3

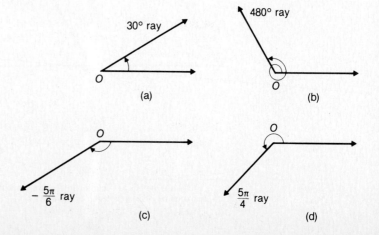

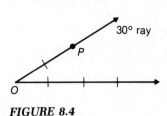

FIGURE 8.4   **FIGURE 8.5**

Solution

  a] 405° ray; −315° ray
  b] 3π ray; −3π ray
  c] $(2.5 + 2\pi)$ ray ≈ 8.78 ray; $(2.5 - 2\pi)$ ray ≈ -3.78 ray
  d] $(-2.5 + 2\pi)$ ray ≈ 3.78 ray; $(-2.5 + 4\pi)$ ray ≈ 10.07 ray

In (c) and (d) the results have been rounded off to two decimal places. The answers are not unique. ∎

EXAMPLE 3   Point P, shown in Fig. 8.4, is on the 30° ray at a distance 2 from the polar origin. Give four different names for P in polar coordinates.

Solution   Any of the following pairs (and many others) can be used as the name of point P:

  $[2, \ 30°]$; $[2, 30° + 360°] = [2, 390°]$;
  $[2, 30° - 360°] = [2, -330°]$; $[-2, 30° + 180°] = [-2, 210°]$. ∎

EXAMPLE 4   Suppose point P is 3 units from the polar origin on the $7\pi/6$ ray. Let Q be the point obtained by reflecting P about the line l perpendicular to the polar axis and passing through the polar origin. Give four different names for Q in polar coordinates.

Solution   From Fig. 8.5 we see that point Q is on the $11\pi/6$ ray and 3 units from O. Therefore Q can be represented by any of the following ordered pairs:

$$\left[3, \frac{11\pi}{6}\right]; \qquad \left[3, -\frac{\pi}{6}\right]; \qquad \left[-3, \frac{5\pi}{6}\right]; \qquad \left[3, -\frac{13\pi}{6}\right]. \quad ∎$$

EXAMPLE 5   In each of the following, draw a sketch to illustrate the point corresponding to the given ordered pairs in polar coordinates.

  a] $[2, 40°]$ **b]** $[-3, 580°]$ **c]** $\left[3, \frac{3\pi}{4}\right]$ **d]** $[-4, -3\pi]$

Solution   See Fig. 8.6. ` ∎

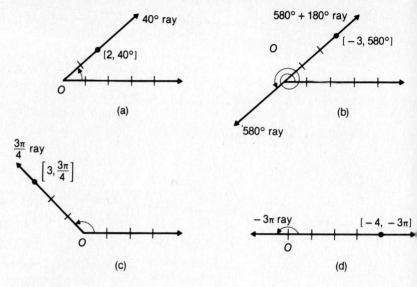

FIGURE 8.6

EXERCISE 8.1

1. In each of the following, a point is described in relation to a polar axis with polar origin O. Draw a diagram showing the given point, and then give four different ordered pairs [r, θ] that name the point in polar coordinates.
   a] P is 3 units from O on the 50° ray.
   b] Q is 4 units from O on the (−60°) ray.
   c] T is 2 units from O on the 540° ray.

2. In Problem 1, suppose each of the points P, Q, and T is reflected about the polar origin O to get new points P_1, Q_1, and T_1, respectively. For each of these points, give an ordered pair [r, θ] that can be used to represent the point in polar coordinates.

3. In Problem 1, suppose that each of the points P, Q, T is reflected about the line containing the polar axis to get new points P_2, Q_2, T_2, respectively. For each of these points, give an ordered pair [r, θ] that corresponds to the point in polar coordinates.

4. In each of the following, a point is described relative to a polar axis with polar origin O. Draw a diagram showing the given point, and then give four different ordered pairs of real numbers [r, θ] that can be used to name the point in polar coordinates.
   a] P is 2 units from O on the 2π/3 ray.
   b] Q is 3 units from O on the −11π/12 ray.
   c] T is 4 units from O on the 17π/6 ray.

5. In Problem 4, suppose that each of the points P, Q, T is reflected about the polar origin to get points P_1, Q_1, T_1, respectively. For each of these

points, give an ordered pair $[r, \theta]$ of real numbers that is a name for the point in polar coordinates.

6. In Problem 5, suppose each of the points P_1, Q_1, T_1 is reflected about the line through O perpendicular to the polar axis to get points P_2, Q_2, T_2, respectively. For each of these points, give an ordered pair $[r, \theta]$ of real numbers that can be used to represent the point in polar coordinates. How are P_2, Q_2, T_2 geometrically related to P, Q, T of Problem 4?

7. In each of the following, draw a diagram that illustrates the point corresponding to the given ordered pairs.

   a] $[3, 60°]$ **b]** $[-4, 45°]$ **c]** $[-2, 180°]$ **d]** $[-3, -450°]$

8. In each of the following, draw a diagram showing the point that corresponds to the given ordered pairs.

   a] $\left[4, \dfrac{4\pi}{3}\right]$ **b]** $\left[-3, \dfrac{5\pi}{12}\right]$ **c]** $[2, 17\pi]$ **d]** $[-2, -2.36]$

9. In each part of Problem 7 the given point is reflected about the polar origin. Give an ordered pair of real numbers $[r, \theta]$ that represents the new point in polar coordinates.

10. In each part of Problem 8 the given point is reflected about the line through the polar axis. Give an ordered pair $[r, \theta]$ of real numbers that can be used to represent the new point.

8.2 Graphs in Polar Coordinates

In earlier parts of this book we encountered a variety of problems in which an equation was given in the form $y = f(x)$, and then by means of a system of rectangular coordinates a graph (curve) corresponding to the given equation was drawn. The analogous problem in polar coordinates is: Given $r = f(\theta)$, draw a curve that corresponds to this equation.

EXAMPLE 1 Sketch the curve whose equation in polar coordinates is $r = 2 \sin \theta$.

Solution We first determine several ordered pairs $[r, \theta]$ that satisfy the given equation. These are shown in the following table. Note that it is not necessary to continue with larger values of θ, since $\sin(\theta + \pi) = -\sin \theta$ is an identity, and so

$$[r, \theta + \pi] = [2 \sin(\theta + \pi), \theta + \pi] = [-2 \sin \theta, \theta + \pi].$$

Therefore

$$[r, \theta + \pi] = [-2 \sin \theta, \theta + \pi] \quad \text{and} \quad [r, \theta] = [2 \sin \theta, \theta]$$

represent the same point.

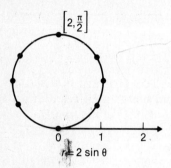

$r = 2 \sin \theta$

FIGURE 8.7

In a similar manner we can show that negative values of θ produce no points that are not already included in the points given by $0 \leq \theta \leq \pi$.

We now plot the points given in the table and draw the curve shown in Fig. 8.7. The curve is a circle (see Problem 11 of Exercise 8.3). ∎

θ	0	$\frac{\pi}{6}$	$\frac{\pi}{4}$	$\frac{\pi}{3}$	$\frac{\pi}{2}$	$\frac{2\pi}{3}$	$\frac{3\pi}{4}$	$\frac{5\pi}{6}$	π
r	0	1	$\sqrt{2}$	$\sqrt{3}$	2	$\sqrt{3}$	$\sqrt{2}$	1	0

EXAMPLE 2 Sketch the curve whose equation in polar coordinates is $r = 1 + \cos \theta$.

Solution As in Example 1, we first make a table giving ordered pairs $[r, \theta]$ that satisfy the given equation. Values of r are given in decimal form to two places. Since $\cos(\theta + 2\pi) = \cos \theta$ is an identity, it is clear that we get no new points by considering values of θ that are outside the interval $0° \leq \theta \leq 360°$. Plot these points and draw the curve, as shown in Fig. 8.8. The curve is an example of a *cardioid*. ∎

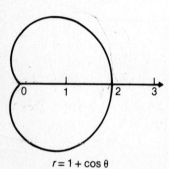

$r = 1 + \cos \theta$

FIGURE 8.8

θ	0°	45°	90°	135°	180°	225°	270°	315°	360°
r	2	1.71	1	0.29	0	0.29	1	1.71	2

EXAMPLE 3 Sketch the curve whose equation in polar coordinates is $r = 3$.

FIGURE 8.9

Solution As in the preceding two examples, first make a table of ordered pairs $[r, \theta]$. The variable θ does not appear explicitly in the given equation; if this causes any problems, we can write the equation in equivalent form as $r = 3 + 0 \cdot \theta$. The value of r is 3 for every value of θ, and so the corresponding points are on a circle with center at the polar origin and radius 3, as shown in Fig. 8.9. ∎

$r = 3$

EXAMPLE 4 Sketch the curve whose equation in polar coordinates is $r = \sin 3\theta$.

Solution First note that $\sin 3(\theta + \pi) = -\sin 3\theta$ is an identity. Thus

$$[r, \theta + \pi] = [\sin 3(\theta + \pi), \theta + \pi] = [-\sin 3\theta, \theta + \pi].$$

Also, $[r, \theta] = [\sin 3\theta, \theta]$. But $[-\sin 3\theta, \theta + \pi]$ and $[\sin 3\theta, \theta]$ represent the same point. Hence it is sufficient to use values of θ in the interval $0 \leq \theta \leq \pi$, as seen in the following table.

θ	0	$\dfrac{\pi}{12}$	$\dfrac{\pi}{6}$	$\dfrac{\pi}{4}$	$\dfrac{\pi}{3}$	$\dfrac{5\pi}{12}$	$\dfrac{\pi}{2}$	$\dfrac{7\pi}{12}$	$\dfrac{2\pi}{3}$	$\dfrac{3\pi}{4}$	$\dfrac{5\pi}{6}$	$\dfrac{11\pi}{12}$	π
r	0	0.71	1	0.71	0	−0.71	−1	−0.71	0	0.71	1	0.71	0

Plotting the points given in this table and connecting them in the appropriate manner gives the *three-leaf* rose shown in Fig. 8.10.

Note in Example 4 that $r = 0$ for values of θ such as $0, \pi/3, 2\pi/3$, and π. In each case the point is the origin, and the curve comes into the origin tangent to the corresponding θ ray (as shown in Fig. 8.10 for $\theta = \pi/3$). This illustrates a general situation: If $r = f(\theta)$ and $f(\theta_1) = 0$, then the curve comes into the origin tangent to the θ_1 ray. ■

EXAMPLE 5 Sketch the curve whose equation in polar coordinates is given by $r = -\theta$, where $\theta \geq 0$.

Solution Note that the given equation implies that radian measure is to be used for θ, since r is a real number. First make a table of ordered pairs $[r, \theta]$ that satisfy the equation; θ is given in exact form, and r is rounded off to two decimal places.

Plotting these points and drawing a curve through them gives a *spiral*, as shown in Fig. 8.11. The curve begins at the polar origin and, as θ increases, winds around in the counterclockwise direction, as illustrated. ■

θ	0	$\dfrac{\pi}{4}$	$\dfrac{\pi}{2}$	$\dfrac{3\pi}{4}$	π	$\dfrac{5\pi}{4}$	$\dfrac{3\pi}{2}$	$\dfrac{7\pi}{4}$	2π
r	0	−0.79	−1.57	−2.36	−3.14	−3.93	−4.71	−5.50	−6.28

FIGURE 8.10

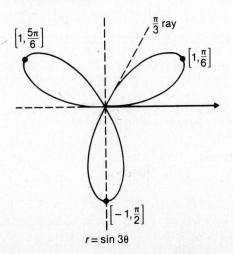

$r = \sin 3\theta$

FIGURE 8.11

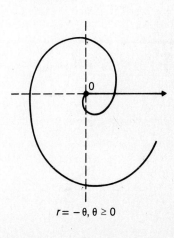

$r = -\theta, \theta \geq 0$

EXERCISE 8.2

In Problems 1–20, sketch the curve that corresponds to the given equation in polar coordinates. A list of some common equations and their graphs appear in the Chapter Summary.

1. $r = \cos\theta$ **2.** $r = 3\cos\theta$ **3.** $r = 2$

4. $r = -2\sin\theta$ **5.** $r = 1 + \sin\theta$ **6.** $r = 1 - \sin\theta$

7. $r = 1 - \cos\theta$ **8.** $r = 3 + \sin^2\theta + \cos^2\theta$

9. $r = \sin 2\theta$ **10.** $r = \cos 3\theta$ **11.** $r = \cos^2\theta - \sin^2\theta$

12. $r^2 = 4$ **13.** $r = \cos\theta\tan\theta$ **14.** $r^2 = \sin^2\theta$

15. $r = \sin\left(\theta + \dfrac{\pi}{4}\right)$ **16.** $r = \cos(\theta + \pi)$ **17.** $r = 1 + 2\cos\theta$

18. $r = 2 - \sin\theta$ **19.** $r = \theta$, where $\theta \geq 0$ **20.** $r = \dfrac{3}{\theta}$, where $\theta \geq 1$

8.3 Relationship between Polar and Rectangular Coordinates

Suppose the polar axis is taken in such a way that it coincides with the positive x-axis, as shown in Fig. 8.12, and let P be any point in the plane. The name of point P is (x, y) relative to the x, y-coordinate system and $[r, \theta]$ relative to the polar coordinate system.

The following equations give the relationship between rectangular and polar coordinates:

$$x = r\cos\theta, \qquad y = r\sin\theta. \tag{8.1}$$

$$r^2 = x^2 + y^2, \qquad \tan\theta = \frac{y}{x}. \tag{8.2}$$

FIGURE 8.12

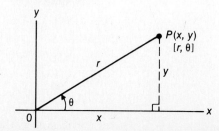

The equations given in (8.1) are *transformation equations from polar to rectangular coordinates.* For each pair $[r, \theta]$ there is precisely one pair (x, y) corresponding to it.

The equations given in (8.2) are known as the *transformation equations from rectangular to polar coordinates.* Note that for a given pair (x, y) we can get multiple pairs $[r, \theta]$, each of which represents the same point. Since r can be taken as $\sqrt{x^2 + y^2}$ or as $-\sqrt{x^2 + y^2}$, and θ satisfying $\tan \theta = y/x$ is multiple-valued, we must be careful to match appropriate values of r and θ. This is illustrated in the following examples.

EXAMPLE 1   In each of the following, find all ordered pairs $[r, \theta]$ that are associated with the given point in rectangular coordinates.

a] $(3, 4)$ **b]** $(-2, -1)$

Solution

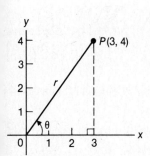

FIGURE 8.13

a] We use Eq. (8.2) as follows (see Fig. 8.13): First find $[r, \theta]$ where $r > 0$, $r = \sqrt{3^2 + 4^2} = 5$, and θ satisfies $\tan \theta = 4/3$, where θ is in the first quadrant. Hence $\theta \approx 53.13°$. This gives the set of ordered pairs

$$A = \{[5, 53.13° + k \cdot 360°] \,|\, k \text{ is any integer}\}.$$

Now find $[r, \theta]$ where $r < 0$, $r = -\sqrt{3^2 + 4^2} = -5$, and θ satisfies $\tan \theta = 4/3$, where θ is in the third quadrant. This gives the set of ordered pairs

$$B = \{[-5, 233.13° + k \cdot 360°] \,|\, k \text{ is any integer}\}.$$

Therefore the name in polar coordinates of the point associated with $(3, 4)$ is given by any one of the ordered pairs in the union of sets A and B, where θ values are rounded off to two decimal places.

FIGURE 8.14

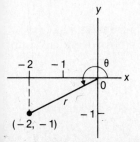

b] In a manner similar to (a) we have: For $r > 0$, $r = \sqrt{(-2)^2 + (-1)^2} = \sqrt{5}$, and θ satisfies $\tan \theta = \frac{1}{2}$, where θ is in the third quadrant (Fig. 8.14). That is, $r = \sqrt{5}$ and $\theta \approx 3.61 + k \cdot 2\pi$. For $r < 0$, $r = -\sqrt{5}$, and θ satisfies $\tan \theta = \frac{1}{2}$, where θ is in the first quadrant. That is, $r = -\sqrt{5}$ and $\theta \approx 0.46 + k \cdot 2\pi$. Therefore the point $(-2, -1)$ is represented in polar coordinates by any of the ordered pairs in the set

$$\{[\sqrt{5}, 3.61 + k \cdot 2\pi] \,|\, k \text{ any integer}\} \cup$$
$$\{[-\sqrt{5}, 0.46 + k \cdot 2\pi] \,|\, k \text{ any integer}\},$$

where θ values are rounded off to two decimal places. ∎

EXAMPLE 2 In each of the following, the given ordered pair names a point P in polar coordinates. Find the corresponding name in rectangular coordinates.

a] $[4, 60°]$ **b]** $[-3, 180°]$ **c]** $\left[4, \dfrac{-3\pi}{4}\right]$ **d]** $[-2, 2.48]$

Solution Use the equations in (8.1), which are valid for all values of θ and r.

a] $x = 4 \cos 60° = 4(1/2) = 2; y = 4 \sin 60° = 4(\sqrt{3}/2) = 2\sqrt{3}$. Therefore the point in rectangular coordinates is given by $(2, 2\sqrt{3})$.

b] $x = -3 \cos 180° = -3(-1) = 3;$ $y = -3 \sin 180° = -3(0) = 0$. Hence the given point is $(3, 0)$ in rectangular coordinates.

c] $x = 4 \cos(-3\pi/4) = -2\sqrt{2}; y = 4 \sin(-3\pi/4) = -2\sqrt{2}$. Thus the given point is denoted by $(-2\sqrt{2}, -2\sqrt{2})$ in rectangular coordinates.

d] $x = -2 \cos 2.48 \approx 1.58; y = -2 \sin 2.48 \approx -1.23$ (to two decimal places). Therefore $[-2, 2.48]$ is represented by $(1.58, -1.23)$ in rectangular coordinates. ∎

EXAMPLE 3 Find an equation in polar coordinates that describes the same set of points (same curve) as $x^2 + y^2 - 2x = 0$ in rectangular coordinates.

Solution Substituting $x = r \cos \theta$ and $y = r \sin \theta$ into the given equation gives

$$(r \cos \theta)^2 + (r \sin \theta)^2 - 2(r \cos \theta) = 0,$$
$$r^2[\cos^2 \theta + \sin^2 \theta] - 2r \cos \theta = 0.$$

This is equivalent to $r^2 - 2r \cos \theta = 0$. Thus $r(r - 2 \cos \theta) = 0$, and so $r = 0$ or $r = 2 \cos \theta$. Since $r = 0$ gives only the polar origin as a point, and from $r = 2 \cos \theta$ we get the point $[0, \pi/2]$, which is also the polar origin, we can ignore $r = 0$ in our solution. That is, $r = 2 \cos \theta$ will describe the same set of points as $x^2 + y^2 - 2x = 0$. ∎

EXAMPLE 4 Find an equation in rectangular coordinates that describes the same set of points in polar coordinates as

$$r = 2 \sin \theta + \cos \theta.$$

Solution Since a direct substitution for r and θ from Eq. (8.2) would involve replacing r by $\sqrt{x^2 + y^2}$, it is simpler to first multiply both sides of the given equation by r:

$$r^2 = 2r \sin \theta + r \cos \theta.$$

Now replacing r^2 by $x^2 + y^2$, $r \sin \theta$ by y, and $r \cos \theta$ by x, we get

$$x^2 + y^2 = 2y + x.$$

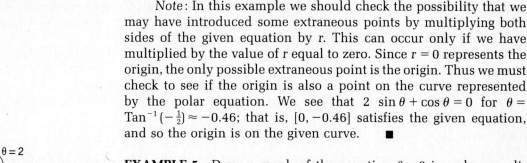

FIGURE 8.15

Note: In this example we should check the possibility that we may have introduced some extraneous points by multiplying both sides of the given equation by r. This can occur only if we have multiplied by the value of r equal to zero. Since r = 0 represents the origin, the only possible extraneous point is the origin. Thus we must check to see if the origin is also a point on the curve represented by the polar equation. We see that $2 \sin \theta + \cos \theta = 0$ for $\theta = \text{Tan}^{-1}(-\frac{1}{2}) \approx -0.46$; that is, $[0, -0.46]$ satisfies the given equation, and so the origin is on the given curve. ■

EXAMPLE 5   Draw a graph of the equation $\theta = 2$ in polar coordinates. Then find an equivalent equation in rectangular coordinates.

Solution   The graph of $\theta = 2$ is a line through the origin, as shown in Fig. 8.15. Since $\tan \theta = y/x$, the corresponding equation in rectangular coordinates is $\tan 2 = y/x$, or $y = x(\tan 2)$. In decimal form this is $y \approx -2.19x$. ■

EXERCISE 8.3

For each answer that is to be expressed in decimal form, give the result correct to two decimal places.

1. In each of the following, a point is given in rectangular coordinates. Find one name of the point in polar coordinates.
    a] $(-1, 1)$ b] $(-1, -\sqrt{3})$ c] $(\pi, 4)$ d] $(-1.57, 2.43)$

2. For each of the points given in Problem 1, give the set of all possible ordered pairs [r, θ] that can be used as polar coordinates for the given points.

3. Express each of the following in polar coordinates with $r \geq 0$ and $0 \leq \theta \leq 2\pi$.

    a] $(-3, 3)$ b] $(1, -3)$ c] $\left(\pi, \dfrac{1 + \sqrt{5}}{2}\right)$

4. Express each of the following in polar coordinates, using the smallest positive angle θ and r < 0.
    a] $(4, -3)$ b] $(-\sqrt{3}, \sqrt{3})$ c] $(2.52, -2\pi)$

5. Express each of the following in rectangular coordinates.

    a] $\left[2, \dfrac{\pi}{2}\right]$ b] $\left[-3, -\dfrac{3\pi}{4}\right]$ c] $[2.24, -0.37]$

6. Express each of the following in rectangular coordinates.
    a] $[0, 30°]$ b] $[4, -630°]$ c] $[-2, 47°37']$

7. In each of the following, determine whether or not the given pair satisfies the equation $r^2 \sin \theta = 1$.

    a] $\left[1, \dfrac{\pi}{2}\right]$ b] $\left[-1, -\dfrac{\pi}{2}\right]$ c] $\left[\sqrt{2}, \dfrac{5\pi}{6}\right]$ d] $[0, 0]$ e] $\left[1, \dfrac{3\pi}{2}\right]$

8. In each of the following, the coordinates of a point P are given in rectangular coordinates. Determine whether or not P lies on the curve whose equation in polar coordinates is $r = 1 + \cos \theta$.

a] $(0, 0)$ b] $(0, 1)$ c] $(2, 0)$ d] $\left(\dfrac{1 + \sqrt{2}}{2}, \dfrac{1 + \sqrt{2}}{2} \right)$

9. Let $[r_1, \theta_1]$ be polar coordinates of point P and $[r_2, \theta_2]$ be polar coordinates of point Q. Let d represent the distance between P and Q. Show that d is given by
$$d = \sqrt{r_1^2 + r_2^2 - 2r_1r_2 \cos(\theta_1 - \theta_2)}.$$

10. Use the result in Problem 9 to find the distance between the given pairs of points.

a] $[3, 0]$, $[\pi, \pi]$ b] $\left[1, \dfrac{\pi}{3} \right]$, $\left[-2, \dfrac{3\pi}{4} \right]$

c] $[-3.4, 32°]$, $[1.6, 47°]$ d] $[-2.4, 3.2]$, $[3.7, -0.64]$

In Problems 11–18, find an equation in rectangular coordinates that describes the same set of points (same curve) as the given equation in polar coordinates.

11. $r = 2 \sin \theta$ 12. $r = 4 \cos \theta$ 13. $3\theta = 4$ 14. $r \cos \theta = 3$

15. $r(1 - \sin \theta) = 2$ 16. $r(1 + \cos \theta) = 2$

17. $r = 2 \cos(\theta + \pi)$ 18. $r = \cos 2\theta$

In Problems 19–22, find an equation in polar coordinates that describes the same set of points (same curve) as the given equation in rectangular coordinates. Then sketch the curve, using the equation either in rectangular or in polar form.

19. $x^2 + y^2 = 1$ 20. $2xy = 3$ 21. $3x - y = 0$

22. $x^2 + y^2 + x = \sqrt{x^2 + y^2}$

23. Are all points on the curve whose equation is $r = \sin \theta$ also on the curve with equation $r \csc \theta = 1$? Give the reason for your answer.

24. Express $r = \sin 2\theta$ as an equation in rectangular coordinates.

25. Suppose P is a point in the plane given in polar coordinates by $[-2, \pi]$. Is P on the curve whose equation is $r = 1 + \cos \theta$? (*Hint*: P is also given by $[2, 0]$.)

Summary

Coordinate systems

Ordered pairs of numbers are used to denote points in a plane. In *rectangular coordinates* the name (x, y) of a point P is relative to two perpendicular lines (the x-axis and y-axis), and there is a one-to-one correspondence between the ordered pairs (x, y) and points P in the plane.

In *polar coordinates* a name $[r, \theta]$ for a point P in the plane is given in relation to a ray (called the polar axis where the endpoint is the polar origin). For each ordered pair $[r, \theta]$ there is associated a unique point P, but with each point P we can associate infinitely many ordered pairs $[r, \theta]$.

Graphs in polar coordinates

The following is a list of some equations in polar coordinates and their corresponding curves. We assume that a is a nonzero number.

1. For $r = a$, $r = a \sin \theta$, or $r = a \cos \theta$ the graph is a *circle*.
2. For $r = a (1 \pm \cos \theta)$ or $r = a (1 \pm \sin \theta)$ the graph is a *cardioid*.
3. For $r = a \sin 3\theta$ or $r = a \cos 3\theta$ the graph is a *three-leaf rose*.
4. For $r = a \sin 2\theta$ or $r = a \cos 2\theta$ the graph is a *four-leaf rose*.
5. For $r = a\theta$ or $r = a/\theta$ the graph is a *spiral*.
6. For $r (1 \pm \cos \theta) = a$ or $r (1 \pm \sin \theta) = a$ the graph is a *parabola*.

Transformation equations

From polar to rectangular coordinates:

$$x = r \cos \theta, \qquad y = r \sin \theta.$$

From rectangular to polar coordinates:

$r = \pm\sqrt{x^2 + y^2}$, and θ is given by $\tan \theta = y/x$.
A specific value of θ is given by $\theta = \mathrm{Tan}^{-1}(y/x)$.

Computer Problems (Optional)

Section 8.1

1. Write a program that will allow you to input the polar coordinates of any point $P[r, \theta]$ where r and θ (in degrees) *are integers* (positive, negative, or zero) to get an output of $Q[r_1, \theta_1]$ such that
   a] $r_1 \geq 0$ and $0° \leq \theta_1 < 360°$
   b] $r_1 \geq 0$ and $-180° < \theta_1 \leq 180°$
   c] $0° \leq \theta_1 < 180°$. Here r_1 will be negative in some cases.

Section 8.2

2. Write a program that will list a table of r, θ values for $\theta = 0, \pi/12, \pi/6, \pi/4, \pi/3, \ldots, 2\pi$, where $r = 1 + \cos \theta$. Have your program include

headings for the r and θ columns and give r rounded off to two decimal places. Now plot the corresponding points given in your table and then draw a smooth curve through them. Compare your graph with that shown in Fig. 8.8.

3. Follow the instructions of Problem 2 for $r = 1/(1 + \cos \theta)$. If your program gives an ERROR response at some point, make the necessary adjustments to take care of the problem.

4. Follow the instructions of Problem 2 for $r = 1 - 2 \cos \theta$. Here you should be careful to connect the points in proper order when drawing the graph.

Section 8.3

5. Write a program that will give polar coordinates $[r, \theta]$ for any point $P(x, y)$ given in rectangular coordinates. Have your program give the values of r and θ to two decimal places where $r \geq 0$ and $0 \leq \theta < 2\pi$.

6. Write a program that will give the distance between two points given in polar coordinates. See Problem 9 of Exercise 8.3 for a formula giving the distance between two points.

Review Exercises

In any problem in which both rectangular and polar coordinates are used, assume that the positive x-axis coincides with the polar axis.

1. In each of the following, the name of a point is given in rectangular coordinates. Give one name of the point in polar coordinates.
   a] $(1, 0)$ b] $(-3, 0)$ c] $(4, 4)$ d] $(-2, 2)$
   e] $(-\sqrt{3}, -1)$ f] $(\sqrt{2}, -\sqrt{2})$ g] $(0, 4)$ h] $(0, -3)$

2. Find the name in polar coordinates for the given points. Give r and θ (in radians) to two decimal places with $r > 0$ and $0 \leq \theta \leq 2\pi$.
   a] $(3, 4)$ b] $(-5, 1)$ c] $(3, -5)$ d] $(-2, -1)$

3. In each of the following, a name of a point is given in polar coordinates. Draw a diagram illustrating the point, and then give the name of the point in rectangular coordinates.
   a] $\left[4, \dfrac{\pi}{3}\right]$ b] $\left[-2, \dfrac{5\pi}{6}\right]$ c] $[4, \pi]$

   d] $\left[-1, \dfrac{9\pi}{4}\right]$ e] $\left[-3, \dfrac{-3\pi}{4}\right]$

4. Follow the instructions of Problem 3. Give answers to two decimal places.
   a] $\left[1, \dfrac{5\pi}{7}\right]$ b] $[-4, 3.47]$ c] $[2.3, 1.35]$

   d] $\left[-2, \dfrac{17\pi}{5}\right]$ e] $[3, -4.32]$

In Problems 5–12 an equation is given in polar coordinates. Draw a graph of the corresponding curve.

5. $r = \sin \theta$ **6.** $r^2 = 16$ **7.** $r = 2 \sin(-\theta)$

8. $r = \cos \theta - 1$ **9.** $r = 3 \sec \theta$ **10.** $r = \cos 2\theta$

11. $2r = \theta$, where $\theta \geq 0$ **12.** $r = \sin\left(\theta + \dfrac{\pi}{2}\right)$

13. Find an equation in polar coordinates that describes the same curve as $x^2 + y^2 = 4$. Draw a graph of the curve.

14. Find an equation in polar coordinates that describes the same curve as $x^2 + y^2 + y = \sqrt{x^2 + y^2}$. Draw a graph of the curve.

15. Draw a graph of $r(1 + \cos \theta) = 1$. Then find an equation in rectangular coordinates that describes the same curve.

16. Draw a graph of $r \sin \theta = 3$. Then find an equation in rectangular coordinates that describes the same curve.

Complex Numbers

9

9.1 Introduction

The system of real numbers is essential in the development of pure mathematics, as well as in applications of mathematics. However, even such a simple problem as finding the roots of the equation $x^2 + 1 = 0$ has no solution in the set of real numbers. To remedy this situation, we introduce a number denoted by i (also written as $\sqrt{-1}$) with the property of $i^2 = -1$. Thus the solutions to $x^2 + 1 = 0$ are i and $-i$. Similarly, the quadratic equation $x^2 - 4x + 5 = 0$, which can be written as $(x - 2)^2 + 1 = 0$, has no solution in the set of real numbers. However, $x - 2 = i$ and $x - 2 = -i$ give solutions $2 + i$ and $2 - i$. The solutions in these examples, called complex numbers, lead us to the following definition.

DEFINITION 9.1   The set **C** given by **C** = $\{u + vi \mid u$ and v are real numbers$\}$ is called the *set of complex numbers*.

Note that if we take $v = 0$ in Definition 9.1, the complex number $u + vi$ becomes simply u, a real number. Thus the set of real numbers **R** is a subset of **C**. If we take $u = 0$ and $v \neq 0$, the resulting complex number is vi. Such a number is called an *imaginary number*. We shall refer to u as the *real part* and v as the *imaginary part* of the complex number $u + vi$.

Basic properties of real numbers related to the four binary operations $(+, -, \times, \div)$ and the order relations ($<$ and $>$) are discussed in algebra courses. Now that the set **R** is extended to the set **C**, it is of interest to define addition, subtraction, multiplication, and division of complex numbers. Since **R** $\subset$ **C**, we want these definitions to be such that when they are applied to real numbers, the properties studied in earlier courses are still valid. In the system of complex numbers it is not possible to define an order relation similar to that of "less than" for the real numbers; that is, we do not talk about one complex number being less than a second unless both are real numbers.

First let us define equality of two complex numbers.

DEFINITION 9.2 Suppose a, b, c, and d are real numbers. We say that the complex numbers $a + bi$ and $c + di$ are equal if and only if $a = c$ and $b = d$.

DEFINITION 9.3 *Binary Operations on Complex Numbers*
Suppose $a + bi$ and $c + di$ are two complex numbers, where a, b, c, and d are real numbers. Their sum, difference, product, and quotient are given by the following.

Addition: $(a + bi) + (c + di) = (a + c) + (b + d)i;$

Subtraction: $(a + bi) - (c + di) = (a - c) + (b - d)i;$

Multiplication: $(a + bi) \cdot (c + di) = (ac - bd) + (ad + bc)i;$

Division: $\dfrac{a + bi}{c + di} = \left(\dfrac{ac + bd}{c^2 + d^2}\right) + \left(\dfrac{bc - ad}{c^2 + d^2}\right)i,$

where c and d are not both zero.

Addition and subtraction as stated in Definition 9.3 appear to be natural, but the definitions of multiplication and division require some explanation. These are motivated by thinking of $a + bi$ and $c + di$ as algebraic expressions to which we can apply the familiar rules of algebra, except that we replace i^2 by -1. Thus for multiplication we have

$$(a + bi) \cdot (c + di) = ac + adi + bci + bdi^2$$
$$= ac + (ad + bc)i + bd(-1)$$
$$= (ac - bd) + (ad + bc)i.$$

For division the first step in the following sequence involves multiplying the numerator and denominator by $c - di$; this gives the real number $c^2 + d^2$ in the denominator, as seen in the third step.

$$\frac{a + bi}{c + di} = \frac{(a + bi)(c - di)}{(c + di)(c - di)} = \frac{ac + bci - adi - bdi^2}{c^2 - d^2i^2}$$
$$= \frac{(ac + bd) + (bc - ad)i}{c^2 + d^2} = \left(\frac{ac + bd}{c^2 + d^2}\right) + \left(\frac{bc - ad}{c^2 + d^2}\right)i.$$

Actually, we shall follow the pattern above for multiplying or dividing two complex numbers, rather than substitute into Definition 9.3.

In the process of division described above, the numerator and denominator were multiplied by $c - di$. We call $c - di$ the *conjugate* of $c + di$.

DEFINITION 9.4   Suppose $z = x + yi$, where x and y are real numbers. The *conjugate* of z, denoted by $\bar{z}$, is given by $\bar{z} = x - yi$.

A complex number is in *standard form* if it is written as $a + bi$, where a and b are real numbers. For instance, $(1 + i)/i$ represents a complex number that can be written in standard form as follows:

$$\frac{1 + i}{i} = \frac{(1 + i)(-i)}{i(-i)} = \frac{-i - i^2}{-i^2} = \frac{-i + 1}{1} = 1 - i.$$

Square roots

The square root of a nonnegative real number b is defined to be a number x satisfying $x^2 = b$. For instance, the square root of 4 is a number x satisfying $x^2 = 4$; there are two such numbers, 2 and -2. We choose 2 as the *principal square root* and write $\sqrt{4} = 2$.

In a similar manner we can talk about the square root of a negative real number. For example, $\sqrt{-4}$ is a number z satisfying $z^2 = -4$. Since $(2i)^2 = 4i^2 = 4(-1) = -4$ and $(-2i)^2 = 4i^2 = 4(-1) = -4$, we see that $z = 2i$ or $z = -2i$. We choose $2i$ as the *principal square root* of -4 and write $\sqrt{-4} = 2i$.

In general, suppose b is a positive real number. Then

$$\boxed{\sqrt{-b} = \sqrt{b}\,i.}$$

Let us now recall the square root property for real numbers: If a and b are nonnegative real numbers, then $\sqrt{a}\sqrt{b} = \sqrt{ab}$. This can be generalized to the following.

Square Root Property
Suppose a and b are real numbers such that *not both are negative*. Then $\sqrt{a}\sqrt{b} = \sqrt{ab}$.

Note that the conclusion stated in the square root property is not valid if both a and b are negative numbers. For instance, if $a = -3$ and $b = -12$, then

$$\sqrt{-3}\sqrt{-12} = (\sqrt{3}i)(\sqrt{12}i) = \sqrt{3}\sqrt{12}i^2 = \sqrt{36}(-1) = -6;$$

whereas $\sqrt{(-3)(-12)} = \sqrt{36} = 6$.

Thus whenever we have $\sqrt{-b}$, where $b > 0$, it is good practice to write it as $\sqrt{b}i$ before performing algebraic manipulations. This is illustrated in the following example.

EXAMPLE 1   Evaluate $(2 + \sqrt{-3})(2 - \sqrt{-3})$.

Solution

$$(2 + \sqrt{-3})(2 - \sqrt{-3}) = (2 + \sqrt{3}i)(2 - \sqrt{3}i) = 2^2 - (\sqrt{3}i)^2$$
$$= 4 - 3i^2 = 4 + 3 = 7. \quad \blacksquare$$

From the above discussion, note that we can determine the *square root of any real number.* We could continue with the investigation of the square root of any complex number $a + bi$ where $b \neq 0$. For instance, it is a simple matter to show that

$$\left(\frac{\sqrt{2}}{2} + \frac{\sqrt{2}}{2}i\right)^2 = i \quad \text{(see Problem 6a)},$$

and so we could define $\sqrt{i}$ as the number $(\sqrt{2}/2) + (\sqrt{2}/2)i$. However, it is not in our interest to pursue this matter further at this point. See Section 9.5 for a discussion of roots of complex numbers.

EXAMPLE 2 Write each of the following as complex numbers in standard form.

a] $(3 + 4i) + (5 - 8i)$ b] $(2 - 3i) - (-4 + i)$
c] $(3 - 4i)(2 + i)$ d] $(1 - 3i) \div (3 + 4i)$

Solution

a] $(3 + 4i) + (5 - 8i) = (3 + 5) + (4 - 8)i = 8 - 4i.$
b] $(2 - 3i) - (-4 + i) = (2 + 4) + (-3 - 1)i = 6 - 4i.$
c] $(3 - 4i)(2 + i) = 6 + 3i - 8i - 4i^2 = 6 - 5i + 4 = 10 - 5i.$
d] $(1 - 3i) \div (3 + 4i) = \dfrac{1 - 3i}{3 + 4i} = \dfrac{(1 - 3i)(3 - 4i)}{(3 + 4i)(3 - 4i)} = \dfrac{3 - 13i + 12i^2}{9 - 16i^2}$

$$= \frac{3 - 13i - 12}{9 + 16} = \frac{-9 - 13i}{25} = \frac{-9}{25} - \frac{13}{25}i. \quad \blacksquare$$

EXAMPLE 3 Given that $f(z) = z^3 + 2z^2 - 3$, find $f(1 + i)$.

Solution

$$f(1 + i) = (1 + i)^3 + 2(1 + i)^2 - 3 = 1 + 3i + 3i^2 + i^3 + 2(1 + 2i + i^2) - 3$$
$$= 1 + 3i - 3 - i + 2 + 4i - 2 - 3 = -5 + 6i.$$

Note that we used the familiar rules of algebra, treating i as though it were a variable and replacing i^2 by -1. $\blacksquare$

EXAMPLE 4 Given that $z = 2 - i$, find the following.

a] $\bar{z}$ b] $z \cdot \bar{z}$ c] $\dfrac{\bar{z}}{z}$

Solution

a] $\bar{z} = 2 + i$.

b] $z \cdot \bar{z} = (2 - i)(2 + i) = 4 - i^2 = 4 + 1 = 5$.

c] $\dfrac{\bar{z}}{z} = \dfrac{2 + i}{2 - i} = \dfrac{(2 + i)(2 + i)}{(2 - i)(2 + i)} = \dfrac{4 + 4i + i^2}{4 - i^2} = \dfrac{4 + 4i - 1}{4 + 1} = \dfrac{3}{5} + \dfrac{4}{5}i$. ◢

EXAMPLE 5   Find the roots of $2z^2 + 2iz - 1 = 0$.

Solution   We apply the quadratic formula* to get

$$z = \frac{-2i \pm \sqrt{(2i)^2 - 4(2)(-1)}}{2(2)} = \frac{-2i \pm \sqrt{-4 + 8}}{4} = -\frac{1}{2}i \pm \frac{1}{2}.$$

Therefore the roots are given by $z = \frac{1}{2} - \frac{1}{2}i$ and $z = -\frac{1}{2} - \frac{1}{2}i$.   ■

EXAMPLE 6   Is $1 + \sqrt{3}i$ a zero of the polynomial $P(z) = z^2 - 2z + 4$?

Solution   To answer this question, we evaluate $P(1 + \sqrt{3}i)$ to see whether the result is equal to zero:

$$\begin{aligned}
P(1 + \sqrt{3}i) &= (1 + \sqrt{3}i)^2 - 2(1 + \sqrt{3}i) + 4 \\
&= (1 + 2\sqrt{3}i + 3i^2) - 2 - 2\sqrt{3}i + 4 \\
&= 1 + 2\sqrt{3}i - 3 - 2 - 2\sqrt{3}i + 4 \\
&= (1 - 3 - 2 + 4) + (2\sqrt{3} - 2\sqrt{3})i \\
&= 0 + 0i = 0.
\end{aligned}$$

Therefore the answer to the question is yes.   ■

EXERCISE 9.1

Express answers in $a + bi$ form, where a and b are real numbers.

1. Evaluate each of the following.
   a] i^3 **b]** i^6 **c]** i^{32} **d]** i^{17}
   e] $(-i)^3$ **f]** $(-i)^5$ **g]** $(-i)^8$ **h]** $(-i)^{17}$

2. Evaluate each of the following.
   a] $\dfrac{1}{i^4}$ **b]** $\dfrac{3 + i}{i^3}$ **c]** $2i^4 - 3i^{20}$ **d]** $\dfrac{1}{i(i - 1)}$

3. Evaluate each of the following.
   a] $\sqrt{9} \cdot \sqrt{16}$ **b]** $\sqrt{9}\sqrt{-16}$ **c]** $\sqrt{-9}\sqrt{-16}$
   d] $\dfrac{\sqrt{9}}{\sqrt{-16}}$ **e]** $\dfrac{\sqrt{-9}}{\sqrt{16}}$ **f]** $\dfrac{\sqrt{-9}}{\sqrt{-16}}$

---

* It can be shown that the quadratic formula is valid for quadratic equations whose coefficients are complex numbers.

4. Evaluate each of the following for $z = 1 - i$.

   a] z^2 b] $\dfrac{1}{z^2}$ c] $3z^2 - 2z^3$

   d] $z \cdot \bar{z}$ e] $(\bar{z})^3$ f] $z \div \bar{z}$

5. Given that $f(z) = 2 - 3z - z^2$, evaluate each of the following.

   a] $f(-2)$ b] $f(1 + i)$ c] $f\left(\dfrac{1}{\sqrt{2}} + \dfrac{1}{\sqrt{2}}i\right)$

6. Show that the following are true.

   a] $\left(\dfrac{1}{\sqrt{2}} + \dfrac{1}{\sqrt{2}}i\right)^2 = i$ b] $\left(\dfrac{1}{\sqrt{2}} - \dfrac{1}{\sqrt{2}}i\right)^2 = -i$

7. Show that the following are true.

   a] $\left(\dfrac{\sqrt{3}}{2} + \dfrac{1}{2}i\right)^3 = i$ b] $\left(\dfrac{1}{2} + \dfrac{\sqrt{3}}{2}i\right)^3 = -1$

8. Given that $f(z) = z^2 + iz - 3$, evaluate the following.
   a] $f(1 + i)$ b] $f(-3i)$

9. Express each of the following in standard $a + bi$ form.

   a] $\sqrt{-4} + (3 - 5\sqrt{-4})$ b] $(\sqrt{-48} + 2) - \sqrt{-27}$

   c] $\sqrt{-8}(2 + \sqrt{-2})$ d] $(1 + \sqrt{-8})(1 - \sqrt{-8})$

   e] $\dfrac{1}{1 - \sqrt{-9}}$ f] $\dfrac{\sqrt{-2}}{3 + \sqrt{-8}}$

10. In each of the following, determine the roots of the given equation.

    a] $z^2 - 3z + 4 = 0$ b] $3z^2 + z - 1 = 0$ c] $z^2 + 16 = 0$

11. Determine the roots of the given equations.

    a] $2z^2 - 3iz + 2 = 0$ b] $z^2 + 2iz + 3 = 0$

    c] $iz^2 - 3z + i = 0$ d] $2iz^2 + z + i = 0$

12. Given that $z = x + iy$, where x and y are real numbers, prove the following.

    a] The real part of z is equal to $\dfrac{z + \bar{z}}{2}$.

    b] The imaginary part of z is equal to $\dfrac{z - \bar{z}}{2i}$.

13. Determine real numbers x and y that satisfy the equation

    $$x - 3y - (3x + y)i = -7 + i.$$

14. Solve the equation $z - 3\bar{z} = 1 + i$ for z. (*Hint*: let $z = x + iy$; then find x and y.)

15. Determine all pairs of real numbers x, y such that $x^2 + 2x + yi = 2 + y + (8 - x)i$.

16. a] Is $1 + i$ a root of the equation $z^2 - z + 1 - i = 0$?
    b] Is $1 - i$ a root of the equation given in (a)?

17. Is $-3i$ a solution of the equation $2z^3 - z^2 + 18z - 9 = 0$?

18. Is $1 - \sqrt{5}i$ a zero of the polynomial given by $f(z) = z^3 - z^2 + 4z + 6$?

19. a] Is $1 - i$ a root of the equation $z^3 - 3z^2 + 2z - 1 - i = 0$?
    b] Is $1 + i$ a solution of the equation given in (a)?

20. a] Is $1 + \sqrt{3}i$ a solution of the equation $z^3 - 3z^2 + 6z - 4 = 0$?

   b] Is $1 - \sqrt{3}i$ a root of the equation given in (a)?

9.2 Geometric Representation of Complex Numbers

The set of complex numbers **C** is given by

$$\mathbf{C} = \{x + iy \mid x \text{ and } y \text{ are real numbers and } i^2 = -1\}.$$

We can establish a correspondence between **C** and the set of points in the plane in a natural way: With each complex number $x + iy$, associate the point (x, y) in the plane, and indicate this correspondence by

$$\boxed{x + iy \leftrightarrow (x, y).}$$

In this setting, the plane is referred to as the *complex plane*, where points are labeled either by (x, y) or by $x + iy$. The real numbers are associated with points on the x-axis $(x \leftrightarrow (x, 0))$, and the imaginary numbers correspond to points on the y-axis $(yi \leftrightarrow (0, y))$. Thus the x-axis is called the *real axis*, and the y-axis is referred to as the *imaginary axis*. Some examples of this correspondence are illustrated in Fig. 9.1.

FIGURE 9.1

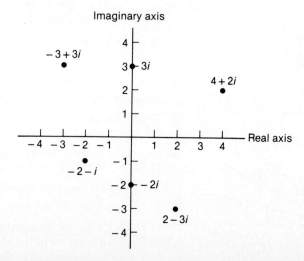

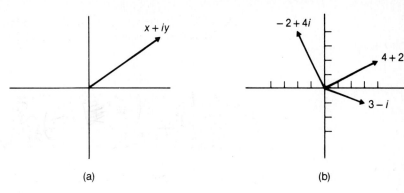

(a)   (b)

FIGURE 9.2

In some problems it is useful to associate each complex number with a *geometric vector*, as shown in Fig. 9.2(a), in which the origin is the initial point and $x + iy$ is the terminal point. Figure 9.2(b) illustrates some examples of this correspondence.

Representation of complex numbers by geometric vectors provides us with a convenient geometric interpretation of the sum of complex numbers. The sum $(a + bi) + (c + di)$ is associated with the geometric vector represented by the diagonal of the parallelogram illustrated in Fig. 9.3.

EXAMPLE 1   For each of the given complex numbers, show the corresponding point (x, y) in the complex plane. Also, draw the corresponding geometric vector.

a] $5 + 3i$ b] $-\frac{5}{2} + 3i$ c] $\pi - 2i$ d] $3i$

Solution   See Fig. 9.4.   ∎

FIGURE 9.3

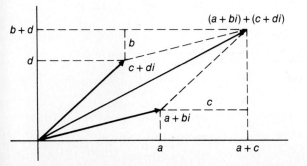

FIGURE 9.4

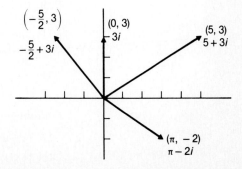

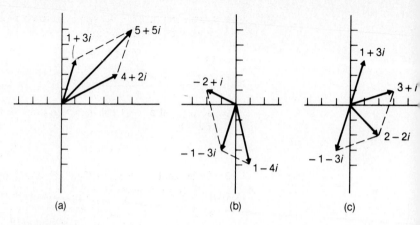

FIGURE 9.5

EXAMPLE 2 Illustrate each of the following by a diagram using geometric vectors.

a] $(4 + 2i) + (1 + 3i)$ **b]** $(1 - 4i) + (-2 + i)$ **c]** $(3 + i) - (1 + 3i)$

Solution   The solutions are shown in Fig. 9.5, where in (c) we use

$$(3 + i) - (1 + 3i) = (3 + i) + (-1 - 3i). \qquad \blacksquare$$

EXERCISE 9.2

In Problems 1–8, give the ordered pair of real numbers associated with the given complex number.

1. $3 + 5i$ **2.** $-3 + i$ **3.** $4i$ **4.** $\sqrt{5}$

5. $-\sqrt{-4} + 2i$ **6.** $1 - \pi i$ **7.** $i(1 - \sqrt{-4})$ **8.** $\dfrac{1}{1 - i}$

In Problems 9–12, give the complex number associated with the given ordered pair.

9. $(0, -4)$ **10.** $(5, 2)$ **11.** $(-4, -3)$ **12.** $(\sqrt{2}, -\sqrt{3})$

In Problems 13–16, illustrate the given complex number by drawing the associated geometric vector.

13. $-1 + 3i$ **14.** $-4 - 5i$ **15.** $-\sqrt{2} + i$ **16.** $\dfrac{1}{1 - 2i}$

In Problems 17–20, illustrate geometrically the given sum or difference.

17. $(2 + 3i) + (5 + i)$ **18.** $(1 - 3i) + (4 + 2i)$

19. $(4 - i) - (3 + 5i)$ **20.** $(2 - 3i) - (5 + 2i)$

21. Given that $z = 3 - 4i$, show on the same set of axes the points associated with the following.

a] z **b]** $-z$ **c]** $\bar{z}$ **d]** $\dfrac{z + \bar{z}}{2}$

e] $\dfrac{z - \bar{z}}{2}$ **f]** $\sqrt{z \cdot \bar{z}}$

22. Given that $z = -1 + i$, give the ordered pairs corresponding to the following.

a] z^2 **b]** $(\bar{z})^2$ **c]** $\dfrac{1}{z}$ **d]** $z^2 + z + 1$

23. Given that $z = -1/2 + (\sqrt{3}/2)i$, draw the geometric vector associated with the following.

a] z **b]** z^2 **c]** $\dfrac{1}{(\bar{z})^2}$ **d]** $\sqrt{z \cdot \bar{z}}$

24. Let $z = 2(1 + \sqrt{3}i)$. Express each of the following in standard form.
a] z^2 **b]** z^3 **c]** z^4 **d]** z^5

25. Suppose point $P(x, y)$ is associated with the complex number $x + iy$. State the conditions on x and y that characterize each of the following.
a] P is on the positive real axis.
b] P is on the imaginary axis.
c] P is in the first quadrant.
d] P is to the right of the imaginary axis.
e] P is below the real axis.

9.3 Trigonometric Form for Complex Numbers

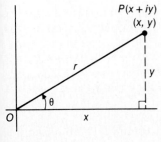

$P(x + iy)$
(x, y)

FIGURE 9.6

We continue the development of the preceding section, in which complex numbers are represented as points in the complex plane or as geometric vectors. Suppose $x + iy$ is associated with point $P(x, y)$ in the complex plane, as shown in Fig. 9.6. Let r denote the distance from the origin O to P, and θ the directed angle that OP makes with the positive real axis. Since $\cos \theta = x/r$ and $\sin \theta = y/r$, we have $x = r \cos \theta$ and $y = r \sin \theta$. Hence

$$x + iy = r \cos \theta + ir \sin \theta = r (\cos \theta + i \sin \theta).$$

The result, $r (\cos \theta + i \sin \theta)$, is called the *trigonometric form* or the *polar form* of the complex number $z = x + iy$. The real number r is given by

$$r = \sqrt{x^2 + y^2}$$

and is called the *absolute value* or the *modulus of* z; it is frequently denoted by |z|. Since r is the length of the geometric vector associated with z, it is sometimes referred to as the *length of* z.

The angle θ is called an *argument of* z and is denoted by $\theta = \arg z$. It is determined by the two equations

$$\sin \theta = \frac{y}{\sqrt{x^2 + y^2}} \quad \text{and} \quad \cos \theta = \frac{x}{\sqrt{x^2 + y^2}}$$

or by $\tan \theta = y/x$, with proper quadrant selection for θ.

Note that θ is not unique, since we can add or subtract any integral multiple of 2π (or 360°) to a given θ, and the resulting angle can be used in place of θ. The smallest nonnegative angle that can be used for θ is sometimes called the *principal argument of* z. Also, note that

$$z \cdot \overline{z} = (x + iy)(x - iy) = x^2 - i^2 y^2 = x^2 + y^2 = r^2,$$

and so

$$\boxed{r = \sqrt{z \cdot \overline{z}}.}$$

In the special case where P is the origin $(0, 0)$, we take $r = 0$ and do not specify any particular corresponding value of θ.

Representing complex numbers in trigonometric form is particularly useful in problems that involve multiplication or division.

Multiplication of complex numbers in polar form

Let $z_1 = r_1 (\cos \theta_1 + i \sin \theta_1)$ and $z_2 = r_2 (\cos \theta_2 + i \sin \theta_2)$ be complex numbers in polar form. Let us consider the product $z_1 \cdot z_2$, using the polar-form expressions:

$$
\begin{aligned}
z_1 \cdot z_2 &= r_1 (\cos \theta_1 + i \sin \theta_1) \cdot r_2 (\cos \theta_2 + i \sin \theta_2) \\
&= r_1 r_2 [(\cos \theta_1 \cos \theta_2 - \sin \theta_1 \sin \theta_2) + i (\sin \theta_1 \cos \theta_2 + \cos \theta_1 \sin \theta_2)] \\
&= r_1 r_2 [\cos (\theta_1 + \theta_2) + i \sin (\theta_1 + \theta_2)],
\end{aligned}
$$

where in the last step we used identities I-10 and I-11 from Chapter 4. Therefore

$$\boxed{z_1 \cdot z_2 = r_1 r_2 [\cos (\theta_1 + \theta_2) + i \sin (\theta_1 + \theta_2)].} \qquad \text{[9.1]}$$

From Eq. (9.1) a geometric interpretation of the product of two complex numbers can be given: $z_1 \cdot z_2$ is a complex number and has length r_1r_2 and argument $\theta_1 + \theta_2$. This is stated as follows:

$$\boxed{|z_1z_2| = |z_1| \cdot |z_2| \quad \text{and} \quad \arg(z_1z_2) = \arg z_1 + \arg z_2 .} \quad \textbf{[9.2]}$$

Note: The addition of arguments in the product of complex numbers suggests that a complex number can be expressed in exponential form. This is indeed true. In advanced mathematics courses, one learns that z can be expressed as $z = r \cdot e^{i\theta}$, where e is the irrational number $2.71828\ldots$ introduced in Section 10.1.

Division of complex numbers in polar form

Let z_1 and z_2 be complex numbers expressed in polar form as above, and suppose that $z_2 \neq 0$. Then

$$\boxed{\frac{z_1}{z_2} = \frac{r_1}{r_2}[\cos(\theta_1 - \theta_2) + i \sin(\theta_1 - \theta_2)].} \quad \textbf{[9.3]}$$

The proof of Eq. (9.3), which is similar to that of (9.1), is left to Problem 1 of Exercise 9.3.

From Eq. (9.3), note that the modulus and argument of z_1/z_2 are given by

$$\boxed{\left|\frac{z_1}{z_2}\right| = \frac{|z_1|}{|z_2|} \quad \text{and} \quad \arg\left(\frac{z_1}{z_2}\right) = \arg z_1 - \arg z_2 .} \quad \textbf{[9.4]}$$

In Examples 1, 2, and 3, complex numbers z_1, z_2, z_3, and z_4 are given by

$$z_1 = 1 + i, \qquad z_2 = \sqrt{3} - i, \qquad z_3 = -2 - 2\sqrt{3}i, \qquad z_4 = -3 + 4i.$$

EXAMPLE 1 Express the following in polar form.

a] z_1 b] z_2 c] z_3 d] z_4

Solution

a] $r_1 = |z_1| = \sqrt{1^2 + 1^2} = \sqrt{2}$, and $\theta_1 = \pi/4 = 45°$ (see Fig. 9.7(a)).
Therefore

$$z_1 = \sqrt{2}\left(\cos\frac{\pi}{4} + i \sin\frac{\pi}{4}\right)$$
$$= \sqrt{2}(\cos 45° + i \sin 45°).$$

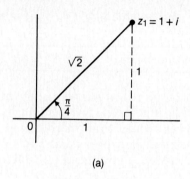

(a)

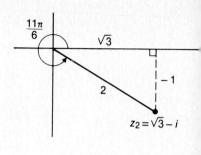

(b)

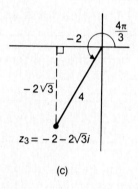

(c)

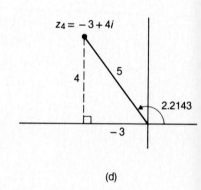

(d)

FIGURE 9.7

b] $r_2 = |z_2| = \sqrt{(\sqrt{3})^2 + (-1)^2} = \sqrt{4} = 2$, and $\theta_2 = 11\pi/6 = 330°$ (see Fig. 9.7(b)). Thus

$$z_2 = 2\left(\cos\frac{11\pi}{6} + i \sin\frac{11\pi}{6}\right)$$
$$= 2\,(\cos 330° + i \sin 330°).$$

c] From Fig. 9.7(c) we see that

$$z_3 = 4\left(\cos\frac{4\pi}{3} + i \sin\frac{4\pi}{3}\right)$$
$$= 4\,(\cos 240° + i \sin 240°).$$

d] From Fig. 9.7(d) we see that $\theta_4 = \mathrm{Cos}^{-1}(-\frac{3}{5}) \approx 2.2143 \approx 126.87°$. Therefore

$$z_4 \approx 5\,(\cos 2.2143 + i \sin 2.2143)$$
$$\approx 5\,(\cos 126.87° + i \sin 126.87°). \qquad \blacksquare$$

EXAMPLE 2   Find the following. Express each answer both in polar form and in rectangular form. Approximate answers to two decimal places.

a] $z_1 \cdot z_2$ **b]** $z_3 \cdot z_4$ **c]** $z_1 \cdot z_2 \cdot z_3$

Solution   In each case the formula given by Eq. (9.1) is used.

a] $z_1 \cdot z_2 = (\sqrt{2})(2)[\cos(45° + 330°) + i \sin(45° + 330°)]$

$= 2\sqrt{2}\,[\cos 375° + i \sin 375°]$

$= 2\sqrt{2}\,(\cos 15° + i \sin 15°)$ (polar form)

$\approx 2.73 + 0.73i$ (rectangular form).

b] $z_3 \cdot z_4 = (4)(5)[\cos(240° + 126.87°) + i \sin(240° + 126.87°)]$

$= 20\,[\cos 366.87° + i \sin 366.87°]$

$= 20\,(\cos 6.87° + i \sin 6.87°)$ (polar form)

$\approx 19.86 + 2.39i$ (rectangular form).

c] $z_1 \cdot z_2 \cdot z_3 = (\sqrt{2})(2)(4)\left[\cos\left(\dfrac{\pi}{4} + \dfrac{11\pi}{6} + \dfrac{4\pi}{3}\right) + i \sin\left(\dfrac{\pi}{4} + \dfrac{11\pi}{6} + \dfrac{4\pi}{3}\right)\right]$

$= 8\sqrt{2}\left[\cos\dfrac{41\pi}{12} + i \sin\dfrac{41\pi}{12}\right]$

$= 8\sqrt{2}\left(\cos\dfrac{17\pi}{12} + i \sin\dfrac{17\pi}{12}\right)$ (polar form)

$\approx -2.93 - 10.93i$ (rectangular form).   ■

EXAMPLE 3   Evaluate the following. Express each answer in both polar form and rectangular form.

a] $\dfrac{z_1}{z_2}$ **b]** $\dfrac{z_3}{z_4}$

Solution   We use Eq. (9.3):

a] $\dfrac{z_1}{z_2} = \dfrac{\sqrt{2}}{2}\,[\cos(45° - 330°) + i \sin(45° - 330°)]$

$= \dfrac{\sqrt{2}}{2}\,[\cos(-285°) + i \sin(-285°)]$ (polar form)

$= \dfrac{\sqrt{2}}{2}\,[\cos 285° - i \sin 285°]$

$\approx 0.18 + 0.68i$ (rectangular form).

b] $\dfrac{z_3}{z_4} \approx \dfrac{4}{5}\left[\cos\left(\dfrac{4\pi}{3} - 2.2143\right) + i \sin\left(\dfrac{4\pi}{3} - 2.2143\right)\right]$

$\approx \dfrac{4}{5}\,[\cos(1.9745) + i \sin(1.9745)]$ (polar form)

$\approx -0.31 + 0.74i$ (rectangular form).   ■

EXAMPLE 4   Express $3(\cos 60° - i \sin 60°)$ in polar form.

Solution

Method 1

For a complex number to be in polar form, it must be expressed as $r(\cos \theta + i \sin \theta)$, where $r \geq 0$. The given number is not in polar form because of the minus sign. However, since $\cos(-60°) = \cos 60°$ and $\sin(-60°) = -\sin 60°$, we can write

$$3(\cos 60° - i \sin 60°) = 3[\cos(-60°) + i \sin(-60°)],$$

which is in polar form. Since $-60°$ and $300°$ are coterminal angles this can also be written as $3(\cos 300° + i \sin 300°)$.

Method 2

Write the given number in rectangular form first.

$$3(\cos 60° - i \sin 60°) = \frac{3}{2}(1 - \sqrt{3}i)$$

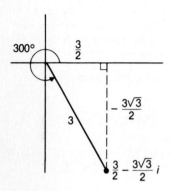

From Fig. 9.8 we see that $3(\cos 300° + i \sin 300°)$ is a polar form of the given number.   ∎

FIGURE 9.8

EXAMPLE 5   Express $-4(\cos 120° + i \sin 120°)$ in polar form.

Solution   The given number is not in polar form, because the -4 is not an acceptable value for r (we require $r \geq 0$). The given number can be written as follows:

$$
\begin{aligned}
-4(\cos 120° + i \sin 120°) &= 4[-\cos 120° + i(-\sin 120°)] \\
&= 4[\cos(180° + 120°) + i \sin(180° + 120°)] \\
&= 4(\cos 300° + i \sin 300°).
\end{aligned}
$$

Note that we used identities $\cos(180° + \theta) = -\cos \theta$ and $\sin(180° + \theta) = -\sin \theta$ with $\theta = 120°$. Thus

$$-4(\cos 120° + i \sin 120°) = 4(\cos 300° + i \sin 300°),$$

which is in polar form.

We encourage the reader to follow Method 2 of Example 4 to get the solution.   ∎

EXAMPLE 6   Express $3(\sin 47° - i \cos 47°)$ in polar form.

Solution

$$
\begin{aligned}
3(\sin 47° - i \cos 47°) &= 3[\cos(270° + 47°) + i \sin(270° + 47°)] \\
&= 3(\cos 317° + i \sin 317°).
\end{aligned}
$$

Here we used identities $\cos(270° + \theta) = \sin\theta$ and $\sin(270° + \theta) = -\cos\theta$ with $\theta = 47°$. Thus the polar form of the given number is

$$3(\cos 317° + i \sin 317°). \quad \blacksquare$$

EXERCISE 9.3

In each of the problems of this exercise, give answers in exact form whenever it is reasonable to do so; otherwise, use a calculator and state the results in decimal form (two places for degree measure, four places for radian measure).

1. Given that z_1 and z_2 are complex numbers expressed in polar form, prove that

$$\frac{z_1}{z_2} = \frac{r_1}{r_2}[\cos(\theta_1 - \theta_2) + i \sin(\theta_1 - \theta_2)].$$

2. Express each of the given numbers in polar form.
   a] -3 b] $1 - i$ c] $-i$ d] $1 + \sqrt{3}i$

3. Express the following in polar form.
   a] $-3i$ b] $3 - 4i$ c] $i^5 - i^4$ d] $12 - 5i$

4. Express the following in polar form.
   a] $-3 - 3i$ b] $5i^2 - 2i - 3$ c] $\frac{1}{i}$ d] $\frac{1}{i - i^2}$

5. Express the following in rectangular form.
   a] $3(\cos 45° + i \sin 45°)$ b] $5(\cos 180° + i \sin 180°)$
   c] $\cos\frac{4\pi}{3} + i \sin\frac{4\pi}{3}$

6. Express the following in rectangular form.
   a] $\cos\left(-\frac{7\pi}{6}\right) + i \sin\left(-\frac{7\pi}{6}\right)$ b] $\cos 450° + i \sin 450°$
   c] $3(\cos 137° + i \sin 137°)$

7. Determine why the given number is not in polar form. Then express it in polar form.
   a] $4(\cos 45° - i \sin 45°)$ b] $-3(\cos 300° + i \sin 300°)$
   c] $-\cos\frac{5\pi}{6} + i \sin\frac{5\pi}{6}$

8. Express the following in polar form.
   a] $3\left(-\cos\frac{\pi}{6} + i \sin\frac{\pi}{6}\right)$ b] $-5(\cos 40° - i \sin 40°)$
   c] $-\cos 120° - i \sin 120°$

In Problems 9–12, perform the indicated operations and express answers in (a) polar form, (b) rectangular form. (*Hint*: Write numbers in polar form first and then use Eq. (9.1) or Eq. (9.3).)

9. $(\cos 15° + i \sin 15°) \cdot (\cos 30° + i \sin 30°)$

10. $4\left(\cos 47° - i \sin 47°\right) \cdot \left(\cos 43° - i \sin 43°\right)$

11. $\dfrac{8\left(\cos 150° + i \sin 150°\right)}{4\left(\cos 30° + i \sin 30°\right)}$

12. $\dfrac{\cos 50° + i \sin 50°}{\cos 80° - i \sin 80°}$

In Problems 13–15, let $z_1 = 3\left(\cos 210° - i \sin 210°\right)$ and $z_2 = 6\left(\sin 60°\right.$ $i \cos 60°)$. Evaluate the given expressions by using Eq. (9.1) or Eq. (9.3).

13. $z_1 \cdot z_2$ **14.** $z_2 \div z_1$ **15.** $\dfrac{1}{z_2}$

In Problems 16–20, let $z_1 = \sqrt{3} + i$ and $z_2 = -2 + 2i$. Express each of the give numbers in polar form.

16. a] z_1 **b]** z_2 **17. a]** $\overline{z}_1$ **b]** $\overline{z}_2$

18. a] $z_1 \cdot z_2$ **b]** $\overline{z}_1 \cdot \overline{z}_2$ **19. a]** $z_1 \div z_2$ **b]** $\overline{z}_1 \div \overline{z}_2$

20. a] $\dfrac{1}{z_1}$ **b]** $\dfrac{1}{z_2}$

21. Given that $z = r\left(\cos \theta + i \sin \theta\right)$ represents a complex number in pol form, show that the following are true.

a] $z^2 = r^2\left(\cos 2\theta + i \sin 2\theta\right)$ **b]** $z^3 = r^3\left(\cos 3\theta + i \sin 3\theta\right)$

22. Given that $z = r\left(\cos \theta + i \sin \theta\right)$ represents a complex number in pol form and $r \neq 0$, show that the following are true.

a] $\dfrac{1}{z} = \dfrac{1}{r}\left[\cos\left(-\theta\right) + i \sin\left(-\theta\right)\right]$ **b]** $\dfrac{1}{z^2} = \dfrac{1}{r^2}\left[\cos\left(-2\theta\right) + i \sin\left(-2\theta\right)\right]$

23. Use Problem 21 to evaluate the following.
a] $\left(\sqrt{2} - \sqrt{2}i\right)^2$ **b]** $\left(1 + \sqrt{3}i\right)^3$

24. Use Problem 22 to evaluate the following.

a] $\dfrac{1}{1 + i}$ **b]** $\dfrac{1}{\left(\sqrt{3} - i\right)^2}$

9.4 DeMoivre's Theorem

Suppose z is a complex number in polar form, $z = r\left(\cos \theta + i \sin \theta\right)$ Applying Eq. (9.1) to the special case where both z_1 and z_2 are taken to be z gives

$$z \cdot z = r \cdot r\left[\cos\left(\theta + \theta\right) + i \sin\left(\theta + \theta\right)\right],$$
$$z^2 = r^2\left(\cos 2\theta + i \sin 2\theta\right).$$

If Eq. (9.1) is applied again with $z_1 = z$ and $z_2 = z^2$, we get

$$z^3 = r^3\left(\cos 3\theta + i \sin 3\theta\right).$$

This suggests that in general

$$z^n = r^n (\cos n\theta + i \sin n\theta)$$ **[9.5]**

for each positive integer n. This is indeed a true statement; the reader is asked to give a formal proof in Problem 16 of Exercise 9.4.

Taking $r = 1$ in Eq. (9.5) gives the special case

$$(\cos \theta + i \sin \theta)^n = \cos n\theta + i \sin n\theta$$

for each positive integer n. This is known as *DeMoivre's theorem*.*

Equation (9.5) is stated for n a positive integer. For exponents that are not positive integers we follow a pattern similar to that already encountered in algebra. We first define z^k where k is zero, then for k a negative integer.

DEFINITION 9.5   *Zero Exponent*
If $z \neq 0$, then $z^0 = 1$.

Negative-integer Exponent

If n is any positive integer and $z \neq 0$, then $z^{-n} = \dfrac{1}{z^n}$.

We now investigate z^{-n}, where n is a positive integer. Let $z = r (\cos \theta + i \sin \theta)$. Then

$$z^{-n} = \frac{1}{z^n} \quad \text{(by Definition 9.5)}$$

$$= \frac{1}{r^n (\cos n\theta + i \sin n\theta)} \quad \text{(by Eq. (9.5))}$$

$$= \frac{1}{r^n} \left(\frac{\cos 0 + i \sin 0}{\cos n\theta + i \sin \theta} \right) \quad \text{(since } 1 = \cos 0 + i \sin 0)$$

$$= r^{-n} [\cos (-n\theta) + i \sin (-n\theta)] \quad \text{(by Eq. 9.3)).}$$

Thus we have

$$z^{-n} = r^{-n} [\cos (-n\theta) + i \sin (-n\theta)].$$

This is precisely Eq. (9.5) for negative integer exponents. Equation (9.5) also holds for $n = 0$, since $z^0 = 1$ and

$$r^0 [\cos (0 \cdot \theta) + i \sin (0 \cdot \theta)] = 1 \cdot (\cos 0 + i \sin 0) = 1.$$

---

* Named after the French-born English mathematician Abraham DeMoivre (1667–1754).

Therefore the result given by Eq. (9.5) is generalized to the following

> If $z = r(\cos\theta + i\sin\theta)$ and n is any integer, then
> $$z^n = r^n(\cos n\theta + i\sin n\theta).$$ [9.6]

EXAMPLE 1   Express each of the following as a complex number in both polar form and rectangular form.

a] $(1 + i)^6$ **b]** $(-1 + \sqrt{3}i)^8$ **c]** $(3 - 4i)^4$

Solution

a] We first express $1 + i$ in polar form and then use the formula given by Eq. (9.6):

$$(1 + i)^6 = [\sqrt{2}(\cos 45° + i\sin 45°)]^6$$
$$= (\sqrt{2})^6[\cos(6 \cdot 45°) + i\sin(6 \cdot 45°)]$$
$$= 8(\cos 270° + i\sin 270°) \quad \text{(polar form)}$$
$$= 8[0 + i(-1)] = -8i \quad \text{(rectangular form)}.$$

b] $(-1 + \sqrt{3}i)^8 = \left[2\left(\cos\dfrac{2\pi}{3} + i\sin\dfrac{2\pi}{3}\right)\right]^8$

$$= 2^8\left[\cos\left(8 \cdot \dfrac{2\pi}{3}\right) + i\sin\left(8 \cdot \dfrac{2\pi}{3}\right)\right]$$

$$= 256\left[\cos\dfrac{16\pi}{3} + i\sin\dfrac{16\pi}{3}\right]$$

$$= 256\left[\cos\left(4\pi + \dfrac{4\pi}{3}\right) + i\sin\left(4\pi + \dfrac{4\pi}{3}\right)\right]$$

$$= 256\left[\cos\dfrac{4\pi}{3} + i\sin\dfrac{4\pi}{3}\right] \quad \text{(polar form)}$$

$$= 256\left[-\dfrac{1}{2} + i\left(-\dfrac{\sqrt{3}}{2}\right)\right]$$

$$= -128 - 128\sqrt{3}i \quad \text{(rectangular form)}.$$

FIGURE 9.9

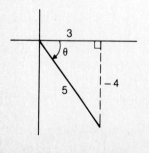

c] $(3 - 4i)^4 = [r(\cos\theta + i\sin\theta)]^4 = r^4(\cos 4\theta + i\sin 4\theta)$, where $r = 5$ and $\theta = \text{Sin}^{-1}(-4/5)$ (see Fig. 9.9). Using a calculator, we evaluate

$$4\theta = 4\,\text{Sin}^{-1}\left(-\tfrac{4}{5}\right) \approx -212.52°.$$

Therefore

$$(3 - 4i)^4 = 625[\cos(-212.52°) + i\sin(-212.52°)] \quad \text{(polar form)}$$
$$= -527 + 336i \quad \text{(rectangular form)}. \quad \blacksquare$$

EXAMPLE 2 Evaluate each of the following, and express answers in both polar form and rectangular form.

a] $[2(\cos 22°30' + i \sin 22°30')]^4$ b] $(\cos 45° - i \sin 45°)^5$

Solution

a] Using Eq. (9.6) gives

$[2(\cos 22°30' + i \sin 22°30')]^4 = 2^4[\cos 4(22°30') + i \sin 4(22°30')]$
$$= 16(\cos 90° + i \sin 90°) \quad \text{(polar form)}$$
$$= 16i \quad \text{(rectangular form)}.$$

b] First express $\cos 45° - i \sin 45°$ in polar form as
$$\cos 45° - i \sin 45° = \cos(-45°) + i \sin(-45°).$$

Applying Eq. (9.6), we have

$(\cos 45° - i \sin 45°)^5 = [\cos(-45°) + i \sin(-45°)]^5$
$$= \cos 5(-45°) + i \sin 5(-45°)$$
$$= \cos(-225°) + i \sin(-225°) \quad \text{(polar form)}$$
$$= -\frac{\sqrt{2}}{2} + \frac{\sqrt{2}}{2}i \quad \text{(rectangular form)}. \quad \blacksquare$$

EXAMPLE 3 Express $\sin 4\theta$ and $\cos 4\theta$ as identities in terms of $\sin \theta$ and $\cos \theta$.

Solution Substituting $n = 4$ into DeMoivre's theorem gives
$$(\cos \theta + i \sin \theta)^4 = \cos 4\theta + i \sin 4\theta.$$

Using the binomial expansion on the left-hand side of this equation, we get

$\cos^4 \theta + 4(\cos^3 \theta \sin \theta)i + 6(\cos^2 \theta \sin^2 \theta)i^2$
$$+ 4(\cos \theta \sin^3 \theta)i^3 + (\sin^4 \theta)i^4 = \cos 4\theta + i \sin 4\theta.$$

Now use $i^2 = -1$, $i^3 = i^2 \cdot i = -i$, and $i^4 = i^2 \cdot i^2 = (-1)(-1) = 1$ and collect real and imaginary terms to get

$[\cos^4 \theta - 6 \cos^2 \theta \sin^2 \theta + \sin^4 \theta] + [4 \cos^3 \theta \sin \theta - 4 \cos \theta \sin^3 \theta]i$
$$= \cos 4\theta + i \sin 4\theta.$$

Using the definition of equality of two complex numbers (see Definition 9.2), we get

$$\sin 4\theta = 4 \cos^3 \theta \sin \theta - 4 \cos \theta \sin^3 \theta,$$
$$\cos 4\theta = \cos^4 \theta - 6 \cos^2 \theta \sin^2 \theta + \sin^4 \theta.$$

These are identities. $\blacksquare$

By using the technique illustrated in Example 3 we can solve the general problem of determining identities in which $\sin n\theta$ and $\cos n\theta$ are expressed in terms of $\sin \theta$ and $\cos \theta$.

EXERCISE 9.4

In the following problems, give answers in exact form whenever it is reasonable to do so; otherwise, state results in decimal form, with numbers rounded off to two decimal places and angles to two places for degree measure and four places for radian measure. Express answers in both polar form and rectangular form.

In Problems 1–8, perform the indicated operations.

1. a] $(\cos 30° + i \sin 30°)^5$ **b]** $[2 (\cos(-45°) + i \sin(-45°))]^4$
   c] $(\cos 40° + i \sin 40°)^{-3}$

2. a] $(\cos 47° + i \sin 47°)^6$ **b]** $\left[3 \left(\cos \dfrac{\pi}{3} + i \sin \dfrac{\pi}{3}\right)\right]^4$
   c] $[\cos(-20°) + i \sin(-20°)]^{-6}$

3. a] $[2 (\cos 150° - i \sin 150°)]^3$ **b]** $\dfrac{16}{[2 (\cos 45° - i \sin 45°)]^4}$

4. a] $[-3 (\cos 20° + i \sin 20°)]^4$ **b]** $\dfrac{81}{[-3 (\cos (\pi/12) + i \sin (\pi/12))]^4}$

5. a] $(-1 + i)^8$ **b]** $(\sqrt{3} - i)^4$ **c]** $(1 + i)^{-3}$

6. a] $(\sqrt{2} + \sqrt{2}i)^4$ **b]** $\dfrac{1}{(1 - \sqrt{3}i)^6}$ **c]** $(2 + i)^6$

7. a] $(-1 + i)^4 \cdot (1 + \sqrt{3}i)^6$ **b]** $\dfrac{(2 + 2i)^4}{(\sqrt{3} + i)^3}$

8. a] $(1 - i)^{-3} \cdot (1 + i)^4$ **b]** $(2 - 3i)^2 \cdot (4 + 3i)^4$

In Problems 9–12, $z = 1 - i$ and $w = -\sqrt{3} + i$. Evaluate the given expression.

9. $z^4 - z$ **10.** $z^3 \cdot w^4$ **11.** $z^4 - w^4$

12. $z^4 + z^3 + z^2 + z + 1$ (*Hint*: The identity $(z - 1)(z^4 + z^3 + z^2 + z + 1) = z^5 - 1$ may be useful.)

13. Given that $f(z) = z^4 - 2z^3 + z$, find the following.
   a] $f(i)$ **b]** $f(-1 + i)$

14. In Eq. (9.6), take $n = 2$, $r = 1$ and get double-angle identities I-25 and I-26 given in Section 4.6.

15. Express $\sin 3\theta$ and $\cos 3\theta$ as identities in terms of $\sin \theta$ and $\cos \theta$ (see Example 3 in Section 9.4).

16. Prove that $z^n = [r (\cos \theta + i \sin \theta)]^n = r^n (\cos n\theta + i \sin n\theta)$ for each positive integer n. (*Hint*: Use mathematical induction.)

9.5 Roots of Complex Numbers

In Section 9.1 we discussed the problem of determining *the square root of any real number* and arrived at the following: Suppose c is a positive real number. Then $\sqrt{c}$ is the positive solution of $x^2 = c$, and $\sqrt{-c}$ is equal to the imaginary number $\sqrt{c}\,i$. In this section we are interested in the general problem of determining the nth roots of any complex number $z = a + bi$, where n is an integer greater than or equal to 2.

DEFINITION 9.6   Suppose z is a given complex number. The *nth roots of z* are the complex numbers w satisfying the equation $w^n = z$.

Although we were able to talk about *the* square root of a real number, we shall make no attempt to define *the* nth root of z in general. Definition 9.6 refers to "the nth roots of z," and we do not select a particular solution of $w^n = z$ and call it the principal value or the nth root of z as we did for the square root of a real number.

Let us proceed with the problem of solving the equation $w^n = z$, where z is a given complex number. For the trivial case of $z = 0$ the solution is $w = 0$. Hence in the following we shall assume that $z \neq 0$. Suppose z and w are expressed in polar form as

$$z = r\,(\cos\theta + i\,\sin\theta),$$
$$w = R\,(\cos\alpha + i\,\sin\alpha).$$

Then $w^n = z$ becomes

$$[R\,(\cos\alpha + i\,\sin\alpha)]^n = r\,(\cos\theta + i\,\sin\theta).$$

Applying Eq. (9.6) to the left-hand side gives

$$R^n\,(\cos n\alpha + i\,\sin n\alpha) = r\,(\cos\theta + i\,\sin\theta).$$

From the definition of equality of two complex numbers (see Definition 9.2) it follows that

$$R^n \cos n\alpha = r\cos\theta \quad\text{and}\quad R^n \sin n\alpha = r\sin\theta.$$

Solving this pair of simultaneous equations for R and α (see Problem 21 of Exercise 9.5) gives

$$R = r^{1/n} = \sqrt[n]{r} \quad\text{and}\quad \alpha = \frac{\theta + k\cdot 2\pi}{n} = \frac{\theta}{n} + \frac{2\pi k}{n},$$

where k is any integer. Therefore $w^n = z$ has solutions given by

$$\boxed{w_k = r^{1/n}\left[\cos\left(\frac{\theta}{n} + \frac{2\pi k}{n}\right) + i\,\sin\left(\frac{\theta}{n} + \frac{2\pi k}{n}\right)\right].}$$ [9.7]

If we let k take on various integral values, we see that w_0, w $w_2, \ldots, w_{n-1}$ will be n *distinct* complex numbers. These are given b

$$
\begin{aligned}
w_0 &= r^{1/n} \left[\cos \frac{\theta}{n} + i \sin \frac{\theta}{n} \right], \\
w_1 &= r^{1/n} \left[\cos \left(\frac{\theta}{n} + \frac{2\pi}{n} \right) + i \sin \left(\frac{\theta}{n} + \frac{2\pi}{n} \right) \right], \\
w_2 &= r^{1/n} \left[\cos \left(\frac{\theta}{n} + \frac{4\pi}{n} \right) + i \sin \left(\frac{\theta}{n} + \frac{4\pi}{n} \right) \right], \\
&\vdots \\
w_{n-1} &= r^{1/n} \left[\cos \left(\frac{\theta}{n} + \frac{2(n-1)\pi}{n} \right) + i \sin \left(\frac{\theta}{n} + \frac{2(n-1)\pi}{n} \right) \right].
\end{aligned}
$$

[9.8

Suppose we evaluate w_n by replacing k by n in Eq. (9.7). This give

$$
\begin{aligned}
w_n &= r^{1/n} \left[\cos \left(\frac{\theta}{n} + \frac{2\pi n}{n} \right) + i \sin \left(\frac{\theta}{n} + \frac{2\pi n}{n} \right) \right] \\
&= r^{1/n} \left[\cos \left(\frac{\theta}{n} + 2\pi \right) + i \sin \left(\frac{\theta}{n} + 2\pi \right) \right] \\
&= r^{1/n} \left[\cos \frac{\theta}{n} + i \sin \frac{\theta}{n} \right] = w_0.
\end{aligned}
$$

Thus $w_n = w_0$.

In a similar manner we can show that each value of k for which $k \geq n$ or $k < 0$ will give a w_k that is already included in Eqs. (9.8).

Geometrically, the n numbers given by (9.8) are located on the circle with center at the origin and radius $\sqrt[n]{r}$; they are equally spaced around the circle with the angle between any two consecutive values being $2\pi/n$. These are shown in Fig. 9.10.

FIGURE 9.10

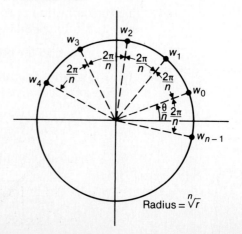

Radius $= \sqrt[n]{r}$

EXAMPLE 1 Find the four roots of $z^4 + 1 = 0$.

Solution We wish to solve for the roots of the equation $z^4 = -1$. First we express -1 in polar form: $-1 = \cos \pi + i \sin \pi$. Substituting $n = 4$, $r = 1$, and $\theta = \pi$ into Eq. (9.7) gives

$$w_k = \cos \left(\frac{\pi}{4} + \frac{2\pi k}{4} \right) + i \sin \left(\frac{\pi}{4} + \frac{2\pi k}{4} \right)$$

$$= \cos \left(\frac{\pi}{4} + k \cdot \frac{\pi}{2} \right) + i \sin \left(\frac{\pi}{4} + k \cdot \frac{\pi}{2} \right).$$

Replacing k by 0, 1, 2, and 3 gives the four roots w_0, w_1, w_2, and w_3, respectively:

$$w_0 = \frac{\sqrt{2}}{2} + \frac{\sqrt{2}}{2} i, \qquad w_1 = -\frac{\sqrt{2}}{2} + \frac{\sqrt{2}}{2} i,$$

$$w_2 = -\frac{\sqrt{2}}{2} - \frac{\sqrt{2}}{2} i, \qquad w_3 = \frac{\sqrt{2}}{2} - \frac{\sqrt{2}}{2} i. \qquad \blacksquare$$

EXAMPLE 2 Solve the equation $z^4 - 2z^2 + 2 = 0$.

Solution The given equation is quadratic in z^2. Solving for z^2 by use of the quadratic formula gives

$$z^2 = \frac{-(-2) \pm \sqrt{(-2)^2 - 4(1)(2)}}{2(1)} = \frac{2 \pm \sqrt{-4}}{2} = 1 \pm i,$$

and

$$1 + i = \sqrt{2} \,(\cos 45° + i \sin 45°),$$
$$1 - i = \sqrt{2} \,(\cos 315° + i \sin 315°).$$

Using (9.8) with $n = 2$, we get the following solutions: $z^2 = 1 + i$ gives

$$w_0 = (\sqrt{2})^{1/2} \left[\cos \frac{45°}{2} + i \sin \frac{45°}{2} \right]$$

$$= \sqrt[4]{2} \,(\cos 22.5° + i \sin 22.5°) \approx 1.10 + 0.46i,$$

$$w_1 = (\sqrt{2})^{1/2} \left[\cos \left(\frac{45°}{2} + \frac{360°}{2} \right) + i \sin \left(\frac{45°}{2} + \frac{360°}{2} \right) \right]$$

$$= \sqrt[4]{2} \,(\cos 202.5° + i \sin 202.5°) \approx -1.10 - 0.46i;$$

$z^2 = 1 - i$ gives

$$w_0' = (\sqrt{2})^{1/2} \left[\cos \frac{315°}{2} + i \sin \frac{315°}{2} \right]$$

$$= \sqrt[4]{2} \,(\cos 157.5° + i \sin 157.5°) \approx -1.10 + 0.46i,$$

$$w_1' = (\sqrt{2})^{1/2} \left[\cos \left(\frac{315°}{2} + \frac{360°}{2} \right) + i \sin \left(\frac{315°}{2} + \frac{360°}{2} \right) \right]$$

$$= \sqrt[4]{2} \,(\cos 337.5° + i \sin 337.5°) \approx 1.10 - 0.46i.$$

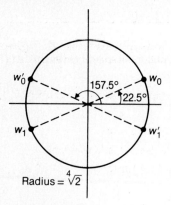

Radius = $\sqrt[4]{2}$

FIGURE 9.11

Therefore the solution set for the given equation is

$$\{1.10 + 0.46i, \quad -1.10 + 0.46i, \quad -1.10 - 0.46i, \quad 1.10 - 0.46i\},$$

where the numbers are given to two decimal places. The numbers in the solution set are shown in Fig. 9.11, in which the radius of the circle is $\sqrt[4]{2}$. ■

EXAMPLE 3 Find the square roots of $-3 - 4i$.

Solution We want to solve the equation $w^2 = -3 - 4i$. First express $-3 - 4i$ in polar form:

$$-3 - 4i = 5(\cos \theta + i \sin \theta),$$

where θ is the angle shown in Fig. 9.12. Substituting into (9.8) with $n = 2$ gives

$$w_0 = 5^{1/2}\left(\cos\frac{\theta}{2} + i \sin\frac{\theta}{2}\right),$$

$$w_1 = 5^{1/2}\left[\cos\left(\frac{\theta}{2} + \frac{2\pi}{2}\right) + i \sin\left(\frac{\theta}{2} + \frac{2\pi}{2}\right)\right]$$

$$= 5^{1/2}\left[\cos\left(\frac{\theta}{2} + \pi\right) + i \sin\left(\frac{\theta}{2} + \pi\right)\right]$$

$$= 5^{1/2}\left[-\cos\frac{\theta}{2} - i \sin\frac{\theta}{2}\right]$$

$$= -5^{1/2}\left(\cos\frac{\theta}{2} + i \sin\frac{\theta}{2}\right) = -w_0.$$

FIGURE 9.12

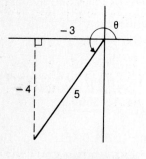

To determine w_0 explicitly, we can use a calculator as follows. From Fig. 9.12, $\text{Sin}^{-1} 4/5$ is the reference angle for θ, and so $\theta = \pi + \text{Sin}^{-1} 4/5$. Evaluate $\theta/2$, store it and then get $w_0 = -1 + 2i$.

However, we can also get w_0 by using half-angle identities as follows. Since $\pi < \theta < 3\pi/2$, then $\pi/2 < \theta/2 < 3\pi/4$, and so $\theta/2$ is an angle in the second quadrant. Therefore $\cos(\theta/2)$ is negative, and $\sin(\theta/2)$ is positive. Since $\cos \theta = -3/5$ (see Fig. 9.12),

$$\cos\frac{\theta}{2} = -\sqrt{\frac{1 + \cos\theta}{2}} = -\sqrt{\frac{1 + (-3/5)}{2}} = -\frac{1}{\sqrt{5}},$$

$$\sin\frac{\theta}{2} = \sqrt{\frac{1 - \cos\theta}{2}} = \sqrt{\frac{1 - (-3/5)}{2}} = \frac{2}{\sqrt{5}}.$$

Thus we have

$$w_0 = \sqrt{5}\left(\cos\frac{\theta}{2} + i \sin\frac{\theta}{2}\right) = \sqrt{5}\left(-\frac{1}{\sqrt{5}} + \frac{2i}{\sqrt{5}}\right) = -1 + 2i,$$

$$w_1 = -w_0 = 1 - 2i.$$

Therefore the square roots of $-3 - 4i$ are $-1 + 2i$ and $1 - 2i$. ■

EXERCISE 9.5

In the problems of this exercise, express answers in polar form. Then give answers in rectangular form as exact numbers when reasonable, otherwise to two decimal places.

1. Find the cube roots of 1.   **2.** Determine the fourth roots of i.

3. Find the fifth roots of $1 - \sqrt{3}i$.

4. Determine the roots of the equation $z^4 + 1 - i = 0$.

5. Find the sixth roots of -1 and show the results in a diagram.

6. Determine the sixth roots of $64 (\cos 126° + i \sin 126°)$.

7. Find the fourth roots of $16 (\sqrt{3} + i)$.

8. Determine the fourth roots of $(\sqrt{3} - i)^3$.

9. Determine the cube roots of $\left(\dfrac{1-i}{\sqrt{2}}\right)^{-2}$.

In Problems 10–13, solve the given quadratic equations.

10. $z^2 - (2 + 3i)z - 1 + 3i = 0$ **11.** $z^2 - 3z + 3 - i = 0$

12. $2z^2 + 2\sqrt{2}(-1 + i)z - 1 - 2i = 0$

13. $z^2 + z + 1 - i = 0$

14. Find the roots of the equation $z^4 + 1 = 0$.

15. Find the roots of the equation $z^3 + z^2 + iz + i = 0$. (*Hint*: Factor first.)

16. Find the roots of the equation $z^5 + 2z^3 - z^2 - 2 = 0$. (*Hint*: Factor first.)

17. Find the square roots of $3 - 4i$.   **18.** Find the square roots of $3 + 4i$.

19. Find the square roots of $-5 + 12i$.

20. Find the roots of $z^2 - iz - 1 + i = 0$.

21. In the derivation of Eq. (9.7) we encountered the problem of solving the following two equations simultaneously for R and α in terms of r and θ:

$$R^n \cos(n\alpha) = r \cos \theta, \qquad R^n \sin(n\alpha) = r \sin \theta.$$

Carry out the solution and show that $R = r^{1/n}$ and $\alpha = (\theta + k \cdot 2\pi)/n$. (*Hint*: First eliminate α by squaring each of the given equations and then adding the resulting equations. After getting R, substitute the result in either of the given equations and then solve for α.)

Summary

Trigonometric form for complex numbers

$$z = x + iy = r (\cos \theta + i \sin \theta)$$

where r and θ are given by $r = \sqrt{x^2 + y^2}$ and $\sin \theta = (y/r)$, $\cos \theta = (x/r)$.

Product and quotient in polar form

$$z_1 \cdot z_2 = r_1 r_2 \left[\cos\left(\theta_1 + \theta_2\right) + i \sin\left(\theta_1 + \theta_2\right)\right],$$

$$\frac{z_1}{z_2} = \left(\frac{r_1}{r_2}\right)\left[\cos\left(\theta_1 - \theta_2\right) + i \sin\left(\theta_1 - \theta_2\right)\right].$$

DeMoivre's theorem

$$z^n = r^n \left(\cos n\theta + i \sin n\theta\right)$$

where n is *any integer.*

Roots of a complex number

For any given complex number z the equation $w^n = z$ (where n is a positive integer) has n solutions (called the nth roots of z) given by

$$w_k = r^{1/n}\left[\cos\left(\frac{\theta}{n} + \frac{2\pi k}{n}\right) + i \sin\left(\frac{\theta}{n} + \frac{2\pi k}{n}\right)\right]$$

where $k = 0, 1, 2, 3, \ldots, n - 1$.

Computer Problems (Optional)

Section 9.1

1. *Evaluating Linear Functions*
   a] Write a program that will allow you to enter complex numbers a, b, and z to get an output of the complex number $az + b$.
   b] Adjust your program to allow you to enter a and b and then compute $az + b$ for several values of z that you enter. Have your program include an output of the type, "are you through?" so that you can then enter other values of a and b.

Section 9.2

2. Write a program that will tell the quadrant in which the terminal point of the vector $w + z$ lies, where w and z are any complex numbers that you enter. Include in your program the possibility that the terminal point may be on one of the coordinate axes.

Section 9.3

3. Write a program that gives you the trigonometric form (the r- and θ-numbers) for any complex number $x + iy$ that you enter. Have the θ values be in the interval $0 \le \theta < 2\pi$. If you use $\mathrm{Tan}^{-1}(y/x)$ in computing θ then include in your program the possibility that x may be zero.

4. Write a program that will give the trigonometric form of $w \cdot z$ where w and z are any complex numbers, say, $w = u + iv$ and $z = x + iy$ where u, v, x, and y are entered. (*Hint:* You may wish to use your program

from Problem 3 to get w and z in trigonometric form and then use the formula given in Eq. (9.1).)

Section 9.4

5. Write a program that will allow you to input the real numbers x and y of $z = x + iy$, then convert to trigonometric form, $z = r\,(\cos\theta + i\sin\theta)$, and then evaluate z^n where n is any positive integer that is also entered. Use the formula given in (9.5). Have the output show z^n both in trigonometric form and in rectangular form. In the final results, round off the numbers in the answers to two decimal places.

6. Continue the program of Problem 5 and have the output also include $1/(x + iy)^n$ (that is, z^{-n}) in both trigonometric and rectangular form.

Section 9.5

7. Write a program that will allow you to input a complex number z in trigonometric form and then determine the n roots of z as given in Eq. (9.8) where n is an integer, $n \geq 2$. Have the answers given in trigonometric form with numbers rounded off to two decimal places.

Review Exercises

In Problems 1–12, evaluate the given expressions and express results in standard form. Give answers in exact form whenever it is reasonable to do so; otherwise, give them correct to two decimal places.

1. $(1 + i)^3$
2. $(3 - 2i)^2$
3. $(1 + 2i)^4$
4. $(\sqrt{3} + i)^6$
5. $(1 + i)^{-2}$
6. $625\,(3 + 4i)^{-4}$
7. $(1 + i)(\sqrt{3} - i)^{-4}$
8. $\dfrac{(3 + 4i)^5}{(4 + 3i)^4}$
9. $\dfrac{(1 + 2i)(3 + 4i)^3}{(1 - i)^4}$
10. $\left(\dfrac{\sqrt{3}}{2} - \dfrac{1}{2}i\right)^6$
11. $(1 + i)^3 - (1 - i)^5$
12. $\left(\dfrac{1}{2} + \dfrac{\sqrt{3}}{2}i\right)^{12}$

In Problems 13–15 the function f is defined on the set of complex numbers and is given by $f(z) = 3 - 4z + z^2$, where z is any complex number. Evaluate the given expressions in exact form, and state answers in standard form.

13. $f(-3)$
14. $f(2 - 2i)$
15. $f(1 - \sqrt{3}i)$

In Problems 16–20, give answers in standard form.

16. Solve the quadratic equation $z^2 + (2 - i)z - i = 0$.

17. Find the cube roots of $\frac{1}{2}(\sqrt{3} - i)$.   18. Find the fourth roots of $\frac{3}{5} - \frac{4}{5}i$.

19. Solve the equation $z^4 + (1 + i)z^2 + i = 0$.

20. Solve the equation $z^2 - 2iz - 2 = 0$.

Exponential and Logarithmic Functions
10

Until recent years, one of the main applications of logarithms has been as an aid in performing cumbersome arithmetic computations.* With the introduction of calculators and computers the importance of logarithms for computational purposes has been essentially eliminated. However, logarithmic functions occur frequently in applications and in the study of pure mathematics. The treatment in this chapter concentrates on the fundamental properties of logarithmic functions. In order to proceed with this development it is necessary to first talk about exponential functions.

10.1 Exponential Functions

We assume that the reader already has some knowledge of working with integer exponents. For instance, if $a = 3^5$, we say that 3 is the *base*, 5 is the *exponent*, and 3^5 means the product of five threes, or 243. Also, if $b = 3^{-2}$, then by definition, $b = 1/3^2$. For the product $a \cdot b$ we add exponents to get

$$a \cdot b = 3^5 \cdot 3^{-2} = 3^{5+(-2)} = 3^3 = 27.$$

The same familiar "rules" that are valid for integer exponents also apply for any real number exponents (rational numbers such as 3/5 or irrational numbers such as $\sqrt{3}$ or π).

We shall make no attempt to give a detailed development of exponents but merely summarize the important properties that allow

---

*"Trigonometric calculations for triangulation had now been simplified as a result of another advance in mathematics. The invention of logarithmic tables, perhaps the most universally useful mathematical discovery of the seventeenth century, cut in half the time it took to calculate distances." from *The Mapmakers*, John Noble Wilford.

us to perform algebraic manipulations with expressions involving exponents. We shall rely on calculators for evaluation of numerical quantities, particularly with noninteger exponents.

EXAMPLE 1   Evaluate the following expressions.

a] $3^{4/7}$ **b]** $4^{-\sqrt{5}}$ **c]** $(-3)^{\sqrt{5}}$

Solution   Scientific calculators have a key labeled $\boxed{y^x}$. The following key sequence will accomplish the desired computations:

a] Alg.: 3, $\boxed{y^x}$, (, 4, $\boxed{\div}$, 7,), $\boxed{=}$
RPN: 3, $\boxed{\text{ENT}}$, 4, $\boxed{\text{ENT}}$, 7, $\boxed{\div}$, $\boxed{y^x}$
The result is $3^{4/7} \approx 1.873444$.

b] Alg.: 4, $\boxed{y^x}$, 5, $\boxed{\sqrt{x}}$, $\boxed{+/-}$, $\boxed{=}$
RPN: 4, $\boxed{\text{ENT}}$, 5, $\boxed{\sqrt{x}}$, $\boxed{\text{CHS}}$, $\boxed{y^x}$
This gives $4^{-\sqrt{5}} \approx 0.0450560$.

c] If we attempt to evaluate $(-3)^{\sqrt{5}}$ with a key sequence similar to that in (b), the calculator will give an ERROR response. The reason is that in this case the $\boxed{y^x}$ key will not accept a negative number as a base. Also, $(-3)^{\sqrt{5}}$ is not a real number. ■

Properties of exponents _____

In the study of exponential functions we require the base to be a *positive* number. For instance, functions given by $y = 3^x$, $y = (2/3)^x$, $y = (\sqrt{5})^x$ are examples of exponential functions, but $y = (-3)^x$ is not. Hence in the following list of properties of exponents we restrict the *base number b to be positive*, and the exponents x and y are *any real numbers*:

(E1) $b^{-x} = \dfrac{1}{b^x}$, (E2) $b^0 = 1$, (E3) $b^x \cdot b^y = b^{x+y}$,

(E4) $b^x/b^y = b^{x-y}$, (E5) $(b^x)^y = b^{xy}$.

If c is also a positive number, then we have

(E6) $(b \cdot c)^x = b^x \cdot c^x$ (E7) $(b/c)^x = b^x/c^x$.

Note that (E6) and (E7) involve a product or quotient to a power. Similar formulas are not valid for a sum or a difference. That is, in general, $(b + c)^n$ is not equal to $b^n + c^n$.

EXAMPLE 2   Simplify each of the following.

a] $3^{5/2} \cdot 3^{-1/2}$ **b]** $\dfrac{5^{3/2}}{5^{7/2}}$ **c]** $(2^{-3/2})^4$

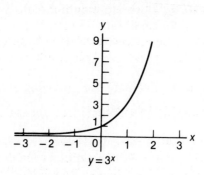

$y = 3^x$

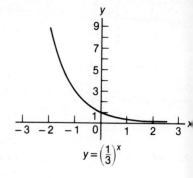

$y = \left(\dfrac{1}{3}\right)^x$

FIGURE 10.1   **FIGURE 10.2**

Solution

a] Apply (E3) to get $3^{5/2} \cdot 3^{-1/2} = 3^{5/2 + (-1/2)} = 3^2 = 9$.

b] Use (E4) and then (E1) to get $5^{3/2}/5^{7/2} = 5^{3/2 - 7/2} = 5^{-2} = \dfrac{1}{5^2} = \dfrac{1}{25}$.

c] Using (E5), we have $(2^{-3/2})^4 = 2^{(-3/2)(4)} = 2^{-6} = \dfrac{1}{2^6} = \dfrac{1}{64}$.   ■

EXAMPLE 3   Draw a graph of $y = 3^x$.

Solution   First complete the following table, then plot the corresponding (x, y)-points and draw a graph, as shown in Fig. 10.1.

x	−3	−2	−1	−0.5	0	0.5	1	1.5	2	3
y	0.04	0.11	0.33	0.58	1	1.73	3	5.20	9	27

From the graph in Fig. 10.1, note that $f(x) = 3^x$ is an *increasing* function; the *domain and range* of f are given by

$$D(f) = \{x \mid x \text{ is a real number}\}, \qquad R(f) = \{y \mid y > 0\}.$$ ■

EXAMPLE 4   Draw a graph of $y = (1/3)^x = 3^{-x}$.

Solution   Following the pattern of Example 3, we make a table of x, y-values, plot the corresponding points, and draw the curve, as shown in Fig. 10.2. It is instructive to compare the x, y-values in this table with those in the table of Example 3.

x	−3	−2	−1.5	−1	−0.5	0	0.5	1	2	3
y	27	9	5.20	3	1.73	1	0.58	0.33	0.11	0.04

From the graph of Fig. 10.2 we see that the function g given by $g(x) = (1/3)^x$ is a *decreasing* function with *domain and range*

$$D(g) = \{x \mid x \text{ is a real number}\}, \qquad R(g) = \{y \mid y > 0\}.$$ ■

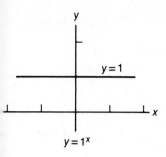

y = 1ˣ

FIGURE 10.3

EXAMPLE 5 Draw a graph of $y = 1^x$.

Solution Since 1^x equals 1 for all values of x, the given equation is equivalent to $y = 1$. The graph is a horizontal line shown in Fig. 10.3. ■

Exponential functions

The functions given in Examples 3 and 4 are called exponential functions, while that in Example 5 is not (its graph is a line and it is a linear function). This suggests the following definition.

DEFINITION 10.1 If b is any given *positive real number* and $b \neq 1$, then the function f given by the formula $f(x) = b^x$ is called the *exponential function* with base b.

For each positive number b, $b \neq 1$, we get an exponential function. For instance, $y = 5^x$, $y = (3/4)^x$, and $y = (1 + \sqrt{3})^x$ are examples of three different exponential functions.

Properties of exponential functions

1. The *domain* of any exponential function is the set of real numbers.
2. The *range* of any exponential function is the set $\{y \mid y > 0\}$.
3. If $b > 1$, then $y = b^x$ is an *increasing function*.
4. If $0 < b < 1$, then $y = b^x$ is a *decreasing function*.

From properties 3 and 4 we conclude that all exponential functions are one-to-one, and so their *inverses are also functions*. See Section 3.3 for a discussion of inverse functions.

The number e

The reader is already familiar with the number π, which happens to be the ratio of the circumference of any circle to its diameter. Another number that is not so well known but is equally important in pure and applied mathematics is denoted by e. We introduce it through the following example.

EXAMPLE 6 Suppose f is a function given by $f(x) = (1 + x)^{1/x}$, where $x > -1$ and $x \neq 0$. Make a table of $f(x)$ values for several values of x near zero. From the table, draw a conclusion regarding the behavior of $f(x)$ as x approaches zero.

Note: The function f given here is not an exponential function, since the base, $1 + x$, is not a constant and also the exponent is $1/x$, not just x.

Solution Use a calculator to complete the following table of $f(x)$ values for the given values of x.

x	$f(x)$
1	2
0.5	2.25
0.2	2.48832
0.1	2.59374
0.01	2.70481
0.001	2.71692
0.0001	2.71815

x	$f(x)$
−0.8	7.47674
−0.5	4
−0.2	3.05167
−0.1	2.86797
−0.01	2.73200
−0.001	2.71964
−0.0001	2.71842

From the values of $f(x)$ in the above table we conclude that $(1 + x)^{1/x}$ appears to be approaching a number (as x approaches 0) that is between 2.71815 and 2.71842. ∎

With tools of calculus we are able to show that the limiting value of $(1 + x)^{1/x}$ as x approaches zero is indeed a number as suggested in Example 6. It is an irrational number (also transcendental, as is π), and to 24 decimal places it is given by

$$e = 2.718281828459045235360287\ldots .$$

The number e is an important number that occurs frequently in applied as well as theoretical problems in mathematics.*

EXAMPLE 7 Draw a graph of $y = e^x$.

Solution First make a table of x, y-values, then plot the corresponding points and draw a curve through these points, as shown in Fig. 10.4. Some calculators have an $\boxed{e^x}$ key, so the value of y can be determined directly by pressing the $\boxed{e^x}$ key after x is entered in the display. For calculators that do not have an $\boxed{e^x}$ key, we suggest using the $\boxed{y^x}$ key, where 2.718281828 (rounded off to calculator

FIGURE 10.4

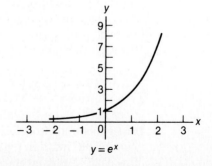

$y = e^x$

---

* The letter e is used in honor of the Swiss mathematician Leonhard Euler (1707–1783), one of the greatest mathematicians of all time.

capacity) is first stored with the ⟨STO⟩ key and recalled with the ⟨RCL⟩ key when needed. In Section 10.3, methods for evaluating e^x without first storing e will be given. In the following table the values of y have been rounded off to two decimal places. ∎

x	−3	−2.5	−2	−1.5	−1	−0.5	0	0.5	1	1.5	2	2.5	3
y	0.05	0.08	0.14	0.22	0.37	0.61	1	1.65	2.72	4.48	7.39	12.18	20.09

EXERCISE 10.1

In Problems 1–8, evaluate the given expressions. Give answers rounded off to three decimal places.

1. $3^{\sqrt{2}}$ **2.** $(\sqrt{2})^3$ **3.** $5^{-\sqrt{3}}$ **4.** $(2/3)^{1/7}$

5. e^2 **6.** $e^{\sqrt{3}}$ **7.** $(1+e)^{-2}$ **8.** $e^{1/3} - e^{1/4}$

In Problems 9–16, apply the rules of exponents to simplify the given expressions.

9. $3^{5/2} \cdot 3^{-3/2}$ **10.** $5^{5/2} \div 5^{1/2}$ **11.** $5^{1/3} \cdot 135^{2/3}$

12. $8 \cdot 16^{-3/4}$ **13.** $(16^{1/3} + 2^{1/3})^3$ **14.** $(7^{5/2} - 63^{3/2}) \div 7^{1/2}$

15. $(4^{1/3} \cdot 2^{1/4})^{12}$ **16.** $(5^{-1/3})^9$

In Problems 17–20, draw a graph of the given exponential function.

17. $y = 4^x$ **18.** $y = (2/5)^x$ **19.** $y = (1/2)^x$ **20.** $y = (e+1)^x$

21. Function f is given by $f(x) = 4 - 5^{-x}$. Evaluate the following expressions (three decimal places).
 a] $f(0)$ **b]** $f(1)$ **c]** $f(\tfrac{1}{2})$ **d]** $f(-0.24)$

22. Function g is given by $g(x) = 4/(1 + e^x)$. Evaluate the following expressions (three decimal places).
 a] $g(0)$ **b]** $g(1)$ **c]** $g(-2)$ **d]** $g(\sqrt{3})$

23. The predicted population of a city is given by the formula

$$P = 45000\,(1.08)^{n/12},$$

where n is the number of years after 1980. Find the predicted population for each of the following years. Give answers rounded off to the nearest thousand.
 a] 1990 **b]** 2000 **c]** 2050

24. A function that occurs frequently in the study of probability and statistics is given by

$$f(x) = \frac{1}{\sqrt{2\pi}}\, e^{-x^2/2},$$

where x is any real number. Evaluate $f(x)$ at $x = 0, 0.2, 0.4, \ldots, 1.8, 2.0$. Draw a graph of the given function. Note that $f(-x) = f(x)$.

25. Is the function $f(x) = 3^x$ additive? That is, is $f(u + v) = f(u) + f(v)$ for all real numbers u and v? (See Section 4.1).

10.2 Logarithmic Functions

In the preceding section we defined exponential functions given by $f(x) = b^x$, where b is a given positive number and $b \neq 1$. We noted that f is an increasing function for $b > 1$ and a decreasing function for $0 < b < 1$. In Section 3.3 inverse functions were reviewed, and it was noted that an increasing or decreasing function is one-to-one, and consequently its inverse relation is a function. Thus the *inverse of any exponential function is also a function*, which we shall call a *logarithmic function*.

In Section 3.3 we illustrated a useful technique for determining a formula for some inverse functions. For instance, suppose g is a function defined by

$$g : y = 3x - 2. \qquad \textbf{[10.1]}$$

To get a formula for g^{-1} we can interchange x and y in Eq. (10.1), obtaining

$$x = 3y - 2.$$

Then solving for y gives

$$g^{-1} : y = \frac{x + 2}{3}.$$

Now suppose we attempt a similar approach to determine a formula for the inverse function of f, given by

$$f : y = 3^x. \qquad \textbf{[10.2]}$$

Since f is an increasing function, we can be certain that its inverse is a function. Interchanging x and y in Eq. (10.2) gives $x = 3^y$. In this case we cannot solve for y in terms of x in a simple manner, as we did in the above example. We could express the inverse function as a set of ordered pairs and denote it by the symbol f^{-1}, as used in general situations:

$$f^{-1} = \{(x, y) \mid x = 3^y\}. \qquad \textbf{[10.3]}$$

However, inverses of exponential functions play an important role in applied as well as theoretical mathematics, and so they deserve special names. These functions are called *logarithmic functions*. The function given in Eq. (10.3) is denoted by $\log_3$, which is read "log base 3 function." For example

$$\log_3 = \{(x, y) \mid x = 3^y\}$$

means that $\log_3$ is a function defined by $y = \log_3 x$ if and only if $x = 3^y$.

Now let us return to inverses of exponential functions in general and state the following definition.

DEFINITION 10.2   Suppose b is a given positive number and $b \neq 1$. The *logarithmic function base b*, denoted by $\log_b$ is defined by

$$y = \log_b x \text{ if and only if } x = b^y.$$

That is, $\log_b$ is the inverse of the exponential function with base b.

The domain and range of the $\log_b$ function are given by

$$D\,(\log_b) = \{x \mid x > 0\}, \qquad R\,(\log_b) = \mathbf{R}.$$

Graphs of logarithmic functions

The following example illustrates a procedure for drawing a graph of a logarithmic function.

EXAMPLE 1   Draw a graph of the function given by

$$y = \log_3 x. \qquad\qquad \textbf{[10.4]}$$

Solution   Since $\log_3$ is the inverse of the exponential function given by $f(x) = 3^x$, we can draw its graph by simply reflecting the graph shown in Fig. 10.1 about the line $y = x$. Or we can get a table of x, y values satisfying Eq. (10.4) by interchanging the x and y values in the table shown in Example 3 of the preceding section. This gives the following table, which can be used to draw the graph of $y = \log_3 x$.

x	0.04	0.11	0.33	0.58	1	1.73	3	5.20	9	27
y	−3	−2	−1	−0.5	0	0.5	1	1.5	2	3

In Fig. 10.5 the broken curve is the graph of $y = 3^x$, and the solid curve is the graph of $y = \log_3 x$. ∎

FIGURE 10.5

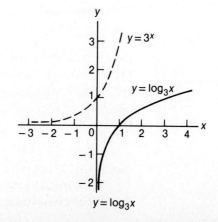

$$y = \log_3 x$$

Properties of logarithmic functions

We now state three useful properties that form a basis for algebraic manipulation of logarithmic functions. They are analogous to corresponding properties (E3), (E4), and (E5) for exponential functions discussed in the preceding section.

> Suppose u and v are positive numbers and t is any real number. Then
>
> (L1) $\quad \log_b (u \cdot v) = \log_b u + \log_b v,$ $\quad$ (L2) $\quad \log_b \left(\dfrac{u}{v}\right) = \log_b u - \log_b v,$
>
> (L3) $\quad \log_b (u^t) = t \, (\log_b u).$

We shall prove the statement given by property (L1); proofs for the other two are similar.

Let $\log_b u = h$ and $\log_b v = k$. Using Definition 10.2, we get

$$u = b^h \quad \text{and} \quad v = b^k.$$

Since (L1) involves the product $u \cdot v$, we multiply and use property (E3) to get

$$u \cdot v = b^h \cdot b^k = b^{h+k}.$$

Applying Definition 10.2 to $u \cdot v = b^{h+k}$ gives

$$\log_b (u \cdot v) = h + k.$$

Replacing h by $\log_b u$ and k by $\log_b u$, we have

$$\log_b (u \cdot v) = \log_b u + \log_b v.$$

Note: Properties (L1), (L2), and (L3) involve logarithms of products, quotients, and powers. We do not give similar formulas for sums and differences because there are no simple results for $\log_b (u + v)$ and $\log_b (u - v)$.

There are some additional properties of logarithmic functions that are worth noting. Let us evaluate $\log_b x$ for $x = 1$ and $x = b$.

Let $\log_b 1 = c$; then $b^c = 1$, and so $c = 0$. Thus $\log_b 1 = 0$.

Let $\log_b b = d$; then $b^d = b$, and so $d = 1$. Thus $\log_b b = 1$.

These two special cases occur frequently, and so we label them as property (L4) for easy reference.

> (L4) $\quad \log_b 1 = 0 \quad$ and $\quad \log_b b = 1.$

Since the $\log_b$ function and the exponential function with base b are inverses of each other, we have the following identities:

$$
\begin{array}{lll}
\text{(L5)} \quad b^{\log_b x} = x & \text{for} & x > 0, \\
\text{(L6)} \quad \log_b(b^x) = x & \text{for} & x \in \mathbf{R}.
\end{array}
$$

Let us consider some examples illustrating the use of properties (L1)–(L6) and Definition 10.2.

EXAMPLE 2 Evaluate each of the following and give answers in exact form.

a] $\log_2 8$ **b]** $\log_{10}(0.0001)$ **c]** $\log_{0.5}(4\sqrt{2})$

Solution

a] Let $\log_2 8 = r$. By Definition 10.2, $2^r = 8 = 2^3$. Thus $r = 3$, and so $\log_2 8 = 3$.

b] Let $\log_{10}(0.0001) = q$. By Definition 10.2, $10^q = 0.0001 = 10^{-4}$, and so $q = -4$. Thus $\log_{10}(0.0001) = -4$.

c] Let $\log_{0.5}(4\sqrt{2}) = m$. By Definition 10.2, $(0.5)^m = 4\sqrt{2} = 2^2 \cdot 2^{1/2} = 2^{5/2}$. Hence $(0.5)^m = 2^{5/2}$. But $(0.5)^m = (1/2)^m = 1/2^m = 2^{-m}$. Thus $2^{-m} = 2^{5/2}$, and so $m = -5/2$. Therefore $\log_{0.5} 4\sqrt{2} = -5/2$. ∎

EXAMPLE 3 Using only the information that $\log_5 3 \approx 0.6826$ and $\log_5 6 \approx 1.1133$ (correct to four decimal places), evaluate the given expressions. Give answers rounded off to three decimal places.

a] $\log_5 2$ **b]** $\log_5(\log_2 8)$

Solution

a] $\log_5 2 = \log_5(6/3) = \log_5 6 - \log_5 3 \approx 1.1133 - 0.6826 = 0.4307$. Here we used property (L2). Thus we have $\log_5 2 \approx 0.431$.

b] From Example 1(a) we have $\log_2 8 = 3$. Therefore $\log_5(\log_2 8) = \log_5 3 \approx 0.6826$. Rounding off to three decimal places, we have $\log_5(\log_2 8) \approx 0.683$. ∎

EXAMPLE 4 Combine $3\log_5 2 + (3/2)\log_5 8 - (1/2)\log_5 32$ and express the result as $\log_5$ of a number.

Solution

$$
3\log_5 2 + (3/2)\log_5 8 - (1/2)\log_5 32 \underset{\substack{\uparrow \\ \text{by (L3)}}}{=} \log_5 2^3 + \log_5 8^{3/2} - \log_5 32^{1/2}
$$

$$
\underset{\substack{\uparrow \\ \text{by (L1), (L2)}}}{=} \log_5\left(\frac{2^3 \cdot 8^{3/2}}{32^{1/2}}\right) \underset{\substack{\uparrow \\ \text{by (E5)}}}{=} \log_5\left(\frac{2^3 \cdot 2^{9/2}}{2^{5/2}}\right) \underset{\substack{\uparrow \\ \text{by (E3), (E4)}}}{=} \log_5(2^5) = \log_5 32.
$$

Therefore the given expression is equal to $\log_5 32$. ∎

EXAMPLE 5   Solve the following equations.

a] $\log_3 (2x + 5) - \log_3 x = 1$ **b]** $\log_3 (2x - 5) - \log_3 x = 1$

Solution

a] Applying (L2) to the given equation, we get $\log_3 [(2x + 5)/x] = 1$. Using Definition 10.2, we get $(2x + 5)/x = 3$. Thus $2x + 5 = 3x$, and so $x = 5$.

As a check we wish to see if 5 actually satisfies the given equation. Replacing x by 5 in the left-hand side gives

$$\text{LHS:}\quad \log_3 (2 \cdot 5 + 5) - \log_3 (5) = \log_3 15 - \log_3 5$$
$$= \log_3 \frac{15}{5} = \log_3 3 = 1.$$

Hence 5 is a solution of the given equation.

b] Following a pattern similar to that in (a), we get $x = -5$. Now substituting -5 for x in the left-hand side of the given equation gives

$$\text{LHS:}\quad \log_3 (-10 - 5) - \log_3 (-5) = \log_3 (-15) - \log_3 (-5).$$

Since -15 and -5 are not in the domain of the $\log_3$ function, that is, $\log_3 (-15)$ and $\log_3 (-5)$ are not defined, we see that -5 is not a solution of the given equation. Thus there is no real number x that satisfies the given equation.   ∎

EXAMPLE 6   Solve for x: $\log_{10} x + \log_{10} (x + 48) = 2$.

Solution   Applying (L1), we can write the given equation as $\log_{10} [x (x + 48)] = 2$. Using Definition 10.2 gives $x (x + 48) = 10^2$. The quadratic equation

$$x^2 + 48x - 100 = 0$$

can now be solved by factoring: $(x + 50)(x - 2) = 0$. Hence -50 and 2 are solutions to the quadratic equation.

We now check to see if these are solutions to the given equation. Replacing x by -50 yields

$$\text{LHS:}\quad \log_{10} (-50) + \log_{10} (-50 + 48) = \log_{10} (-50) + \log_{10} (-2).$$

This gives an undefined result, since -50 and -2 are not in the domain of $\log_{10}$, and so -50 is not a solution.

If we replace x by 2 in the original equation, it is easy to verify that 2 is a solution. Therefore the solution set for the given equation is $\{2\}$.   ∎

Need to check answers

In Examples 5 and 6 we indicated a need to check answers resulting from intermediate steps to see if they actually are solutions to the given equations. To see the reason for this let us take a closer look at the steps involved in the solution of Example 6.

Let f represent the function given by the left-hand side of the given equation:

$$f(x) = \log_{10} x + \log_{10} (x + 48).$$

The domain of f is given by

$$D(f) = \{x \mid x > 0 \text{ and } x + 48 > 0\} = \{x \mid x > 0\}.$$

The first step in our solution involves the function g given by

$$g(x) = \log x \, (x + 48).$$

The domain of g is given by

$$D(g) = \{x \mid x \, (x + 48) > 0\} = \{x \mid x > 0 \text{ or } x < -48\}.$$

Since $D(g) \neq D(f)$, functions f and g are not equal. In fact, $D(f) \subset D(g)$, and so there may be values of x that are solutions to an equation involving $g(x)$ but are not solutions to the corresponding equation involving $f(x)$. This is so in Example 6.

EXAMPLE 7 Determine the domain of function f given by

$$f(x) = \log_3 (x^2 - 5x + 6).$$

Solution Here we use the fact that $D(\log_3) = \{u \mid u > 0\}$:

$$D(f) = \{x \mid x^2 - 5x + 6 > 0\} = \{x \mid (x - 2)(x - 3) > 0\}$$
$$= \{x \mid x < 2 \text{ or } x > 3\}. \quad \blacksquare$$

EXERCISE 10.2

In Problems 1–15, evaluate the given expressions and give answers in exact form. If the given expression is not defined, tell why.

1. $\log_2 (32)$
2. $\log_3 (1/27)$
3. $\log_5 (125/\sqrt{5})$
4. $\log_7 (49/\sqrt{7})$
5. $\log_{10} 100$
6. $\log_{10} 1000$
7. $\log_{10} (0.0001/\sqrt{0.0001})$
8. $\log_e (1/e)$
9. $\log_{0.3} (10/3)$
10. $\log_5 (\log_5 5)$
11. $\log_7 (\log_7 1)$
12. $\log_8 (\log_3 3)$
13. $\log_3 (\log_5 (1/5))$
14. $\log_{10} (\log_{10} 0.1)$
15. $\log_2 (4\sqrt{2})$

In Problems 16–23, use the following to evaluate the given expressions and give answers rounded off to four decimal places (see Example 3):

$$\log_5 2 = 0.43068, \qquad \log_5 3 = 0.68261, \qquad \log_5 7 = 1.20906,$$
$$\log_3 11 = 2.18266, \qquad \log_3 22 = 2.81359.$$

16. $\log_5 6$ **17.** $\log_5 63$ **18.** $\log_5 75$ **19.** $\log_3 2$

20. $\log_3 66$ **21.** $\log_3 \sqrt{44}$ **22.** $\log_3 \sqrt{54}$ **23.** $\log_5 (\log_3 9)$

In Problems 24–26, write each of the given expressions as $\log_b$ of a number for the given b (see Example 4).

24. $\log_3 5 + \log_3 20$

25. $2 \log_3 5 - \log_3 4$

26. $\frac{1}{2} \log_7 4 + \frac{2}{3} \log_7 27 - \frac{1}{6} \log_7 64$

In Problems 27–34, solve for the indicated letter. When necessary, be certain to check to see that your solution satisfies the given equation.

27. If $\log_3 x = 4$, then $x = $ _____ .

28. If $\log_5 \left(\frac{1}{25}\right) = y$, then $y = $ _____ .

29. If $\log_5 (3x - 1) = 1$, then $x = $ _____ .

30. If $\log_5 (4x) - \log_5 (2x - 1) = 2$, then $x = $ _____ .

31. If $\log_3 (2x) + \log_3 (5x) = \log_3 10$, then $x = $ _____ .

32. If $\log_b \left(\frac{1}{27}\right) = -3$, then $b = $ _____ .

33. If $\log_5 25 + \log_3 27 = 2x + 1$, then $x = $ _____ .

34. If $\log_{10} x + \log_{10} (x + 3) = 1$, then $x = $ _____ .

In Problems 35–40, determine whether the given statement is true, false, or meaningless. A statement is meaningless if any part of it is undefined. Give reasons for your answers.

35. $\log_3 9 - \log_3 2 = \log_3 (4.5)$

36. $\log_7 (3^2 + 4^2) = 2 \log_7 3 + 2 \log_7 4$

37. $\log_3 \left(\dfrac{1 - \sqrt{3}}{2}\right) = \log_3 (1 - \sqrt{3}) - \log_3 2$

38. $\log_{10} 100 - \log_{10} 0.01 = 4$

39. $\log_5 \left(\dfrac{4}{\sqrt{5} + 1}\right) = \log_5 (\sqrt{5} - 1)$

40. $\log_2 (\log_2 \frac{1}{2}) = -1$

In Problems 41–45, state the domain of the given functions.

41. $f(x) = \log_{10} (1 + x)$ **42.** $f(x) = \log_5 \sqrt{25 - x^2}$

43. $g(x) = \log_3 (-x)$ **44.** $f(x) = \log_3 (x - 4) + \log_3 x$

45. $g(x) = \log_3 [(x - 4)x]$

10.3 Using a Calculator to Evaluate Logarithmic Functions

In the examples of the preceding section we were able to evaluate logarithms by converting to exponential form. For example, to evaluate $\log_3 \sqrt{27}$, we let $\log_3 \sqrt{27} = y$. This is equivalent to $3^y = \sqrt{27} = 3^{3/2}$. Thus $y = 3/2$, and so $\log_3 \sqrt{27} = 3/2$. However, attempting a similar procedure to evaluate $\log_3 6.4 = x$, we have $3^x = 6.4$. Since 6.4 cannot be expressed as a simple power of 3, we are unable to complete the solution as we did in the first example. In this section we introduce techniques by which a calculator can be used to solve such problems.

Common and natural logarithms

For computational purposes the base of logarithms that is frequently used is $b = 10$. Since it is cumbersome to write the subscript 10 in $\log_{10}$ each time, we shall write log, and understand that the base is 10. For theoretical as well as computational purposes it is an interesting fact that the transcendental number $e = 2.718281828\ldots$ introduced in Section 10.1 occurs naturally as a base of logarithms in the study of calculus. To avoid writing $\log_e$ each time, we replace it by ln. Thus we have the following notation.

$$\log_{10} x \text{ is written as } \log x;$$
$$\log_e x \text{ is written as } \ln x.$$

The notation adopted here is consistent with that appearing on scientific calculators.

Logarithms with base 10 are called *common logarithms*, whereas those with base e are called *natural logarithms*.

Logarithms with calculators

Most scientific calculators have both $\boxed{\log}$ and $\boxed{\ln}$ keys. We shall consider several examples that will illustrate the use of these keys. Some calculators have the $\boxed{\ln}$ key but not the $\boxed{\log}$ key; we shall see that this is sufficient for our purposes.

The $\boxed{\log}$ and $\boxed{\ln}$ keys represent functions of one variable. If a positive number x is entered into the display of the calculator and then the $\boxed{\ln}$ key is pressed, the result $\ln x$ will appear almost immediately in the display; it is not necessary to press the $\boxed{=}$ key on algebraic calculators. Similarly the $\boxed{\log}$ key gives $\log x$ for any positive number x.

EXAMPLE 1   Evaluate each of the following, correct to four decimal places.

a] $\ln 2$ **b]** $\log 0.0037$

Solution

a] Pressing the keys ⊂2⊃ and ⊂ln⊃ gives $\ln 2 \approx 0.6931$.
b] If the calculator has a ⊂log⊃ key, then entering 0.0037 into the display and pressing ⊂log⊃ gives $\log 0.0037 \approx -2.4318$. If there is no ⊂log⊃ key, log 0.0037 can be evaluated by using Eq. (10.5), given below, with $b = 10$ and $u = 0.0037$.   ∎

Change of base

In using a calculator to evaluate $\log_b u$, where b is a positive number and $b \neq 1$, it is necessary to convert to logarithms with base e or base 10. This can be done as follows:

Let $\log_b u = t$, which is equivalent to $b^t = u$. Taking ln of both sides of this equation gives $\ln b^t = \ln u$, which is equivalent to saying $t (\ln b) = \ln u$. Thus $t = \ln u / \ln b$. Therefore we have the following formula, which expresses $\log_b u$ in terms of $\ln u$ and $\ln b$:

$$\log_b u = \frac{\ln u}{\ln b}.$$ **[10.5]**

Similarly, using log in place of ln in the above discussion, we get

$$\log_b u = \frac{\log u}{\log b}.$$ **[10.6]**

EXAMPLE 2   Evaluate each of the following, and give answers rounded off to four decimal places.

a] $\log_3 7.5$ **b]** $\log_5 (0.0348)$

Solution   The formula given in Eq. (10.5) or (10.6) can be used in each of these problems. We choose Eq. (10.5), since some calculators have a ⊂ln⊃ key but not a ⊂log⊃ key.

a] $\log_3 7.5 = \dfrac{\ln 7.5}{\ln 3} \approx 1.8340$

b] $\log_5 (0.0348) = \dfrac{\ln 0.0348}{\ln 5} \approx -2.0865.$ ∎

Inverse logarithms

In the above examples, all the problems were of the following type: Given a positive number u, find log u or ln u. We are now interested in the inverse problem: Given the value of log u or ln u, determine u. For example, given that log u = 0.4735, we wish to find u. The notation that has been traditionally used is u = Antilog 0.4735. However, since this actually involves the inverse of the log function, we shall denote it by u = log^{-1} 0.4735. This is read, "u is the inverse log of 0.4735."

As another example of notation, if ln v = 1.2654, then we can write that v = ln^{-1} 1.2654, and say, "v is the inverse ln of 1.2654."*

So far, in the two examples being considered here, we merely introduced some notation. Let us proceed to actually determine u and v. Since the log function is defined as the inverse of the function given by $f(x) = 10^x$, the inverse of the log function must be this exponential function. Therefore if log u = 0.4735, then

$$u = \log^{-1} 0.4735 = 10^{0.4735}.$$

This is precisely what Definition 10.2 tells us; log u = 0.4735 implies $u = 10^{0.4735}$. We can now determine u by using a calculator, as follows:†

1. If your calculator has a $\boxed{10^x}$ key, then evaluate $10^{0.4735}$ by pressing $\boxed{10^x}$ after entering 0.4735 into the display. This gives u ≈ 2.9751 (to four places).

2. If your calculator does not have a $\boxed{10^x}$ but has an $\boxed{\text{INV}}$ key, then with 0.4735 in the display, pressing the $\boxed{\text{INV}}$ and $\boxed{\log}$ keys gives u ≈ 2.9751.

Similarly, the ln function and the function given by $f(x) = e^x$ are inverses of each other, so the solution of ln v = 1.2654 is v = ln^{-1} 1.2654 = $e^{1.2654}$. Thus v can be found by pressing the $\boxed{\text{INV}}$ and $\boxed{\ln}$ keys or by using the $\boxed{e^x}$ key after entering 1.2654. Therefore v ≈ 3.5445 to four decimal places.

The discussion above illustrates the following:

$\boxed{10^x}$ and $\boxed{\log}$ keys are inverses of each other;

$\boxed{e^x}$ and $\boxed{\ln}$ keys are inverses of each other.

---

* The notation adopted here is consistent with that used for inverse functions in general (see Section 3.3).

† If your calculator does not have $\boxed{\log}$ and $\boxed{10^x}$ keys but has $\boxed{\ln}$ and $\boxed{e^x}$ keys, proceed as follows: Express the original problem, log u = 0.4735, in equivalent ln form by using the change-of-base formula given in Eq. (10.5). That is, with b = 10, ln u = (ln 10) log u. Therefore ln u = (ln 10) log u ≈ 2.30259 log u = (2.30259)(0.4735) ≈ 1.0903. Thus $u ≈ e^{1.0903}$, which can be evaluated by using the $\boxed{e^x}$ key, or the $\boxed{\text{INV}}$ and $\boxed{\ln}$ keys.

Thus we have the following special cases of properties (L5) and (L6) (which are really inverse identities):

(L7) $10^{\log x} = x$ for all $x > 0$ and $\log(10^x) = x$ for $x \in \mathbf{R}$.
(L8) $e^{\ln x} = x$ for all $x > 0$ and $\ln(e^x) = x$ for $x \in \mathbf{R}$.

Evaluation of 10^x and e^x by calculator

To evaluate 10^u or e^u, first enter u into the display. Pressing

$\boxed{10^x}$ or $\boxed{\text{INV}}$ and $\boxed{\log}$ gives 10^u in the display;   **[10.7]**

$\boxed{e^x}$ or $\boxed{\text{INV}}$ and $\boxed{\ln}$ gives e^u in the display.   **[10.8]**

EXAMPLE 3   In each of the following, solve for v, correct to four decimal places.

a] $\ln v = 1.345$ **b]** $\log v = -1.4382$
c] $e^v = 0.456$ **d]** $\ln(2v + 1) - \ln 3 = 1.48$

Solution

a] $\ln v = 1.345$ is equivalent to $v = \ln^{-1} 1.345$, or $v = e^{1.345}$. Following (10.8) gives $v \approx 3.8382$.

b] $\log v = -1.4382$ is equivalent to $v = \log^{-1}(-1.4382)$, or $v = 10^{-1.4382}$. Using (10.7) gives $v \approx 0.0365$.

c] $e^v = 0.456$ is equivalent to $v = \ln 0.456$. Enter 0.456 and press the $\boxed{\ln}$ key to get $v \approx -0.7853$.

d] The given equation is equivalent to $\ln[(2v + 1)/3] = 1.48$. Thus $(2v + 1)/3 = e^{1.48}$, and so $v = (3e^{1.48} - 1)/2$. Now use (10.8) to find $e^{1.48}$, and then continue with the remaining arithmetic operations. This gives $v \approx 6.0894$.   ∎

EXAMPLE 4   Evaluate each of the following. Give answers in exact form.

a] $e^{\ln 5}$ **b]** $10^{-\log 5}$

Solution

a] By property (L8), $e^{\ln 5} = 5$.

b] By properties (L3) and (L7), $10^{-\log 5} = 10^{\log 5^{-1}} = 5^{-1} = \frac{1}{5}$.   ∎

EXAMPLE 5   Solve the equation

$$\ln(2v - 1) + \ln v = 1. \qquad\qquad \textbf{[10.9]}$$

Solution   By using property (L1) the given equation can be written as $\ln[v(2v - 1)] = 1$. Apply Definition 10.2 to get $v(2v - 1) = e$. We now have a quadratic equation to solve:

$$2v^2 - v - e = 0. \qquad\qquad \textbf{[10.10]}$$

Applying the quadratic formula gives

$$v = \frac{1 \pm \sqrt{1 + 8e}}{4}.$$

Evaluating by calculator, we get 1.4423 and -0.9423 as solutions to Eq. (10.10). It is necessary to check these to see if they are solutions to the given equation. Replacing v in Eq. (10.9) by each of these values, we can easily see that 1.4423 is a solution but that -0.9423 is not, since $\ln(-0.9423)$ is undefined.   ■

EXAMPLE 6   Solve the equation

$$5^x = 3 \cdot 4^{1-x}. \qquad\qquad \textbf{[10.11]}$$

Give answers correct to three decimal places.

Solution   Taking ln of both sides of Eq. (10.11) gives $\ln 5^x = \ln(3 \cdot 4^{1-x})$. Applying properties (L1) and (L3), we get

$$x \ln 5 = \ln 3 + (1 - x) \ln 4,$$
$$x \ln 5 = \ln 3 + \ln 4 - x \ln 4,$$
$$x \ln 5 + x \ln 4 = \ln 3 + \ln 4,$$
$$x(\ln 5 + \ln 4) = \ln 3 + \ln 4,$$
$$x = \frac{\ln 3 + \ln 4}{\ln 5 + \ln 4}.$$

We can now evaluate this expression by using a calculator. However, we can simplify slightly by using (L1) to get $x = \ln 12/\ln 20$. Using a calculator gives $x \approx 0.829$. Substituting 0.829 for x in Eq. (10.11), we can check to see that 0.829 is a solution.   ■

In the preceding example the solution is given by $x = (\ln 12)/(\ln 20) \approx 0.829$. Note that $(\ln 12)/(\ln 20) \neq \ln(12/20) \approx -0.511$; also $(\ln 12)/(\ln 20) \neq \ln 12 - \ln 20 \approx -0.511$.

Warning: $\dfrac{\ln a}{\ln b}$ is equal to neither $\ln\left(\dfrac{a}{b}\right)$ nor $\ln a - \ln b$.

EXERCISE 10.3

If your calculator should indicate ERROR while you are solving any problem in this set, determine the reason.

In Problems 1–12, evaluate the given expression and give answers rounded off to four decimal places.

1. $\ln 5$ **2.** $\ln 0.47$ **3.** $\log 1.87$

4. $\log 0.0435$ **5.** $\ln (1.56^2 + 2.73^2)$ **6.** $\log (2.43\sqrt{5.75})$

7. $\ln (2 - \sqrt{5.43})$ **8.** $\log [(2 - \sqrt{6})/5]$ **9.** $\log [(1 + \sqrt{3})/8]$

10. $\log_3 6$ **11.** $\log_5 3.47$ **12.** $\log_7 (\sqrt{3} - 1)$

In Problems 13–18, evaluate by applying (L7) and (L8).

13. $e^{\ln (1.43)}$ **14.** $10^{\log (2.54)}$ **15.** $\log (10^{-0.42})$

16. $\ln (e^{3.2})$ **17.** $e^{-\ln 2}$ **18.** $e^{-3 \ln 2}$

In Problems 19–31, determine the value of v correct to four decimal places.

19. $\ln v = 1.532$ **20.** $\log v = -0.372$ **21.** $\ln v = 1 - \sqrt{3}$

22. $10^v = -0.473$ **23.** $e^v = 0.875$ **24.** $e^{-v} = 1.238$

25. $e^{-v} = -0.471$ **26.** $e^{2v-1} = 1.362$

27. $\ln (2v - 5) - \ln 7 = 2.43$ **28.** $\ln (v - 5) + \ln 2.43 = 1.56$

29. $\ln (e^{v-1}) = e^{-1.6}$ **30.** $\ln (e^{1-3v}) = 4$ **31.** $\ln (2v + 1) + \ln v = 1$

In Problems 32 and 33, find the domains of the functions.

32. $f(x) = \ln (x - 2)$ **33.** $f(x) = \log (x + 1) + \log (x - 1)$

In Problems 34–38, draw graphs of the given functions. In each case label the coordinate intercept points.

34. $f(x) = \ln (x - 2)$ **35.** $f(x) = \log (x^2 - 1)$

36. $f(x) = \log (-x)$ **37.** $g(x) = 1 + \ln x$

38. $f(x) = \ln e^{x-1}$

In Problems 39 and 40, solve the given equations. Give solutions rounded off to two decimal places. See Example 6.

39. $8^x = 3 \cdot 5^x$ **40.** $5^x = 3 \cdot 8^{1-x}$

Summary

Properties of exponential functions

For b any *positive number*, $b \neq 1$, and x, y any *real numbers*, $f(x) = b^x$ is an *exponential function* with the following properties

$$\boxed{b^x \cdot b^y = b^{x+y},} \qquad \boxed{b^x/b^y = b^{x-y},} \qquad \boxed{(b^x)^y = b^{x \cdot y}.}$$

If $0 < b < 1$, then $y = b^x$ is a *decreasing function*.
If $b > 1$, then $y = b^x$ is an *increasing function*.

All exponential functions are one-to-one and have domain D and range R given by

$$D = \mathbf{R}, \qquad R = \{y \mid y > 0\}.$$

The inverse of $y = b^x$ is also a function, denoted by $\log_b$.

$$\boxed{y = \log_b x \qquad \text{if and only if} \qquad x = b^y.}$$

Properties of $\log_b$ functions

$$\text{Domain} = \{x \mid x > 0\}, \qquad \text{Range} = \mathbf{R}$$

If u and v are any positive numbers and t is any number, then properties of logarithmic functions are

$$\boxed{\log_b (uv) = \log_b u + \log_b v,} \qquad \boxed{\log_b (u/v) = \log_b u - \log_b v,}$$

$$\boxed{\log_b u^t = t \,(\log_b u).}$$

Inverse identities

$$\boxed{\log_b (b^t) = t,} \qquad\qquad \boxed{b^{\log_b u} = u, \; u > 0.}$$

Special bases

If $b = 10$, then $\log_{10}$ is written as $\log$. If $b = e$, then $\log_e$ is written as $\ln$.

Change of base formula

$$\boxed{\log_b u = \frac{\ln u}{\ln b},} \qquad\qquad \boxed{\log_b u = \frac{\log u}{\log b}.}$$

Computer Problems (Optional)

Section 10.3

1. **a]** Write a program that will evaluate $y = \ln x$ for $x = 0.2, 0.4, 0.6, \ldots$, 3.8, 4.0. Use the corresponding (x, y)-values and draw a graph of $y = \ln x$.

b] Follow a procedure similar to that in part (a) for the function $y = \log_{10} x$.

(*Note:* In many computer languages the ln function is given by LOG.)

2. a] Write a program that will give the roots of

$$\ln x + \ln (ax + b) = c,$$

where a, b, and c are numbers to be entered. Include in your program a check to see whether the solution actually satisfies the given equation.

b] Follow a procedure similar to that in part (a) for the equation $\ln [x (ax + b)] = c$.

3. Write a program that will allow you to solve any equation of the type $b^x = a (c^x)$, where a, b, and c are positive numbers to be entered.

Review Exercises

In each of the problems give answers in exact form whenever it is reasonable to do so. Otherwise, express results in decimal form rounded off to three decimal places. In problems involving undefined quantities, give reasons for an "undefined" answer.

In Problems 1–15, evaluate the given expression.

1. $\log 8$ **2.** $\log \sqrt{43}$ **3.** $\ln 23$ **4.** $\log (\sqrt{2} + \sqrt{3})$

5. $\ln (36^3)$ **6.** $\log (\ln 48)$ **7.** $\ln (\log 48)$ **8.** $\ln \left(\dfrac{\sqrt{2} + \sqrt{6}}{3} \right)$

9. $\log_5 8$ **10.** $\log_3 (\sqrt{5} + \sqrt{12})$ **11.** $\log_7 (\log 24)$

12. $\log_8 (e^3)$ **13.** $\log (\ln 0.6)$ **14.** $\log_5 (1 - \sqrt{2})$ **15.** $\log_3 (27\sqrt{3})$

In Problems 16–26, solve the given equations.

16. $\ln e^x = 3$ **17.** $\log e^x = 3$ **18.** $1 - \ln (2x + 1) = 3$

19. $\log (\ln x) = 1$ **20.** $e^{2x-1} = 4$ **21.** $e^{3x} = 10^{1-x}$

22. $\log 10^{4-3x} = 1$ **23.** $3^{x-1} = 4$ **24.** $5^x = 3 (7^x)$

25. $2e^x + 1 = 0$ **26.** $3e^x - 1 = 0$

27. Plot a graph of $y = e^{-x}$.

28. Plot a graph of $y = 4^x$.

29. Plot a graph of $y = 1 - 3^x$.

30. Find the domain of $f(x) = \ln (x - 3) + \ln x$.

31. Find the roots of $\ln (x - 1) + \ln x = 1$.

32. Draw a graph of $y = 1 + \ln (x - 1)$.

Introduction to the Use of Calculators

A

APPENDIX

Much of mathematics deals with numbers and the ways in which they are combined to get other numbers. Calculators and computers can be very helpful in evaluating complicated numerical expressions that may be difficult and time consuming when done by hand. Calculators will do the arithmetic (the computations), but they cannot do the algebra for you. It is important that you first master the algebraic techniques that are required to get the result in the form to which calculators can be applied. Then the computational part of the problem is easy—let the calculator do it.

For study of the course presented in this book, one needs a scientific calculator. This means it should have keys labeled $\boxed{\sin}$, $\boxed{\cos}$, and $\boxed{\tan}$. Basically, there are only two different kinds of calculators. One type is called *algebraic*, and the other is referred to as RPN (Reverse Polish Notation). The difference between them is the order in which number and function keys are pressed. You can easily determine whether your calculator is algebraic or RPN. Look at the keyboard; if you find a key labeled $\boxed{=}$, then it is algebraic. Otherwise, it is RPN. RPN calculators are frequently referred to as HP, since Hewlett-Packard is the major producer.

Some of the more sophisticated scientific calculators also have programmable capability, but it is not necessary to have such a calculator for this course. Perhaps the best suggestion we can offer is to learn to use *your calculator* efficiently and accurately and do not concern yourself too much about the other types.

In this appendix we given an introduction to the use of both algebraic and RPN calculators. This is done in separate sections. Also, since *hand-held computers* are becoming readily available (at low costs), we are including an introduction to the calculator features of such machines. These can also be programmed in the BASIC language of computers. If you should have a hand-held computer or a microcomputer available, we encourage you to write programs for the computer problems given at the end of each chapter of this book.

Algebraic Calculators

In general, the order in which keys of an algebraic calculator are pressed is the same as the order in which you read the numbers and arithmetic operations from left to right in any given problem. For instance, to evaluate $2 + 3 \times 5$, you press keys in the following sequence:

$$2 \; \boxed{+} \; 3 \; \boxed{\times} \; 5 \; \boxed{=} .$$

The calculator display will then show the number 17. Note that your calculator is clever enough to know that it is supposed to do the multiplication (3×5) before performing the addition, even though you pressed $\boxed{+}$ before the $\boxed{\times}$ key.

Calculators with AOS (algebraic operating system) are designed to follow the "My Dear Aunt Sally" (MDAS) rule that you learned in elementary school:

> Multiplication and Division before Addition and Subtraction*

In all of our discussion related to algebraic calculators we shall assume that we have an AOS calculator along with parenthesis keys, $\boxed{(}$ and $\boxed{)}$.

Using the keys
$\boxed{+}$ $\boxed{-}$ $\boxed{\times}$ $\boxed{\div}$ $\boxed{=}$ $\boxed{+/-}$ $\boxed{(}$ $\boxed{)}$ $\boxed{x^2}$

In order to use the calculator efficiently it is helpful to know something about the operation of the machine. The series of examples given below is designed to help the reader make some important observations involving the order in which pending operations are carried out in an AOS calculator.[†]

EXAMPLE 1 Calculate $5 - 7 + 4$.

Solution Press the calculator keys corresponding to the numbers and operations, as written from left to right, carefully watching the display to see when a given command is executed. Press

$$5 \; \boxed{-} \; 7 \; \boxed{+} \; 4 \; \boxed{=} . \qquad \blacksquare$$

---

* Some of the earlier models of algebraic calculators are not designed to follow the MDAS rule. On such calculators the key sequence suggested above gives 25 as the result of $2 + 3 \times 5$. Here the 2 and 3 are added first, and the result is multiplied by 5. Actually, this is the evaluation of $(2 + 3) \times 5$.

[†] We are intentionally using simple integer numbers in the examples so that you can mentally follow the calculator computations as the sequence of keys is pressed.

EXAMPLE 2   Calculate $5 - 7 + 4 \cdot 3$.

Solution   Press

$$5 \boxed{-} 7 \boxed{+} 4 \boxed{\times} 3 \boxed{=}.$$

Observe how all pending operations are executed when the $\boxed{=}$ key is pressed.   ∎

EXAMPLE 3   Calculate $\dfrac{5 - 7 + 4}{3}$.

Solution   Press

$$\boxed{(} 5 \boxed{-} 7 \boxed{+} 4 \boxed{)} \boxed{\div} 3 \boxed{=}.$$

Note that the numerator is evaluated after the right-parenthesis key $\boxed{)}$ is pressed. As an alternative solution, press $5 \boxed{-} 7 \boxed{+} 4 \boxed{=}$ $\boxed{\div} 3 \boxed{=}$. Thus when the left-parenthesis key $\boxed{(}$ is not entered, one can use the $\boxed{=}$ key to compute the numerator before dividing by 3.   ∎

EXAMPLE 4   Calculate $5 - 7 + 4 \cdot 3^2$.

Solution   Press

$$5 \boxed{-} 7 \boxed{+} 4 \boxed{\times} 3 \boxed{x^2} \boxed{=}.$$

Note that pressing $\boxed{x^2}$ squares only the contents of the display. Pressing $\boxed{=}$ executes all pending operations.   ∎

EXAMPLE 5   Calculate $5 - 7 + (4 \cdot 3)^2$.

Solution   Press

$$5 \boxed{-} 7 \boxed{+} \boxed{(} 4 \boxed{\times} 3 \boxed{)} \boxed{x^2} \boxed{=}.$$

The problem requires that $4 \cdot 3$ be multiplied before squaring. Parenthesis keys are used here to accomplish this.   ∎

EXAMPLE 6   Calculate $5 \div (-7 + 4 \cdot 3)$.

Solution   Press

$$5 \boxed{\div} \boxed{(} 7 \boxed{+/-} \boxed{+} 4 \boxed{\times} 3 \boxed{)} \boxed{=}.$$

The parentheses serve to compute the divisor before the division is carried out. Special note should be taken of the use of the change-sign key $\boxed{+/-}$. This key changes the sign of the number in the display. The calculator will not accept the sequence $5 \boxed{\div} \boxed{(} \boxed{-} 7 \ldots$ Such a sequence treats the $\boxed{-} 7$ command as subtraction rather than a negative number, but the algebraic calculator cannot accept two operation commands in sequence (such as ⌈÷⌉ and ⌈−⌉). ■

Clearing the calculator

If the last key pressed is ⌈=⌉, all pending operations have been executed, and the calculator is ready for a new problem without pressing the clear key. Some calculators have a clear-entry key that clears only the number in the display, while a separate key is used to clear all pending operations. Other calculators have a key labeled ⌈ON/C⌉ that serves three purposes. It is used to turn the calculator on; then, during computations, if it is pressed once, the number in the display *only* will be cleared, while if it is pressed twice in succession, all pending operations are also cleared.

The clear-entry feature is especially useful, since one of the most frequent mistakes is to key in an incorrect number after the calculator already has several pending operations. We illustrate this in the following example, where a 7 rather than an 8 was entered and this mistake is corrected by using the clear-entry key.

EXAMPLE 7   Evaluate $2 + 3 \cdot 5 - 24 \div 6 + 8$.

Solution   Press

$$2 \;⌈+⌉\; 3 \;⌈\times⌉\; 5 \;⌈-⌉\; 24 \;⌈÷⌉\; 6 \;⌈+⌉\; 7 \;⌈ON/C⌉\; 8 \;⌈=⌉.\qquad ■$$

EXERCISE A.1

Calculations in Problems 1–15 involve integers only. This is intended to allow the student to mentally follow the arithmetic and observe when the pending operations are performed by the calculator. Some important features of the calculator are illustrated in these problems; therefore, the student is encouraged to consider each calculation carefully.

1. $5 + 3 \cdot 7$
2. $(5 + 3) \cdot 7$
3. $(5 + 3) - 7$
4. $(5 + 3)(-7)$
5. $2 + 12 \div 3 - 7$
6. $2 + 12 \div (3 - 7)$
7. $\dfrac{(15 - 4) \cdot 5}{2} + 3 \cdot 5 - 7$
8. $\dfrac{(15 - 4) \cdot 5}{2 + 3 \cdot 5 - 7}$
9. $\dfrac{(1/2) - 3}{4}$
10. $(1/2) - (3/4)$
11. $2 \cdot 3^2 + 4 \cdot 5^2$
12. $(2 \cdot 3)^2 + (4 \cdot 5)^2$
13. $(2 \cdot 3 + 4 \cdot 5)^2$
14. $\left(\dfrac{3 \cdot 4^2}{2}\right) \cdot 5^2$
15. $(3 \cdot 4^2) \div (2 \cdot 5^2)$

Use your calculator to solve Problems 16–30. Answers correct to three decimal places are provided for a quick check.

		Answers
16.	$(1.87)(34.61) + 3.872$	68.593
17.	$(45.9 - 29.76)^2 + 52.86$	313.360
18.	$45.9 - 29.76^2 + 52.86$	-786.898
19.	$\dfrac{563 + 284}{18.7}$	45.294
20.	$563 + \dfrac{284}{18.7}$	578.187
21.	$\dfrac{52.9 \cdot 0.3876}{21.3}$	0.963
22.	$12^2 + 5^2 - 2 \cdot 5 \cdot 12 \cdot 0.9848$	50.824
23.	$(12^2 + 5^2 - 2 \cdot 5 \cdot 12)(0.9848)$	48.255
24.	$(-37.48 + 59.32)^2 - 31.97$	445.016
25.	$(37.48 - 59.32)^2 - 31.97$	445.016
26.	$\dfrac{(15.39 - 4.72) \cdot 5}{2.3} + 3.78 \cdot 5.43$	43.721
27.	$\dfrac{(15.39 - 4.72) \cdot 5}{2.3 + 3.78 \cdot 5.43}$	2.337
28.	$\dfrac{21.8 + 4.32^2}{5.12} - 5.39^2$	229.593
29.	$\dfrac{2}{3} + \dfrac{3}{4} - \dfrac{7}{8}$	0.542
30.	$\dfrac{(2/7) + (3/8)}{(1/6) + (1/7)}$	2.135

Using the keys $\boxed{1/x}$ $\boxed{\sqrt{x}}$ $\boxed{y^x}$ $\boxed{\text{STO}}$ $\boxed{\text{RCL}}$ _____

Scientific calculators have several keys in addition to the basic keys described in the preceding section. Here we shall consider the use of five more keys and defer discussion of others until the appropriate places in the text. The $\boxed{1/x}$ and $\boxed{\sqrt{x}}$ keys give the reciprocal and the square root, respectively, of the number in the display. The $\boxed{y^x}$ key operates by entering a positive number y, followed by $\boxed{y^x}$, then the number x, followed by $\boxed{=}$. For example, to evaluate 7^3, keys are pressed in the following order: 7 $\boxed{y^x}$ 3 $\boxed{=}$ and the result 343 appears in the display. Similarly, to find $\sqrt[3]{7}$, we evaluate $7^{1/3}$ by pressing the following keys: 7 $\boxed{y^x}$ 3 $\boxed{1/x}$ $\boxed{=}$, which gives $\sqrt[3]{7} = 1.9129$ (to four decimal places).

A lengthy computation frequently involves the evaluation of intermediate numbers that must be recorded and used later to complete the calculation. Scientific calculators allow the user to store a number with the $\boxed{\text{STO}}$ key* and recall it when needed with the $\boxed{\text{RCL}}$ key, thus avoiding the necessity of recording intermediate steps. This feature will be illustrated in examples given in this section.

EXAMPLE 8 Calculate $\sqrt{3.9^2 + 7.3^2}$.

Solution Press

$$\boxed{(}\ 3.9\ \boxed{x^2}\ \boxed{+}\ 7.3\ \boxed{x^2}\ \boxed{)}\ \boxed{\sqrt{x}}.$$

The display shows 8.2764727.

Alternative Solution Press

$$3.9\ \boxed{x^2}\ \boxed{+}\ 7.3\ \boxed{x^2}\ \boxed{=}\ \boxed{\sqrt{x}}.$$

This method uses the $\boxed{=}$ key to calculate the radicand before taking the square root. ■

EXAMPLE 9 Calculate $12^3 - 4^5$.

Solution Press

$$12\ \boxed{y^x}\ 3\ \boxed{-}\ 4\ \boxed{y^x}\ 5\ \boxed{=}.$$

The display shows 704. ■

EXAMPLE 10 Calculate $\sqrt[3]{24.3} \cdot \sqrt[5]{32.7}$.

Solution The problem can be rewritten as $(24.3)^{1/3} \cdot (32.7)^{1/5}$; then press

$$24.3\ \boxed{y^x}\ 3\ \boxed{1/x}\ \boxed{\times}\ 32.7\ \boxed{y^x}\ 5\ \boxed{1/x}\ \boxed{=}.$$

The display shows 5.8180615. Note that when the $\boxed{\times}$ key is pressed in this sequence, at that point the calculator evaluates $(24.3)^{1/3}$; in this computation it is not necessary to press the $\boxed{=}$ key before the $\boxed{\times}$ key. ■

EXAMPLE 11 Calculate $\sqrt{1.3^2 + 2.8^2 - 2(1.3)(2.8)(0.3215)}$.

Solution Press

$$1.3\ \boxed{x^2}\ \boxed{+}\ 2.8\ \boxed{x^2}\ \boxed{-}\ 2\ \boxed{\times}\ 1.3\ \boxed{\times}\ 2.8\ \boxed{\times}\ 0.3215\ \boxed{=}\ \boxed{\sqrt{x}}.$$

The display shows 2.6813206. ■

---

* Some calculators have multiple storage capacity and require a number address to follow the $\boxed{\text{STO}}$ key. The owner's manual that accompanies such a calculator gives details.

EXAMPLE 12 Calculate $\dfrac{1}{\sqrt{5.61 + 24.93}}$.

Solution Press

$$\boxed{(} \;\; 5.61 \;\; \boxed{+} \;\; 24.93 \;\; \boxed{)} \;\; \boxed{\sqrt{x}} \;\; \boxed{1/x}.$$

The display shows 0.18095287. ■

EXAMPLE 13 Calculate $\dfrac{1}{5.2^3 + 3.8^4} + \sqrt{4.2^2 + 3.97}$.

Solution Press

$$5.2 \;\; \boxed{y^x} \;\; 3 \;\; \boxed{+} \;\; 3.8 \;\; \boxed{y^x} \;\; 4 \;\; \boxed{=} \;\; \boxed{1/x} \;\; \boxed{STO} \;\; 4.2 \;\; \boxed{x^2} \;\; \boxed{+} \;\; 3.97 \;\; \boxed{=}$$
$$\boxed{\sqrt{x}} \;\; \boxed{+} \;\; \boxed{RCL} \;\; \boxed{=}.$$

The display shows 4.6515201. Storage is used to store the first part while the second part is being calculated. ■

EXAMPLE 14 Calculate $(5.873)^3 + 3(5.873)^2 - 9(5.873) + 4$.

Solution Press

$$5.873 \;\; \boxed{STO} \;\; \boxed{y^x} \;\; 3 \;\; \boxed{+} \;\; 3 \;\; \boxed{\times} \;\; \boxed{RCL} \;\; \boxed{x^2} \;\; \boxed{-} \;\; 9 \;\; \boxed{\times} \;\; \boxed{RCL} \;\; \boxed{+} \;\; 4 \;\; \boxed{=}.$$

The display shows 257.19166. Use of the $\boxed{STO}$ key eliminates the need to key in the four-digit number 5.873 three separate times. (Note: The $\boxed{y^x}$ key will function only when the base is positive. The calculator will indicate an ERROR if the base is negative.) ■

EXAMPLE 15 Use the calculator to evaluate the following:

a] $\sqrt{5.3 - 9.7}$ **b]** $\sqrt[3]{-12.97}$ **c]** $(-3.1)^4$ **d]** $(-3.1)^5$

Solution

a] Press $\boxed{(} \;\; 5.3 \;\; \boxed{-} \;\; 9.7 \;\; \boxed{)} \;\; \boxed{\sqrt{x}}$. The display will indicate an ERROR. This is predictable, since $5.3 - 9.7 = -4.3$ and the square root of a negative number is not a real number.

b] Rewrite $\sqrt[3]{-12.97}$ as $(-12.97)^{1/3}$ and press

$$12.97 \;\; \boxed{+/-} \;\; \boxed{y^x} \;\; 3 \;\; \boxed{1/x} \;\; \boxed{=};$$

the result indicates an ERROR. This is because the calculator will not accept a negative base y when the $\boxed{y^x}$ key is used. However, $\sqrt[3]{-12.97}$ is a real number equal to $-\sqrt[3]{12.97}$. We therefore calculate $\sqrt[3]{12.97}$ by pressing $12.97 \;\; \boxed{y^x} \;\; 3 \;\; \boxed{1/x} \;\; \boxed{=}$. The display shows 2.3495. Therefore we have $\sqrt[3]{-12.97} \approx -2.3495$.

c] When evaluating $(-3.1)^4$, the calculator will indicate an ERROR if we press 3.1 $\boxed{+/-}$ $\boxed{y^x}$ 4 $\boxed{=}$, but we know that $(-3.1)^4 =$ $(3.1)^4$, and this can be calculated by using the $\boxed{y^x}$ key. Press 3.1 $\boxed{y^x}$ 4 $\boxed{=}$. The display shows 92.3521. Thus $(-3.1)^4 =$ 92.3521.

d] Since $(-3.1)^5 = -(3.1)^5$, we first evaluate $(3.1)^5$ by pressing 3. $\boxed{y^x}$ 5 $\boxed{=}$. The display shows 286.29151, so we conclude that $(-3.1)^5 = -286.29151$. ∎

EXERCISE A.2

Use a calculator to solve problems 1–20. Answers rounded off to three decimal places are given as a check.

Answers

1. $\sqrt{47.23 + 52.18}$ 9.970

2. $\sqrt{39.4 + (5.8)(7.3)}$ 9.041

3. $\sqrt{54.6 - 31.93}$ 4.761

4. $\sqrt{(9.1)(3.6) - (7.28)(5.97)}$ Imaginary number

5. $\sqrt{9.2^2 + 4.1^2}$ 10.072

6. $\sqrt{(3.87 + 9.4) \cdot 4.83^2}$ 17.595

7. $\sqrt[3]{12.96}$ 2.349

8. $\sqrt[3]{-243.78}$ -6.247

9. $\sqrt[5]{32.786}$ 2.010

10. $\sqrt[4]{17.39}$ 2.042

11. $\dfrac{1}{2} + \dfrac{1}{3} + \dfrac{1}{4} + \dfrac{1}{5}$ 1.283

12. $\dfrac{2}{3} + \dfrac{3}{4} + \dfrac{5}{6}$ 2.250

13. $\dfrac{1}{\sqrt{2}} + \dfrac{1}{\sqrt{3}} + \dfrac{1}{\sqrt{4}}$ 1.784

14. $\dfrac{5}{\sqrt{12}} + \dfrac{7}{\sqrt{3}}$ 5.485

15. $\sqrt[3]{3.47^5 + 29.3^3}$ 29.494

16. $(-4.3)^2 + (-5.9)^3$ -186.889

17. $(-4.1)^3 + (-5.9)^4$ 1142.815

18. $\sqrt{11.9^2 + 13.2^2 - 2(11.9)(13.2)(0.4937)}$ 12.679

19. $\sqrt{[11.9^2 + 13.2^2 - 2(11.9)(13.2)](0.4937)}$ 0.913

20. $\sqrt{4 - \sqrt{2}}$ 1.608

The problems given in Exercise A.4 (pp. 374–375) provide the opportunity for additional practice in using AOS calculators. The student is urged to do most of them.

RPN Calculators

Calculators using Reverse Polish Notation (RPN) can easily be identified by the presence of the (ENT) key (and the absence of the (=) key).
The manufacturer of RPN calculators is Hewlett-Packard (HP). In the
following discussion we shall describe the operation of RPN calculators consistent with HP scientific calculators.

Registers and use of stack

The only external means of communication between the calculator
and its user is through the keyboard and the numbers appearing in
the display. At any time there is only one number in the display;
however, the calculator accepts several numbers and stores them for
recall on keyboard command. The places used to store the numbers
are called registers and may be thought of as physical places inside
the machine where a number is kept until needed. HP machines have
four such registers. The content of one register is displayed by the
machine. This is called the X register. Registers not visible to the user
are called Y, Z, and T. These four registers form the *stack* or *automatic memory* of the machine. In order to use RPN calculators efficiently it is essential to understand the operation of the stack.

If we represent the stack as a mailboxlike set of compartments
$\boxed{X \mid Y \mid Z \mid T}$, where X, Y, Z, and T are the addresses for the boxes,
then we can visualize what is happening inside the calculator. When
a sequence of digit keys is pressed, the corresponding number appears in the X register. Pressing the (ENT) key shifts the number into
the Y register, and the machine is ready to accept a second number.
For example, pressing 2 gives $\boxed{2 \mid Y \mid Z \mid T}$; when we follow this with
(ENT) we get $\boxed{2 \mid 2 \mid Z \mid T}$. If we now press 3, the 2 in the X register is
replaced by 3 and the 2 in the Y register remains. Pressing (ENT) shifts
the contents as shown: $X \to Y \to Z \to T \to$ lost, retaining the number
entered in the X register as well as in the Y register. The series of key
strokes

$$2 \ (ENT) \ 3 \ (ENT) \ 1 \ 5 \ (ENT) \ 4$$

provides us with this arrangement of numbers in the stack: $\boxed{4 \mid 15 \mid 3 \mid 2}$.

Observe that the 15 in the Y register was accomplished without
pressing key (ENT) between 1 and 5. This feature best describes the
purpose of the (ENT) key; that is, to separate the numbers entered into
the machine. Pressing the (ENT) key after 4 will give $\boxed{4 \mid 4 \mid 15 \mid 3}$, losing
the 2 (and the calculator is now ready to accept a new number in the
X register). It may appear that having only a four-stack capacity is a
serious limitation; but this is not the case, since we can perform most

of our computations without any additional registers, as will be demonstrated in the following examples. In fact, some RPN calculators have only three register stacks, and they perform adequately in most problems.

For arithmetic operations, only the numbers in the X and Y registers are used directly. If x is in X and y is in Y, then pressing any one of the keys (+) (−) (×) or (÷) gives the corresponding result $y + x$, $y − x$, $y \times x$ or $y \div x$ in the display.

For example, to evaluate $2 + 3$, press 2 (ENT) 3 to get (3 2); then pressing the (+) key gives (5). To evaluate $15 − 4$, press 15 (ENT) 4 (−); the result will show 11 in the display. Similar steps are followed in the operations of multiplication and division.

In the following examples we use grids to show the contents of each register of the stack after the indicated key has been pressed. A blank register does not necessarily mean an empty register (containing 0) but rather that we are not concerned with its content, since it is not used in our computations. The solutions given do not necessarily include the most efficient key sequences. Here we are interested in illustrating important features of RPN calculators.

In the examples and exercises we use small integer numbers so that you can follow the computations as they appear in the display.

EXAMPLE 1   Calculate $7 + 6 \times 4$.

Solution   Press

7 (ENT) 6 (ENT) 4 (×) (+).

At each entry the contents of the stack registers are shown in the following grid:

T							
Z				7	7		
Y		7	7	6	6	7	
X	7	7	6	6	4	24	31
Key	7	(ENT)	6	(ENT)	4	(×)	(+)

Thus $7 + 6 \times 4 = 31$.   ∎

EXAMPLE 2   Evaluate $7 + 3(4 + 6)$.

Solution   Press the following sequence of keys:

7 (ENT) 3 (ENT) 4 (ENT) 6 (+) (×) (+).

The following grid shows the contents of the stack registers at each stage:

T							7	7	7	7	7
Z				7	7	3	3	7	7	7
Y		7	7	3	3	4	4	3	7	7
X	7	7	3	3	4	4	6	10	30	37
Key	7	(ENT)	3	(ENT)	4	(ENT)	6	(+)	(×)	(+)

Thus the result is 37. ■

In the above solution we first entered all four numbers into the stack and then performed appropriate operations on the contents of the X and Y registers. A more efficient method would be to evaluate $(4 + 6) \times 3 + 7$ with the key sequence

$$4 \ (ENT) \ 6 \ (+) \ 3 \ (\times) \ 7 \ (+).$$

However, the solution given above illustrates an important feature of RPN calculators. When a number, such as 7 in the above grid, reaches the T register and the lower numbers are used up in operations, then that number remains in the upper registers (just as 7 did in the grid above). This is a useful feature that will be illustrated in Example 4.

The contents of the stack registers can be displayed in the X register by using the roll key, (R↓). For example, suppose the stack contains 6, 4, 3, and 7, as it did at one stage in Example 2. Pressing (R↓) repeatedly "rolls" these numbers in the stack as illustrated in the following grid. Also note that in the final step, when we press (ENT), the 7 is lost.

T	7	6	4	3	7	3
Z	3	7	6	4	3	4
Y	4	3	7	6	4	6
X	6	4	3	7	6	6
Key		(R↓)	(R↓)	(R↓)	(R↓)	(ENT)

The (CHS) and (x↔y) keys

The (CHS) key changes the sign of the contents of the X register *only* and must be used to enter a negative number into the machine. The (CHS) key does not shift the content of the X register to Y; hence it is necessary to use the (ENT) key to separate numbers after the (CHS) key is pressed and before a new number is entered. The (x↔y) key interchanges the contents of the X and Y registers and leaves the contents of Z and T undisturbed. This key is frequently used when performing lengthy calculations involving subtraction and/or division.

EXAMPLE 3   Evaluate $\dfrac{-50}{5 + 4(8 - 3)}$.

Solution If we attempt to enter all of the numbers (-50, 5, 4, 8, and 3) into the stack first, the -50 will be lost. However, suppose we first evaluate the denominator with the sequence of keys

$$5 \;\boxed{\text{ENT}}\; 4 \;\boxed{\text{ENT}}\; 8 \;\boxed{\text{ENT}}\; 3 \;\boxed{-}\;\boxed{\times}\;\boxed{+}.$$

At this point we see 25 in the display (the X register). Now we can press 50, $\boxed{\text{CHS}}$. This gives -50 in X, and the 25 moves to the Y register. These are in the wrong order for the division operation. So press the $\boxed{x \leftrightarrow y}$ key, and this interchanges the contents of the X and Y registers. Now press $\boxed{\div}$ to get -2 as the answer.

Another key sequence that can be used is as follows: After the denominator, 25, is in the display, press $\boxed{1/x}$, 50, $\boxed{\text{CHS}}$, $\boxed{\times}$. ∎

EXAMPLE 4 Evaluate $f(-3)$ where $f(x) = 4x^2 + 3x - 6$.

Solution Write $f(x)$ as $f(x) = (4x + 3)x - 6$. Suppose we first "fill the stack" with -3 by pressing $\boxed{\text{ENT}}$ several times. Then we can proceed as follows:

T					-3	-3	-3	-3	-3	-3	-3	-3
Z				-3	-3	-3	-3	-3	-3	-3	-3	-3
Y			-3	-3	-3	-3	-3	-12	-3	-3	27	-3
X	3	-3	-3	-3	-3	4	-12	3	-9	27	6	21
Key	3	CHS	ENT	ENT	ENT	4	$\times$	3	$+$	$\times$	6	$-$

Thus $f(-3) = 21$. ∎

Clearing the calculator

Calculators have various keys for clearing parts of the machine. One key that clears the display only (that is, the X register) is generally labeled $\boxed{\text{CLX}}$ and is especially useful in correcting an error when a wrong number is entered into the display. Some of the more sophisticated calculators have special keys for clearing only the storage registers, or the prefix, or the program in programmable calculators. The owner's manual explains how these keys operate in a particular calculator. In fact, the reader is urged to consult the owner's manual whenever there is a question concerning the operation of any key.

If one wishes to clear the entire machine, turning the calculator off and then on will do it, except for the sophisticated calculators with a continuous memory. It is not always necessary to clear the stack (or even the display) before beginning a new computation, since only the numbers entered for a given calculation are used and the content of the other registers is irrelevant.

EXERCISE A.3

The problems in this exercise can be solved by using the keys $\boxed{+}$, $\boxed{-}$, $\boxed{\times}$, $\boxed{\div}$, $\boxed{\text{CHS}}$, $\boxed{x \leftrightarrow y}$; however, the more experienced student may prefer other keys, such as $\boxed{1/x}$ and $\boxed{x^2}$.

1. For each indicated calculation, two keying methods are given. In each key sequence, fill in a grid giving the content of the X, Y, Z, and T registers after each command has been executed by the calculator. Determine which method evaluates the given calculation correctly.
   a] Evaluate $8 \times 4 - 5$.

   Key sequence 1:   8 $\boxed{\text{ENT}}$ 4 $\boxed{\text{ENT}}$ 5 $\boxed{-}$ $\boxed{\times}$
   Key sequence 2:   8 $\boxed{\text{ENT}}$ 4 $\boxed{\times}$ 5 $\boxed{-}$

   b] Evaluate $(7 + 4) \times 8$.

   Key sequence 1:   7 $\boxed{\text{ENT}}$ 4 $\boxed{+}$ 8 $\boxed{\times}$
   Key sequence 2:   7 $\boxed{\text{ENT}}$ 4 $\boxed{\text{ENT}}$ 8 $\boxed{\times}$ $\boxed{+}$

2. Determine what numerical expression is evaluated by each of the given key sequences.
   a] 2 $\boxed{\text{ENT}}$ 4 $\boxed{\text{ENT}}$ 1 $\boxed{-}$ $\boxed{\times}$ 3 $\boxed{+}$
   b] 5 $\boxed{\text{ENT}}$ 4 $\boxed{\times}$ 2 $\boxed{-}$ 3 $\boxed{+}$
   c] 5 $\boxed{\text{ENT}}$ 4 $\boxed{+}$ 3 $\boxed{\div}$ 2 $\boxed{x \leftrightarrow y}$ $\boxed{-}$

3. Give a sequence of keys that will correctly evaluate each of the given expressions. In each case, make a grid showing the content of all stack registers after each key has been pressed:
   a] $2 + 3 + 4 - 6$ **b]** $2 - 4 + 5 \cdot 7$ **c]** $4 \div 2 + 6 \div 3$

   d] $\dfrac{4 + 6}{2 + 3}$ **e]** $3(2 - 6) + 4(5 - 2)$

In Problems 4–20, evaluate the given expression using a calculator. Make a grid whenever necessary to get a sequence of keys giving the correct answer. Your computations can be checked with the answers given to four decimal places.

	Answers
4. $(1.4 + 3.6)(2.1)$	10.5000
5. $(3.8 - 4.3)(6.3)$	-3.1500
6. $2.9 + 1.6 \div 3$	3.4333
7. $\dfrac{1.96 + 2.3}{4.2 - 3.1}$	3.8727
8. $14.98 - \dfrac{4.3 + 2.6}{5.7}$	13.7695
9. $\dfrac{5.4(6.9 - 1.2) + 4}{7 + 4.3}$	3.0779
10. $\dfrac{1}{4} + \dfrac{1}{5} + \dfrac{1}{7}$	0.5929

11. $\dfrac{3}{4} + \dfrac{4}{5} + \dfrac{2}{7}$ 1.8357

12. $5^2 + 7 \cdot 5 - 3$ 57

13. $\dfrac{2 \cdot 4^2 - 5 \cdot 4 - 3}{2 \cdot 4 + 1}$ 1

14. $\left(\dfrac{3.8}{5.1}\right)^2 + \dfrac{9.6}{4.3}$ 2.7877

15. $5(-1.32)^4 + 4(-1.32)^3$ 5.9799

16. $\dfrac{3.48 - (1.23)(4.75)}{8.41 - 2.54(3.57 - 6.75)}$ -0.1433

17. If $f(x) = 1.47x - 5.36$, find $f(3.4)$ -0.3620

18. If $f(x) = \dfrac{1.56 - 2.36x}{1.57x}$, find $f(-5.7)$ -1.6775

19. If $f(x) = 7.3x^2 - 4.1x + 3.5$, find $f(3.78)$ 92.3073

20. If $f(x) = \dfrac{2.4x^2 - 3.5x - 1.8}{3.2 - 1.5x}$, find $f(-4.3)$.   5.9716

The keys $\boxed{x^2}$ $\boxed{1/x}$ $\boxed{STO}$ $\boxed{RCL}$ $\boxed{y^x}$ $\boxed{\sqrt{x}}$

There is no one correct way to perform a given calculation, although some methods of key entry may be more efficient than others. In the preceding section we considered an example in which we evaluated $f(x) = 4x^2 + 3x - 6$ at $x = -3$. This can also be done by using the $\boxed{x^2}$ key. Pressing the $\boxed{x^2}$ key squares the content of the X register, while the content of the other registers remains unchanged. This is illustrated in the following grid, where we evaluate $f(-3)$ for $f(x) = 4x^2 + 3x - 6$.

EXAMPLE 5   Evaluate $f(-3)$ where $f(x) = 4x^2 + 3x - 6$.

Solution   Here we illustrate the use of the $\boxed{x^2}$ key. Press

4 $\boxed{ENT}$ 3 $\boxed{CHS}$ $\boxed{x^2}$ $\boxed{\times}$ 3 $\boxed{ENT}$ 3 $\boxed{CHS}$ $\boxed{\times}$ $\boxed{+}$ 6 $\boxed{-}$.

The result is 21.   ∎

The $\boxed{\sqrt{x}}$ and $\boxed{1/x}$ keys operate in a manner similar to that of $\boxed{x^2}$; pressing $\boxed{\sqrt{x}}$ takes the square root of the number in the X register and displays the result, while $\boxed{1/x}$ takes the reciprocal of the number appearing in the X register and displays it. Each of these keys leaves the content of the Y, Z, and T registers unchanged.

All scientific calculators have at least one memory storage, and some have several. When the $\boxed{STO}$ key is pressed, the content of the X register is placed in a memory storage separate from any of the stack registers. Pressing the recall key (RCL) will return that number to the X register whenever it is needed and also retain the number in the memory.

If a calculator has more than one memory storage, it is necessary to tell the machine the address of the particular memory to be used. For instance, if the calculator has eight memories numbered 0 through 7, the storage command consists of (STO) followed by one of the numbers 0 through 7. Similarly, for recall, press (RCL) followed by the number 0 through 7 corresponding to the address where the number is stored.

EXAMPLE 6 Evaluate $\dfrac{1}{\sqrt{2}} - \dfrac{1}{\sqrt{3}}$.

Solution We use the (√x̄) to evaluate $\sqrt{2}$ and $\sqrt{3}$, and then the (1/x) key to get $1/\sqrt{2}$ and $1/\sqrt{3}$, since this is simpler than using the (÷) to calculate $1 \div \sqrt{2}$ and $1 \div \sqrt{3}$. The following sequence of keys gives the answer:

$$2 \ (\sqrt{x}) \ (1/x) \ 3 \ (\sqrt{x}) \ (1/x) \ (-).$$

The result is 0.1298 to four decimal places. ∎

The (STO) and (RCL) keys are particularly useful when the same number occurs more than once in a computation. The following example illustrates the use of these keys and the (yˣ) key.

EXAMPLE 7 Evaluate $f(x) = 2x^3 + 5x^2 - 7x + 4$ at $x = 2.5$.

Solution We wish to evaluate

$$f(2.5) = 2(2.5)^3 + 5(2.5)^2 - 7(2.5) + 4.$$

First we store 2.5 in one of the eight memories that are numbered $0, 1, 2, \ldots, 7$, say in memory 4, by pressing 2.4 (STO) 4. Then we can recall the contents of memory 4 by pressing (RCL) 4. We can evaluate $(2.5)^3$ by 2.5 (ENT) 3 (yˣ). However, when 2.5 is stored in or recalled from memory, it is not necessary to press (ENT) to separate 2.5 from the next number to be placed in the display. That is, (RCL) 4, 3 (yˣ) will give $(2.5)^3$.

Returning to the problem, we can find $f(2.5)$ with the following sequence of keys:

$$2.5 \ (\text{STO}) \ 4, 3 \ (y^x) \ 2 \ (\times) \ (\text{RCL}) \ 4 \ (x^2) \ 5 \ (\times) \ (+)$$
$$(\text{RCL}) \ 4, 7 \ (\times) \ (-) \ 4 \ (+).$$

The final answer is 49. ∎

EXERCISE A.4

Evaluate the following expressions to three decimal places. Check your answers; in case of disagreement, complete a grid to determine whether your answer or the authors' (or neither) is correct. Answers to all of the following problems are given at the end of the exercise set.

1. If $f(x) = 3x^2 - 2x + 1$, find $f(2.13)$.

2. Evaluate $f(x) = 1.6x^2 - 2.4x + 4.1$ at $x = 2.46$.

3. Find the value of $g(x) = 5x^2 + \dfrac{1}{x}$ at $x = -1.57$.

4. Evaluate $\dfrac{1}{2} + \dfrac{1}{3} + \dfrac{1}{4} + \dfrac{1}{5} + \dfrac{1}{6}$.

5. If $f(x) = 1 + \dfrac{1}{1 + \dfrac{1}{1 + 1/x}}$, find: a] $f(2)$ b] $f(-1.48)$

6. Evaluate the following expressions by using the ⬭$_\pi$ key on your calculator:

   a] $(24.67)\left(64 + \dfrac{27}{60}\right)\dfrac{\pi}{180}$ b] $\dfrac{1}{2}(24.67)^2\left(64 + \dfrac{27}{60}\right)\left(\dfrac{\pi}{180}\right)$

7. If $u = 2.21$, $v = \dfrac{7\pi}{10}$, $t = 126.43\left(\dfrac{\pi}{180}\right)$, order these three numbers from smallest to largest.

8. Evaluate:

   a] $(34.63)\left(\dfrac{\pi}{180}\right)\sqrt{\dfrac{2(35.61)(180)}{34.63\pi}}$ b] $\sqrt{\dfrac{2(35.61)(34.63)\pi}{180}}$

9. Evaluate:

   a] $\dfrac{1 + \sqrt{7}}{3}$ b] $\left(\dfrac{1 + \sqrt{7}}{3}\right)^2$. c] $\left(\dfrac{1 + \sqrt{7}}{3}\right)^3$

10. The following numbers may be used as rational approximations of π. Calculate each number and use the ⬭$_\pi$ key on your calculator to determine the decimal-place accuracy:

    a] $\dfrac{22}{7}$ b] $\dfrac{333}{106}$ c] $\dfrac{355}{113}$ d] $\dfrac{208341}{66317}$

11. Evaluate:
    a] $(\sqrt{5.38})^3$ b] $\sqrt{5.38^3}$

12. Evaluate:
    a] $\sqrt{24.3 + 36.8}$ b] $\sqrt{24.3} + \sqrt{36.8}$

13. Evaluate:
    a] $\dfrac{\sqrt{3} - 1}{\sqrt{3} + 1}$. b] $2 - \sqrt{3}$

14. If $f(x) = 3x^4 - 8x^2 + 12$, find $f(1.43)$.

15. If $f(x) = \dfrac{x^6 - 1}{x - 1}$, find
    a] $f(3)$ b] $f(2.3)$ c] $f(-1.8)$ d] $f(1)$

16. If $g(x) = x^5 + x^4 + x^3 + x^2 + x + 1$, find:

    a] $g(3)$ **b]** $g(2.3)$ **c]** $g(-1.8)$ **d]** $g(1)$

Compare these results with the answers in Problem 15. What conclusions can you draw about the functions f and g?

17. Evaluate: **a]** $\sqrt{24.7} - \sqrt{36.8}$ **b]** $\sqrt{24.7 - 36.8}$

18. Evaluate: $\left(\dfrac{1 - \sqrt{5}}{2}\right)^3$ **19.** Evaluate: $\sqrt{(1 - \sqrt{3})^2 - 1}$

20. If $f(x) = 3x^4 - 4x^3 + x - 5$, find:

    a] $f(3)$ **b]** $f(-1.2)$ **c]** $f(\pi)$ **d]** $f\left(\dfrac{1 + \sqrt{5}}{2}\right)$

Answers to Exercise A.4

1. 10.351   **2.** 7.879   **3.** 11.688   **4.** 1.450

5. a] 1.600   **b]** 1.245   **6. a]** 27.750   **b]** 342.301

7. $u = 2.210$, $v \approx 2.199$, $t \approx 2.207$; thus $v < t < u$

8. a] 6.561   **b]** 6.561

9. a] 1.215   **b]** 1.477   **c]** 1.795

10. The number agrees with π through

    a] two decimal places   **b]** four decimal places

    c] six decimal places   **d]** at least eight decimal places

11. a] 12.479   **b]** 12.479   **12. a]** 7.817   **b]** 10.996

13. a] 0.268   **b]** 0.268   **14.** 8.186

15. a] 364   **b]** 113.105

    c] -11.790; use $(-1.8)^6 = 1.8^6$ **d]** Calculator indicates ERROR. Why?

16. a] 364   **b]** 113.105   **c]** -11.790 **d]** 6

17. a] -1.096 **b]** Calculator indicates ERROR. Why?

18. -0.236, since $\dfrac{1 - \sqrt{5}}{2}$ is negative; use $\left(\dfrac{1 - \sqrt{5}}{2}\right)^3 = -\left(\dfrac{\sqrt{5} - 1}{2}\right)^3$

19. Calculator indicates ERROR. Why?

20. a] 133   **b]** 6.933; use $3(-1.2)^4 - 4(-1.2)^3 + (-1.2) - 5 = 3(1.2)^4 + 4(1.2)^3 - 1.2 - 5$

    c] 166.344   **d]** 0.236

Hand-Held Computers

A hand-held computer performs both manual calculations and program calculations. The most important feature is that it uses BASIC program language, and this is what distinguishes it from programmable calculators. Here we shall concern ourselves with its function as a calculator. However, we strongly urge you to learn the programmable features and practice with the suggested computer problems at the end of each section of this book.

Several makes of hand-held computers are currently on the market, and their modes of operation are very similar. The instructions given here are in reference to the CASIO FX700P model. There should be no difficulty adapting to other makes such as those produced by Radio Shack and Sharp.

We first note that most of the keys on the keyboard serve two or three operations and are labeled on top, below each key (in blue), and above each key (in red). For instance, the key in the lower left-hand corner is labeled as (Z) on top, SQR below, and RETURN above. The labelings in red (such as RETURN) are commonly used words when programming in BASIC, while the labels in blue (such as SQR) refer to mathematical functions. To activate the SQR (square root) function, we first place the machine in function mode by pressing function key labeled with the letter F in blue in the top row of keys. If your machine does not have the blue labelings, such as the SQR functions, then you can simply type in the appropriate letters from the keyboard (S, Q, R in this case). In the following, we shall refer to the square root function as a single key denoted by (SQR).

Hand-held computers perform manual calculations in a manner very similar to that of *algebraic* calculators discussed at the beginning of this appendix. We consider now the key sequences for a variety of numerical problems in which the machine serves as a calculator.

Algebraic operating system

The hand-held computer performs arithmetic operations according to the MDAS rule: multiplication and division before addition and subtraction.

EXAMPLE 1   Evaluate $2 + 3 \times 7$.

Solution   Press keys as you read the given expression from left to right.

$$2 \; (+) \; 3 \; (*) \; 7 \; (\text{EXE}).$$

The answer 23 appears in the display. Note that (EXE) serves as the equals key. The key labeled (=) is used for other purposes such as in programming or assigning numerical values to variables as illustrated in Example 3.   ■

Entry corrections

The *cursor keys* (→) and (←) are used to move the cursor and are especially useful to correct mistakes that occur when a wrong key is pressed. For instance, if you wish to enter the number 24.876 into the display and, after entry, you check and see that the display shows 24.856, then you can correct this without having to reenter the entire number. Press (←) (←), and the flashing cursor will move under the digit 5; now press 7, which types over the 5. Now you can press (→) to move the cursor to the right of the digit 6, and then continue with the evaluation.

The *delete* (DEL) and *insert* (INS) keys are useful in making corrections. Suppose you have the number 24.856 in the display and you want to change it to 24.56. Move the cursor as described above to appear under the digit 8, then press (DEL), and this will give 24.56 in the display. Suppose you have 24.56 in the display and you wish to change it to 24.856. Move the cursor to the position under the digit 5 and then press (INS). This moves the digits 56 to the right one place and leaves a space for you to press 8.

The *all clear key* labeled (AC) is used to clear the entire display. *It has to be pressed to clear an error message.*

Square root function

EXAMPLE 2   Evaluate $\dfrac{1}{\sqrt{2}} + \dfrac{1}{\sqrt{3}}$.

Solution   The following key sequence can be used to perform this calculation:

$$1 \ (/) \ (SQR) \ 2 \ (+) \ 1 \ (/) \ (SQR) \ 3 \ (EXE).$$

The display now shows the answer 1.28445705.   ■

Assigning numbers to variables, storing numbers

If a number occurs several times in a given problem, you may wish to "store" it in a memory and avoid having to type it in separately each time. This is easily done by using one of the letter keys and the (=) key. For example, to store the number 4, we can *assign* 4 to one of the letters, say A, by pressing (A) (=) 4 (EXE).

Now if we press (A) (*) (A) (EXE), we see 16 appear in the display. The number 4 is assigned to A (even after the machine is turned off and on again) until a new number is given to A. For instance, press Ⓐ ⊜ 0 (EXE), and 0 is now assigned to A. Let us now illustrate how the idea of assigning a number to a letter can be used in the next example.

EXAMPLE 3 Evaluate the function $f(x) = x^2 + 2x + \sqrt{x}$ at $x = 2.48$

Solution We want to evaluate

$$f(2.48) = (2.48)^2 + 2(2.48) + \sqrt{2.48}.$$

Since 2.48 occurs three times, let us first assign it to a letter, say X by pressing Ⓧ ⊜ 2.48 (EXE). Now to perform the desired calculation, press the following key sequence:

Ⓧ ✳ Ⓧ ⊕ 2 ✳ Ⓧ ⊕ (SQR) Ⓧ (EXE). ∎

The result 12.68520157 appears in the display. ∎

Evaluating exponential expressions

The key that corresponds to the function ⓨˣ on a calculator is ⓣ. It is above the decimal ⊙ key and is activated by first pressing the shift key, labeled Ⓢ in red, and then ⊙. We shall denote this combination of keys by ⓣ. Thus to evaluate 4^3, press 4 ⓣ 3 (EXE), and the answer 64 appears in the display.

Care must be taken when evaluating a negative number to a power. If the exponent is *not an integer*, then the ⓣ operation will give ERROR in the display.

For instance, if we try to evaluate $(-8)^{1/3}$ by the key sequence

⊙ ⊖ 8 ⊙ ⓣ ⊙ 1 ⊘ 3 ⊙ (EXE),

we would see ERROR in the display. You can use $(-8)^{1/3} = -(8)^{1/3}$ and evaluate this expression by the sequence

⊖ 8 ⓣ ⊙ 1 ⊘ 3 ⊙ (EXE).

If the exponent is an integer (positive or negative), we can use the ⓣ key. For instance, we can evaluate $(-2)^{-4}$ with the following key sequence:

⊙ ⊖ 2 ⊙ ⓣ ⊖ 4 (EXE).

The answer 0.0625 appears in the display.

EXAMPLE 4 Evaluate $(2.56)^3 - (1.84)^4$.

Solution Press

2.56 ⓣ 3 ⊖ 1.84 ⓣ 4 (EXE).

The answer 5.31492864 appears in the display. ∎

EXAMPLE 5 Evaluate $\sqrt[3]{5.73}$.

Solution $\sqrt[3]{5.73}$ can be written in exponential form as $(5.73)^{1/3}$. To evaluate this, press the following key sequence:

5.73 $\boxed{\uparrow}$ $\boxed{(}$ 1 $\boxed{/}$ 3 $\boxed{)}$ $\boxed{\text{EXE}}$.

The answer 1.789444393 appears in the display.   ■

Logarithmic and exponential functions _____

The keys labeled $\boxed{\text{LOG}}$ and $\boxed{\text{LN}}$ (in blue) correspond to the logarithmic functions with base 10 and e, respectively. The key denoted by $\boxed{\text{EXP}}$ (in blue) is the exponential function e^x. If these labelings do not appear on your machine, you can type in the appropriate letters from the keyboard.

EXAMPLE 6 Evaluate:

   a] $\log 15.63$ **b]** $5 \ln 2.53$ **c]** $e^{-0.48}$

Solution

   a] Press $\boxed{\text{LOG}}$ 15.63 $\boxed{\text{EXE}}$. The result is 1.193958978.
   b] Press 5 $\boxed{*}$ $\boxed{\text{LN}}$ 2.53 $\boxed{\text{EXE}}$. The answer 4.641096514 appears in the display. Note that it is necessary to have the $\boxed{*}$ key between 5 and $\boxed{\text{LN}}$.
   c] Press $\boxed{\text{EXP}}$ $\boxed{-}$ .48 $\boxed{\text{EXE}}$. This gives $e^{-0.48} \approx 0.6187833918$.   ■

Trigonometric functions _____

When evaluating trigonometric functions, *it is important to have the calculator in the proper mode*. Note the $\boxed{\text{MODE}}$ key in the top row of the keyboard. When the machine is turned on, it is in degree mode. Or pressing $\boxed{\text{MODE}}$ 4 places it in degree mode. If you press $\boxed{\text{MODE}}$ 5, the calculator is placed in radian mode. The display indicates the mode with DEG and RAD in small print.

There are three trigonometric function keys labeled $\boxed{\text{SIN}}$, $\boxed{\text{COS}}$, $\boxed{\text{TAN}}$ in blue. Also the corresponding three inverse trigonometric functions are available and are labeled $\boxed{\text{ASN}}$, $\boxed{\text{ACS}}$, $\boxed{\text{ATN}}$. For instance, $\boxed{\text{ASN}}$ refers to arcsine, which is another name for inverse sine.

Also, you should note that there is a pi key labeled $\boxed{\pi}$ in red.

EXAMPLE 7 Evaluate:

   a] $\sin 48.3°$ **b]** $3 \tan 65°$ **c]** $\sec 32.4°$

Solution   Since the given angles are in degrees, be certain that the calculator is in degree mode. You can check this by looking for DEG in small print in the display.

a] (SIN) 48.3 (EXE). The answer is 0.7466381823.

b] Press 3 (✱) (TAN) 65 (EXE). This gives 6.433520762. Note that it is necessary to have the (✱) between 3 and (TAN).

c] Here we first write sec $32.4° = 1/\cos 32.4°$ and evaluate the right-hand side with the key sequence, 1 (/) (cos) 32.4 (EXE). We get 1.184373950.   ∎

EXAMPLE 8   Evaluate:

a] $\cos 1.45$ b] $\sin\left(\dfrac{3\pi}{7}\right)$

Solution   The given angles are in radian measure, and so we first place the calculator in radian mode by pressing (MODE) 5.

a] Press (cos) 1.45 (EXE). This gives 0.1205027694 in the display.

b] With the calculator in radian mode, press

(SIN) (() 3 (✱) (π) (/) 7 ()) (EXE).

The answer is 0.9749279122.   ∎

EXAMPLE 9   Evaluate:

a] $\text{Sin}^{-1} 0.43$ b] $\text{Cos}^{-1}(-0.75)$

Give answers in degrees.

Solution   Place the calculator in degree mode and then press the following key sequences.

a] (ASN) .43 (EXE).   The display shows that $\text{Sin}^{-1} 0.43 \approx$ 25.46756014°.

b] (ACS) (−) .75 (EXE). This gives $\text{Cos}^{-1}(-0.75) \approx 138.5903779°$.   ∎

EXAMPLE 10   Evaluate $\text{Arctan}\left(\dfrac{x}{\sqrt{x^2 + 1}}\right)$ for $x = 1.4582$. Give your answer in radians.

Solution   Place the calculator in radian mode. We wish to compute the number

$$\text{Arctan}\left(\frac{1.4582}{\sqrt{(1.4582)^2 + 1}}\right).$$

Since 1.4582 occurs twice and we would prefer not to enter it a second time, we first assign it to, say, N by pressing Ⓝ ⊜ 1.4582 (EXE). Now press the key sequence

(ATN) (() (N) (/) (SQR) (() (N) (✱) (N) (+) 1 ()) ()) (EXE).

The answer is 0.6896243309.   ■

We conclude with one final suggestion. Your hand-held computer will do considerably more than we have outlined above. Study the manual that accompanies your machine and take full advantage of its computing capabilities.

EXERCISE A.5

In Problems 1–25, evaluate the given numerical expressions. In case your calculator gives "ERROR" as a response, explain. Answers rounded off to three decimal places are provided so that you can get an immediate check.

Answers

1. $\dfrac{3.47 - (8.56)(2.43)}{1.56}$ $\qquad$ -11.109

2. $\dfrac{21.56}{5.87 - (8.56)(2.83)}$ $\qquad$ -1.175

3. $\sqrt{3} + \dfrac{1}{\sqrt{5}}$ $\qquad$ 2.179

4. $\sqrt[3]{4.36} - \sqrt{4.36}$ $\qquad$ -0.454

5. $\sqrt{(2.47)^2 + (3.56)^2}$ $\qquad$ 4.333

6. $\sqrt{(7.23)^2 - (1.56)^2}$ $\qquad$ 7.060

7. $42\left(\dfrac{\pi}{180}\right)$ $\qquad$ 0.733

8. $1.27\left(\dfrac{180}{\pi}\right)$ $\qquad$ 72.766

9. $(2.6)(3.2) \div \left(\dfrac{3}{4.3} - \dfrac{2}{6.5}\right)$ $\qquad$ 21.334

10. $(3.6)^{4/7}$ $\qquad$ 2.079

11. $(3.6 + \sqrt{4.7})^{3/7}$ $\qquad$ 2.119

12. $\ln\left(\dfrac{1 + \sqrt{3}}{2}\right)$ $\qquad$ 0.312

13. $\log\left(\dfrac{2 + \sqrt{3}}{2}\right)$ $\qquad$ 0.271

14. $\ln\left(\dfrac{2 - \sqrt{5}}{3}\right)$ $\qquad$ ERROR, $\dfrac{2 - \sqrt{5}}{3}$ is a negative number.

15. $\sqrt{3 - \sqrt{15}}$ 　　　　　　　ERROR, $3 - \sqrt{15}$ is a negative number.

16. $e^{\sqrt{3}}$ 　　　　　　　　　　　　　　　　5.652

17. $\sin 153°$ 　　　　　　　　　　　　　　0.454

18. $\cos \dfrac{5\pi}{8}$ 　　　　　　　　　　　　-0.383

19. $\text{Sin}^{-1}(-0.475)$, in radians 　　　-0.495

20. $\text{Arccos}\left(\dfrac{1 - \sqrt{3}}{5}\right)$, in degrees 　　　$98.419°$

21. $\dfrac{43.6 \sin 37°}{\sin 16°}$ 　　　　　　　　　　95.194

22. $\sqrt{6.3^2 + 2.4^2 - 2(6.3)(2.4) \cos 49°}$ 　　5.061

23. $\text{Sin}^{-1}\left(\dfrac{63 \sin 43.5°}{55}\right)$, in degrees 　　$52.044°$

24. $(\cos 1.3)^2 - (\sin 1.3)^2$ 　　　　　　-0.857

25. $\tan\left(\dfrac{1 + \sqrt{3}}{5}\right)$ 　　　　　　　　　　0.608

In Problems 26–30, a sequence of keys is given. What numerical expression is evaluated when the sequence is executed?

26. (SQR) (⫐ 2.8 (✳) 2.8 (＋) 1.4 (✳) 1.4 (⫐ (EXE)

27. (SQR) 6.37 (＋) 2.73 (EXE)

28. (SQR) (⫐ 6.37 (＋) 2.73 (⫐ (EXE)

29. 5.48 (↑) -0.41 (EXE)

30. (SQR) 2 (＋) (SQR) (⫐ 1 (＋) (SQR) 3 (⫐ (EXE)

In Problems 31 and 32 we are given four different key sequences. In each case, what expression is being evaluated?

31. a] 2 (✳) (LN) 5 (EXE) 　　　　　　**b]** (LN) 5 (✳) 2 (EXE)
　c] (⫐ (LN) 5 (⫐ (✳) 2 (EXE)
　d] (LN) (⫐ 5 (✳) 2 (⫐ (EXE)

32. a] 3 (✳) (EXP) 1 (EXE) 　　　　　　**b]** (EXP) 1 (✳) 3 (EXE)
　c] (⫐ (EXP) 1 (⫐ (✳) 3 (EXE)
　d] (EXP) (⫐ 1 (✳) 3 (⫐ (EXE)

Approximate Numbers

B

APPENDIX

In most applications of mathematics to real-life problems we encounter two types of numbers: exact and approximate. Examples of exact numbers are 1/2, 4/13, and π. However, when these numbers are expressed in decimal form, we have

$$\frac{1}{2} = 0.5; \qquad \frac{4}{13} = 0.307692307\ldots; \qquad \pi = 3.141592\ldots$$

The decimal representation of 1/2 is finite, while the decimal representations of 4/13 and π are infinite. There is no problem in replacing 1/2 by 0.5, but when the decimal representation of 4/13 or of π is required, it becomes necessary to round off and use only an approximate decimal value. This is one source of approximate numbers.

Another source of approximate numbers comes from applications involving measurements, and in almost all cases the results are expressed as approximate numbers (limited to the degree of accuracy of the measuring instruments). Approximate numbers are then used in formulas to compute other quantities, and so the final numbers are, of necessity, also approximate. In the following discussion our primary goal is to establish rules that can be used in problems involving computations with approximate numbers. In order to do this we first discuss significant digits, scientific notation, and rounding off of numbers.

Significant Digits and Scientific Notation

For a better understanding of approximate numbers it may be helpful to consider some examples first. Suppose that four different objects are measured and their lengths are determined as

$a \approx 24.3$ cm, $\qquad b \approx 0.00407$ m, $\qquad c \approx 832.0$ cm, $\qquad d \approx 34{,}700$ cm.

This means that a is an approximate number representing a length that is actually somewhere between 24.25 and 24.35 cm. Similarly the exact value of b is somewhere between 0.004065 and 0.004075 m while that of c is between 831.95 and 832.05 cm.

In the case of d it is not clear what accuracy is implied. For example, d might have been measured as 347 meters, in which case the exact value is somewhere between 346.5 and 347.5 m (that is, d is actually between 34,650 and 34,750 cm). It is possible that d was measured to the nearest tenth of a meter (nearest 10 cm), in which case we would write $d \approx 347.0$ m. This implies that d is somewhere between 346.95 and 347.05 m (that is, d is between 34,695 and 34,705 cm). Similarly, if d has been measured accurately to the nearest centimeter, then $d \approx 34,700$ means that $34,699.5 < d < 34,700.5$ cm.

Thus the above examples lead to the following question: When a number is represented in decimal form, which of the digits are significant?

For $a \approx 24.3$ cm, all three digits 2, 4, 3 are meaningful in expressing the accuracy of the measurement; thus we say that a has three significant digits.

For $b \approx 0.00407$ m, the zero before the decimal and the two zeros after the decimal merely serve the purpose of telling us where the decimal is located, while the remaining digits 4, 0, 7 give information about the accuracy of measurement. If b were expressed in centimeters, then $b \approx 0.407$ cm, and we would not even encounter the two zeros immediately after the decimal point. Thus b has three significant digits.

In the case of $c \approx 832.0$ cm, the zero after the decimal tells us that the measurement was made to the nearest tenth of a centimeter, and we do not need to be told where the decimal is located. Therefore all four digits 8, 3, 2, 0 are significant.

In the case of $d \approx 34,700$ cm, the two zeros are certainly necessary to locate the decimal point, but it is not clear whether they give us any information about the accuracy of measurement or not. Thus we would say that 3, 4, 7 are significant digits and an additional statement is required concerning the significance of the two zeros. A convenient way to give this information is to use scientific notation. Thus if d is accurate to the nearest meter (nearest 100 cm), then we write $d \approx 3.47 \times 10^4$ cm, and this indicates that only the 3, 4, 7 are significant digits. If d is accurate to the nearest 10 cm, then we write $d \approx 3.470 \times 10^4$ cm and 3, 4, 7, 0 are significant digits. In a similar way, $d \approx 3.4700 \times 10^4$ cm implies that d is measured to the nearest centimeter, and so all of the digits 3, 4, 7, 0, 0 are significant.

The above discussion leads us to the following *general statement concerning significant digits*:

When a number is written in decimal form, its significant digits begin with the first nonzero digit on the left and end with the last digit on the right that definitely gives information about the accuracy of the number.

That is, all nonzero digits are significant, while zeros that merely serve the purpose of locating the decimal point are not, but all other zeros are. In cases in which it is not clear whether a zero merely indicates the place of the decimal point (as in d above), scientific notation is useful. To represent a number in *scientific notation*, we write it as a product of a number between 1 and 10 and a power of 10; all digits of the factor between 1 and 10 are significant.

EXAMPLE 1   Determine which digits are significant in the following numbers:

a] 37.543   b] 136.1030   c] 240.00
d] 0.0048   e] 0.00480   f] 70,400

Solution

a] All five digits are significant.
b] All seven digits are significant (including the zero at the end).
c] The three zeros are significant, and so the number has five significant digits.
d] Only the 4 and 8 are significant digits.
e] The 4, 8, and the final 0 are significant digits.
f] The digits 7, 0, 4 are significant but we cannot say without further information whether the last two zeros are significant. ∎

EXAMPLE 2   Write each of the numbers given in Example 1 in scientific notation.

Solution

a] $37.543 = 3.7543 \times 10$
b] $136.1030 = 1.361030 \times 10^2$
c] $240.00 = 2.4000 \times 10^2$
d] $0.0048 = 4.8 \times 10^{-3}$
e] $0.00480 = 4.80 \times 10^{-3}$
f] $70,400 = 7.04 \times 10^4$ would indicate that only 7, 0, 4 are significant digits. $70,400 = 7.040 \times 10^4$ would imply that 7, 0, 4, 0 are significant digits. $70,400 = 7.0400 \times 10^4$ would tell us that all five digits are significant. ∎

EXAMPLE 3   The following numbers are expressed in scientific notation. Write them in ordinary decimal form:

a] 2.78×10^4 b] 3.47×10^{-4} c] 3.40×10^3 d] 4.800×10^{-1}

Solution

a] 27,800   b] 0.000347   c] 3400   d] 0.4800   ∎

Rounding Off Numbers

When a number is given in decimal form, it is frequently necessary to express it as an approximate number with fewer significant digits. We describe this as the process of *rounding off a number* and illustrate with the following examples.

EXAMPLE 4 Round off the following numbers to three significant digits:

a] 3476   **b]** 24.74   **c]** 73.80
d] 0.473501   **e]** 2435   **f]** 69.95
g] π **h]** $\pi/2$

Solution

a] The number $3480 = 3.48 \times 10^3$ has three significant digits, and it is an approximation to a number between 3475 and 3485. Since the given number 3476 is in this range, we say that 3476 rounded off to three significant digits is 3.48×10^3. Similarly, for (b), (c), and (d) we get

b] 24.7   **c]** 73.8   **d]** 0.474

e] Here we encounter a borderline case in which it is not clear whether we should round off to 2430 or 2440. Both appear to be equally good, and so we shall adopt the rule that we round *up* and use $2440 = 2.44 \times 10^3$ as the answer.*

f] This is similar to (e), and so 70.0 is the approximation of 69.95 with three significant digits.

g] Since $\pi = 3.14159\ldots$, we round off to 3.14.

h] $\pi/2 = 1.57079\ldots$ rounded off to three significant digits is 1.57. ∎

Computations With Approximate Numbers

When approximate numbers are used in computations, it is natural to ask: "How many significant digits should we retain in the final result?" To give an answer, it is helpful to consider some examples.

---

* Some textbooks give a slightly different rule in which the number is sometimes rounded up and other times it is rounded down.

We first take the problem of multiplying or dividing two approximate numbers, and then we study addition and subtraction of such numbers.[†]

Multiplication and division of approximate numbers

Suppose the length and width of a rectangular object are measured with a ruler marked in millimeters and are found to be $\ell \approx 16.4$ cm, $w \approx 8.6$ cm. We wish to find the area of the rectangle. Since Area = $\ell \times w$, we get

$$\text{Area} \approx (16.4 \times 8.6) \text{ cm}^2 \approx 141.04 \text{ cm}^2.$$

This is a computed value based upon the measurements of ℓ and w expressed as approximate numbers. How many of the five digits in 141.04 are really meaningful and not misleading in terms of stating the actual area of the object?

On the basis of the given information about ℓ and w, all we can say is that

$$16.35 < \ell < 16.45 \text{ cm} \qquad \text{and} \qquad 8.55 < w < 8.65 \text{ cm}.$$

This implies that

$$16.35 \times 8.55 < A < 16.45 \times 8.65 \text{ cm}^2.$$

That is, all we can really say about the actual area is

$$139.7925 < A < 142.2925 \text{ cm}^2. \qquad \text{[B.1]}$$

This is the best claim we can make about the area on the basis of the given measurements.

Our computed value of $A \approx 141.04$ cm^2 is certainly in the range given by expression (B.1), but stating that $A \approx 141.04$ cm^2 implies that we know $141.035 < A < 141.045$ cm^2. This says considerably more than what we actually do know.

Suppose we round off the computed value to three significant digits: $A \approx 141$ cm^2. This implies that $140.5 < A < 141.5$ cm^2, and clearly this still claims more than the inequality given in (B.1). Therefore we try rounding off to two significant digits: $A \approx 140$ cm^2 = 1.4×10^2 cm^2. This means that $135 < A < 145$ cm^2, and making such a statement is consistent with the inequality given by Eq. (B.1).

In conclusion, rounding off the computed value of the area to two significant digits results in the best statement we can make that is consistent with what the given measurements tell us about the actual area. Since ℓ was measured to three significant digits and w to

---

[†] The general problem of accuracy in computations involving other operations (such as square root, logarithm, etc.) is a topic for numerical analysis courses.

two significant digits, this suggests that we should round off the product to the smaller number of significant digits of the measured values.

The problem of dividing two approximate numbers is similar. Suppose $a \approx 34.6$ and $b \approx 8.4$ are approximate numbers and we wish to determine $c = a \div b$. Using a calculator to evaluate c, we get

$$c = \frac{34.6}{8.4} = 4.1190\ldots.$$

How many digits should we retain in the answer? Since $34.55 < a < 34.65$ and $8.35 < b < 8.45$, we obtain

$$\frac{34.55}{8.45} < \frac{a}{b} < \frac{34.65}{8.35}.$$

Thus all we know about c is that

$$4.0888 < c < 4.1497 \qquad \text{(to four decimal places).} \qquad \textbf{[B.2]}$$

If we round off c to three significant digits ($c \approx 4.12$), then we are saying that $4.155 < c < 4.125$, and this is not consistent with what we know about c as given by Eq. (B.2). If we round off to two significant digits ($c \approx 4.1$), then we imply that $4.05 < c < 4.15$, which is in agreement with Eq. (B.2). Since $a = 34.6$ has three significant digits and $b = 8.4$ has two signficiant digits, this example suggests that the quotient of two approximate numbers should be rounded off to the smaller number of significant digits of the two measured values.

The above examples suggest the following rule for multiplying and dividing approximate numbers:

> In the multiplication and division of approximate numbers the result should be rounded off to the least number of significant digits in the data used.

For example, suppose $x \approx 47.36$, $y \approx 17.5$, $z \approx 5.2$ and we wish to evaluate $u = (xy) \div z$. Since the numbers of significant digits in x, y, z are four, three, and two, respectively, we should retain two significant digits for u. Thus

$$u \approx (47.36 \times 17.5) \div 5.2 = 159.3846\ldots,$$

and so we have $u \approx 160 = 1.6 \times 10^2$. If this value is to be used in subsequent computations, then we should use one more significant digit ($u \approx 159$) for that purpose, but we must remember that in the final roundoff, u is accurate to only two significant digits.

Addition and subtraction of approximate numbers

In adding or subtracting approximate numbers the situation is a little different from that of multiplying or dividing. For example, suppose a bank reports that a certain fund has $248,000 in it, where this is accurate to the nearest thousand dollars. Now suppose that $72.35 is added to this fund. It would be misleading to say that the fund now has $248,072.35 in it. We would say that the fund still has $248,000 in it to the nearest thousand dollars (based on the given information). That is, we would write $248,000 + 72.35 \approx 248,000$.

It is clear from this example that when we add two approximate numbers, we are not interested in the number of significant digits each has, but we are primarily interested in the *level of precision* of each number. We say that the level of precision of 248,000 is the nearest thousand while that of 72.35 is the nearest hundredth; thus the level of precision of 72.35 is greater than that of 248,000.

As another example, suppose x, y, and z are approximate numbers given by $x \approx 24.65$, $y \approx 0.036$, $z \approx 132.4$. The levels of precision of x, y, z are hundredths, thousandths, and tenths, respectively. Common sense would suggest that the sum

$$x + y + z \approx 24.65 + 0.036 + 132.4 = 157.086$$

should be rounded off to the nearest tenth, since z is no more accurate than the nearest tenth and we cannot expect $x + y + z$ to be more accurate. Thus

$$x + y + z \approx 157.1.$$

The above examples lead us to the following common-sense rule.

> **Rule for adding and subtracting approximate numbers**
> In the addition and subtraction of approximate numbers the result should be rounded off to the least level of precision in the data used.

Linear and angle measurements

In solving triangles the angle and length measurements are usually given as approximate numbers. Therefore, it is desirable to have a guide that can be used to determine the angle measurements with an accuracy corresponding to that of the length measurements. For angles that are not too close to 0° or 90° the following table provides a satisfactory rule:

Lengths accurate to	Corresponding angles accurate to
Two significant digits	Nearest degree
Three significant digits	Nearest 10′ or 0.1°
Four significant digits	Nearest minute or 0.01°
Five significant digits	Nearest tenth of a minute or 0.001°

In the following examples, suppose x, y, z, u, v, t are approximate numbers given by

$$x \approx 3.48, \quad y \approx 0.0360, \quad z \approx 3251, \quad u \approx 5.004,$$
$$v \approx 84{,}000 \quad \text{(only 8 and 4 are significant)},$$
$$t \approx 24{,}800 \quad \text{(the tens 0 is significant)}.$$

EXAMPLE 5 Write the above numbers in scientific notation.

Solution

$$x \approx 3.48 \times 10^0, \quad y \approx 3.60 \times 10^{-2}, \quad z \approx 3.251 \times 10^3,$$
$$u \approx 5.004 \times 10^0, \quad v \approx 8.4 \times 10^4, \quad t \approx 2.480 \times 10^4. \quad \blacksquare$$

EXAMPLE 6 Give the number of significant digits in each of the above numbers.

Solution x has three; y has three; z has four; u has four; v has two; t has four. $\blacksquare$

EXAMPLE 7 State the level of precision of the given numbers.

Solution The level of precision of x is hundredths, of y is ten thousandths, of z is units, of u is thousandths, of v is thousands, and of t is tens. $\blacksquare$

EXAMPLE 8 Using the rule for multiplication and division of approximate numbers, evaluate the following:

a] $x \cdot z$

b] $\dfrac{yv}{x}$

c] $u \cdot t$

Solution

a] $x \cdot z \approx (3.48)(3251) = 11313.48$. Since x has three and z has four significant digits, the result should be rounded off to three significant digits. Thus

$$x \cdot z \approx 11300 = 1.13 \times 10^4.$$

b] $\dfrac{y \cdot v}{x} \approx \dfrac{(0.0360)(84000)}{3.48} = 868.9655\ldots.$

The smallest number of significant digits of x, y, and v is two, and so the answer should be rounded off to two significant digits. That is,

$$\frac{y \cdot v}{x} \approx 870 = 8.7 \times 10^2.$$

c] Both u and t have four significant digits, and so $u \cdot t$ should be rounded off to four significant digits:

$$u \cdot t \approx (5.004)(24800) \approx 1.241 \times 10^5. \qquad \blacksquare$$

EXAMPLE 9   Using the rule for addition and subtraction of approximate numbers, evaluate the following:

a] $x + y$ **b]** $z + t$ **c]** $u - x$ **d]** $v + t$ **e]** $x + z - u$

Solution

a] $x + y \approx 3.48 + 0.0360 = 3.516.$
Since the level of precision of x is hundredths and that of y is ten thousandths, we round off the sum to hundredths:

$$x + y \approx 3.52.$$

b] $z + t \approx 3251 + 24{,}800 = 28{,}051.$
The level of precision of z is units and that of t is tens, and so we round off the sum to tens:

$$z + t \approx 28050 = 2.805 \times 10^4.$$

c] $u - x \approx 5.004 - 3.48 = 1.524.$
The result should be rounded off to the nearest hundredth, and so we have

$$u - x \approx 1.52.$$

d] $v + t \approx 84{,}000 + 24{,}800 = 108{,}800.$
Since v is correct to the nearest thousand and t is accurate to the nearest tens, we round off the sum to the nearest thousand:

$$v + t \approx 109{,}000 = 1.09 \times 10^5.$$

e] $x + z - u \approx 3.48 + 3251 - 5.004 = 3249.476.$
Since the least precise of x, z, u is z (to the nearest unit), we round off the result to the nearest unit:

$$x + z - u \approx 3249. \qquad \blacksquare$$

EXAMPLE 10   Using the rules for computation with approximate numbers, evaluate the following:

a] $z - xu$

b] $\dfrac{v - t}{x}$

Solution

a] We first evaluate xu:

$$xu \approx (3.48)(5.004) = 17.41392 \approx 17.41.$$

Therefore

$$z - xu \approx 3251 - 17.41 = 3233.59 \approx 3234.$$

Note that in the final computation we used an extra digit for xu.

b] We first evaluate $v - t$:

$$v - t \approx 84{,}000 - 24{,}800 = 59{,}200 \approx 59{,}000 = 5.9 \times 10^4.$$

Thus

$$\frac{v - t}{x} \approx \frac{59200}{3.48} = 17011.494 \ldots \approx 17{,}000 = 1.7 \times 10^4.$$

Note that in the final computation we used $v - t = 59{,}200$ (an extra significant digit), but we rounded off the final result to two significant digits.   ■

EXAMPLE 11   The radius of a circle is measured as $r \approx 6.41$ cm. Find the area of the circle.

Solution   We use the formula Area $= \pi r^2$. Since r is measured to three significant digits, the result should be rounded off to three significant digits. We use π as given by the calculator and find that

$$\text{Area} \approx \pi (6.41)^2 = 129.082 \ldots \text{cm}^2 \approx 129 \text{ cm}^2. \quad ■$$

EXERCISE B

In Problems 1–7, suppose x, y, z, u, v, t are approximate numbers given by

$$x \approx 64.75, \qquad y \approx 4830, \qquad z \approx 0.0045, \qquad u \approx 0.0370,$$
$$v \approx 3005.2, \qquad t \approx 3100 \text{ (the tens 0 is significant and the units 0 is not).}$$

1. Write each of the above numbers in scientific notation.

2. Determine the number of significant digits in each of the above numbers.

3. State the level of precision of each of the above numbers.

4. Round off the above numbers to two significant digits.

Using the rules for computing with approximate numbers, evaluate the expressions given in Problems 5–7.

5. a] xu   **b]** vz   **c]** $t \div y$ **d]** $(uy) \div z$

6. a] $x + y$ **b]** $u - z$ **c]** $y - t$ **d]** $y - x - v$

7. a] $xz - u$ **b]** $\dfrac{y - v}{t}$ **c]** $y + ut$

8. The radius of a circle (measured accurately to the nearest millimeter) is found to be $r \approx 2.476$ m. Find the circumference and area of the circle.

9. The radius of a sphere is measured as $r \approx 3.47$ cm. Find the surface area and volume of the sphere.

10. The lengths of the edges of a rectangular box are measured to the nearest millimeter and found to be

$$a \approx 23.4 \text{ cm}, \quad b \approx 12.8 \text{ cm}, \quad c \approx 8.4 \text{ cm}.$$

Determine the volume and the total surface area of the box.

11. The speed of light is approximately 3×10^5 km/sec. A light-year is defined as the distance traveled by light in one year. Assuming 365 days in a year, find the number of kilometers in a light-year. Express your answer in scientific notation.

12. The hypotenuse and an angle of a right triangle are measured and found to be 32.4 cm and 23°40′, respectively. Calculate the area and the perimeter of the triangle.

Tables and Linear Interpolation

C

While the availability of calculators has virtually eliminated the need for tables in trigonometry, it may still be useful to be able to read tables and to interpolate to make educated guesses about information not included in the table.

Trigonometric tables are usually organized in essentially the same way as our Table C.2 on page 396, although the information we have combined for angles measured in degrees and radians often appears in separate tables. Tables including more detailed information or giving more decimal places may be much longer. Books of tables with seven-place accuracy for angles in degrees and minutes may have many, many pages devoted to trigonometric function tables.

Much economy of information is possible through the use of the complementary angle identities. What this means is that since $\sin(90° - x) = \cos x$, one entry in the table serves for both. In Table C.2, for angles between 0° and 45°, we read *down* from the column headings at the top of the page; for angles between 45° and 90°, we read *upward* from the functions named at the bottom of the page. As illustrated in Table C.1, each entry in Table C.2 carries four names:

$$\sin 13° = \cos 77° \approx \sin(0.2269) \approx \cos(1.3439) \approx 0.2250.$$

To go beyond the information displayed in a table, we use an approximation technique called *linear interpolation*. Intuitively, we feel that $\sin(13.5°)$ should be about halfway between the values of $\sin(13°)$ and $\sin(14°)$. This intuitive assumption is equivalent to assuming that the graph of the function is a straight line. Since the graphs of the trigonometric functions are never straight lines, we know that there will always be some error. Under the circumstances it may be surprising that this approximation is ever very good, but it is often quite accurate. A calculator is always more accurate and is incomparably more convenient, but interpolation is still a valuable tool to have. The technique is best illustrated by examples.

Table C.1

	Degrees	Radians	Sin	Tan	Cot	Cos			
For these angles, read down from top headings.	11	0.1920							For these angles, read up from functions across bottom.
	12	0.2094					1.3614	78	
	13	0.2269	0.2250				1.3439	77	
	14	0.2443					1.3265	76	
	15						1.3090	75	
			Cos	Cot	Tan	Sin	Radians	Degrees	

EXAMPLE 1 Use linear interpolation to evaluate $\tan 28°25'$ and compare the interpolated approximation to the calculator value.

Solution From Table C.2 we may read $\tan 28° \approx 0.5317$ and $\tan(29°) \approx 0.5543$. The number we want must lie between these two values. In Fig. C.1 we will find the y-coordinate of the point on the dotted line, hoping that the line is near enough to the actual graph of $y = \tan x$ to give us a useful approximation. The y-coordinate y_0 is just the value of $\tan 28°$ plus the number a in the figure. From similar triangles we have $a/b = c/d$, from which $a = (b/d)c$. This computation is more conveniently organized in the form:

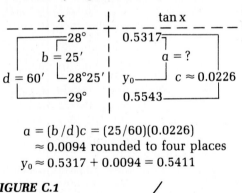

$$a = (b/d)c = (25/60)(0.0226)$$
$$\approx 0.0094 \text{ rounded to four places}$$
$$y_0 \approx 0.5317 + 0.0094 = 0.5411$$

FIGURE C.1

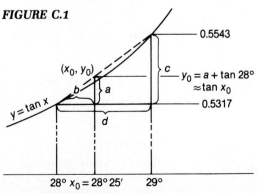

Table C.2
Trigonometric Functions

Degrees	Radians	Sine	Tangent	Cotangent	Cosine		
0	0	0	0	—	1.0000	1.5708	90
1	0.0175	0.0175	0.0175	57.290	0.9998	1.5533	89
2	0.0349	0.0349	0.0349	28.636	0.9994	1.5359	88
3	0.0524	0.0523	0.0524	19.081	0.9986	1.5184	87
4	0.0698	0.0698	0.0699	14.301	0.9976	1.5010	86
5	0.0873	0.0872	0.0875	11.430	0.9962	1.4835	85
6	0.1047	0.1045	0.1051	9.5144	0.9945	1.4661	84
7	0.1222	0.1219	0.1228	8.1443	0.9925	1.4486	83
8	0.1396	0.1392	0.1405	7.1154	0.9903	1.4312	82
9	0.1571	0.1564	0.1584	6.3138	0.9877	1.4137	81
10	0.1745	0.1736	0.1763	5.6713	0.9848	1.3963	80
11	0.1920	0.1908	0.1944	5.1446	0.9816	1.3788	79
12	0.2094	0.2079	0.2126	4.7046	0.9781	1.3614	78
13	0.2269	0.2250	0.2309	4.3315	0.9744	1.3439	77
14	0.2443	0.2419	0.2493	4.0108	0.9703	1.3265	76
15	0.2618	0.2588	0.2679	3.7321	0.9659	1.3090	75
16	0.2793	0.2756	0.2867	3.4874	0.9613	1.2915	74
17	0.2967	0.2924	0.3057	3.2709	0.9563	1.2741	73
18	0.3142	0.3090	0.3249	3.0777	0.9511	1.2566	72
19	0.3316	0.3256	0.3443	2.9042	0.9455	1.2392	71
20	0.3491	0.3420	0.3640	2.7475	0.9397	1.2217	70
21	0.3665	0.3584	0.3839	2.6051	0.9336	1.2043	69
22	0.3840	0.3746	0.4040	2.4751	0.9272	1.1868	68
23	0.4014	0.3907	0.4245	2.3559	0.9205	1.1694	67
24	0.4189	0.4067	0.4452	2.2460	0.9135	1.1519	66
25	0.4363	0.4226	0.4663	2.1445	0.9063	1.1345	65
26	0.4538	0.4384	0.4877	2.0503	0.8988	1.1170	64
27	0.4712	0.4540	0.5095	1.9626	0.8910	1.0996	63
28	0.4887	0.4695	0.5317	1.8807	0.8829	1.0821	62
29	0.5061	0.4848	0.5543	1.8040	0.8746	1.0647	61
30	0.5236	0.5000	0.5774	1.7321	0.8660	1.0472	60
31	0.5411	0.5150	0.6009	1.6643	0.8572	1.0297	59
32	0.5585	0.5299	0.6249	1.6003	0.8480	1.0123	58
33	0.5760	0.5446	0.6494	1.5399	0.8387	0.9948	57
34	0.5934	0.5592	0.6745	1.4826	0.8290	0.9774	56
35	0.6109	0.5736	0.7002	1.4281	0.8192	0.9599	55
36	0.6283	0.5878	0.7265	1.3764	0.8090	0.9425	54
37	0.6458	0.6018	0.7536	1.3270	0.7986	0.9250	53
38	0.6632	0.6157	0.7813	1.2799	0.7880	0.9076	52
39	0.6807	0.6293	0.8098	1.2349	0.7771	0.8901	51
40	0.6981	0.6428	0.8391	1.1918	0.7660	0.8727	50
41	0.7156	0.6561	0.8693	1.1504	0.7547	0.8552	49
42	0.7330	0.6691	0.9004	1.1106	0.7431	0.8378	48
43	0.7505	0.6820	0.9325	1.0724	0.7314	0.8203	47
44	0.7679	0.6947	0.9657	1.0355	0.7193	0.8029	46
45	0.7854	0.7071	1.0000	1.0000	0.7071	0.7854	45
		Cosine	Cotangent	Tangent	Sine	Radians	Degrees

Thus linear interpolation gives an approximate value of 0.5411, which may be compared with the calculator value of 0.5410739727. In this case the linear interpolation is correct when rounded off to four decimal places. ■

EXAMPLE 2 Use linear interpolation to evaluate $\sin(0.2500)$ and compare with the calculator value.

Solution From Table C.2 we set up the information just as we did above:

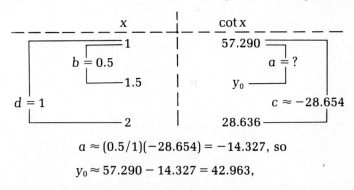

x		$\sin x$
0.2443		0.2419
0.0057		$a = ?$
0.2500		y_0
0.0175		0.0169
0.2618		0.2588

$$a \approx (57/175)(0.0169) \approx 0.0055$$
$$y_0 \approx 0.2419 + 0.0055 \approx 0.2474$$

$\sin(0.2500) \approx 0.2474$ by interpolation;
$\sin(0.2500) \approx 0.24740396$ by calculator. ■

EXAMPLE 3 Use linear interpolation to evaluate $\cot(1.5°)$ and compare with the calculator value.

Solution Since the cotangent is a *decreasing* function, the changes in the y-coordinates (a and c) will be *negative*. See Fig. C.2.

x		$\cot x$
1		57.290
$b = 0.5$		$a = ?$
1.5		y_0
$d = 1$		$c \approx -28.654$
2		28.636

$$a \approx (0.5/1)(-28.654) = -14.327, \text{ so}$$
$$y_0 \approx 57.290 - 14.327 = 42.963,$$

compared with 38.188459 by calculator. This interpolated value is *not* a good approximation, simply because the line is just not close enough to the curve.

In general, linear approximation is not very accurate for angles near 0° or near 90°. ■

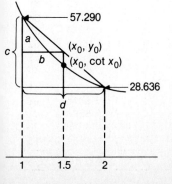

FIGURE C.2

EXERCISE C

Use linear interpolation to evaluate each of the following. Compare the interpolated approximation to the calculator value.

1. $\sin 18°25'$
2. $\cos 6°15'$
3. $\tan 30°37'$
4. $\cot 42°10'$
5. $\cot 47°40'$
6. $\sin 56°18'$
7. $\tan 88°30'$
8. $\cos 45.4°$
9. $\cos 89.8°$
10. $\sin 44.6°$
11. $\sin (0.3660)$
12. $\cos 1$
13. $\cot 1$
14. $\tan (0.0333)$
15. $\sin (1.3)$
16. $\tan (1.502)$
17. $\cos (0.3021)$
18. $\sin (-0.2500)$
19. $\cos (-0.2500)$
20. $\cot (0.4636)$

Answers to odd-numbered exercises
(Calculator values are rounded to six decimal places)

1. 0.3159; 0.315925
3. 0.5919; 0.591791
5. 0.9111; 0.910994
7. 42.963; 38.188459
9. 0.0035; 0.003491
11. 0.3579; 0.357883
13. 0.6421; 0.642093
15. 0.9635; 0.963558
17. 0.9547; 0.954714
19. 0.9689; 0.968912

Answers

In all problems involving approximate decimal values, the answers were obtained by using maximum accuracy in computations with a calculator. Intermediate results (rounded off) *were not used* in subsequent calculations.

Exercise 1.1 (page 11)

1. a]

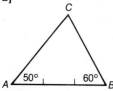

b]

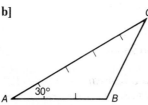

3. $\sqrt{89}$ **5.** $2\sqrt{2}$ **7.** 24 **9.** 3784

11. a] 10 **13.** $45°, 4\sqrt{2}, 4\sqrt{2}$ **15.** $60°, 2.5\sqrt{3}, 2.5$ **17.** $C = 3.2\pi \approx 10.1$
  b] $5\sqrt{3}$ $A = 2.56\pi \approx 8.04$

19. $d = 2\sqrt{\dfrac{3.48}{\pi}} \approx 2.10$ **21.** $|\overline{AB}| = 8, |\overline{BC}| = \sqrt{55}$ **23.** $\dfrac{\sqrt{119}}{2} \approx 5.45$

25. $\dfrac{5.64}{3 + \sqrt{3}} \approx 1.19, \dfrac{5.64\sqrt{3}}{3 + \sqrt{3}} \approx 2.06, \dfrac{11.28}{3 + \sqrt{3}} \approx 2.38$ **27.** $6, 6\sqrt{3}, 12$

29. $\sqrt{37} \approx 6.1$ **31.** $\sqrt{33.13} \approx 5.8$ **33.** $\sqrt{19} \approx 4.4$ **35.** 9.8, 8.1, 3.2

37. Sides are $3\sqrt{2}, 3\sqrt{2}, 6$; right triangle   **39.** Sides are 5, 5, $\sqrt{10}$; isosceles triangle

41. Equilateral triangle; all sides are 4   **43.** All on same line   **45.** $(3.5, -0.5)$

47. a] 3, 4, 5; 5, 12, 13;   **b]** Choose b so that $2b + 1$ is a perfect square; for
  7, 24, 25; 9, 40, 41   instance, $2b + 1 = 81$ gives $b = 40, a = 9, c = 41$

Exercise 1.2 (page 21)

1. a]

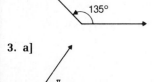
135°

b]
$-60°$

c]
540°

d]

67° 30′

3. a]

$\dfrac{\pi}{3}$

b]

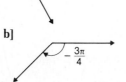
$-\dfrac{3\pi}{4}$

c]

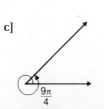
$\dfrac{9\pi}{4}$

399

400 ANSWERS

5. 100°, obtuse   **7.** 84°19′50″, acute

9. a]

$$90° < \alpha < 135°$$

b]

$$\frac{5\pi}{4} < \beta < \frac{3\pi}{2}$$

c]

$$-\pi < \theta < -\frac{\pi}{2}$$

11. a] 156.62°   **b]** 247.38°   **13. a]** 24°21′50″   **b]** 149°22′30″

15. $\frac{\pi}{6}$, 45°, $\frac{\pi}{2}$, $\frac{2\pi}{3}$, 165°, 180°, $\frac{7\pi}{6}$, $\frac{3\pi}{2}$, $\frac{5\pi}{3}$, 330°, 360°, $\frac{\pi x}{180}$, $\left(\frac{180y}{\pi}\right)°$

17. a] $\frac{7\pi}{6} \approx 3.665$ **b]** $-\frac{3\pi}{8} \approx -1.178$ **c]** $\frac{7\pi}{3} \approx 7.330$

19. a] 84.80°, 84°48′   **b]** −197.67°, −197°40′

21. 1.13, 4.21, 140.4°, 205.1°, 7.91, 370.7°

23. $\theta < \beta < \alpha$ **25.** $\beta < \theta < \alpha$ **27.** 69°38′

29. 3.141509434 (agrees to fourth decimal place), three places rounded off

Exercise 1.3 (page 28)

1. a] 29 cm   **3. a]** 19.3 cm   **5. a]** 0.765
b] 4.42 m   **b]** 1.07 m   **b]** 1.55

7. 11.7 cm, 0.747 rad, 0.987 m   **9. a]** 16.0 m   **11. a]** 63.1 cm²
1.15 rad, 13.1 cm   **b]** 50.5 m   **b]** 98.8 cm²
   c] 191 m   **c]** 188 cm²

13. 3.83 cm   **15.** 1040 cm²   **17. a]** $\frac{2\pi}{5}$ **19.** 20.5
   b] 1.3

21. 270,000 miles   **23.** π ft, 3π ft²; 1.5 cm, 2.7 cm²; 3 rad, 3.375 m²; 2.752 rad,
   6.88 cm; 5 cm, 0.8 rad; $\frac{y}{x}$ cm, $\frac{y^2}{2x}$ cm²

25. a] $1.8\pi \approx 5.65$ cm   **27.** Front wheel: 30.0 rev, 189 rad   **29. a]** 339 mi
b] $3.3\pi \approx 10.4$ cm   Rear wheel: 65.7 rev, 413 rad   **b]** 294 NM

31. a] 3330 mi (3 significant digits)   **33.** 8.0°N
b] 2890 NM (3 significant digits)   **35.** 810,000 mi (2 significant digits)

Exercise 1.4 (page 36)

1. a] 130 cm/sec   **b]** 970 cm/sec   **c]** 50 cm/sec (2 sig. digits)
3. a] 1 rev/hr   **b]** 6 deg/min   **c]** $\pi/30 \approx 0.105$ rad/min

5. a] $13\pi \approx 41$ cm/hr   **b]** $\frac{13\pi}{60} \approx 0.68$ cm/min

7. a] $1471.68\pi \approx 4600$ m   **b]** $90.52\pi \approx 280$ m (2 sig. digits)

9. $3200\pi \approx 10,000$ km/hr   **11. a]** $\frac{176}{3} \approx 59$ rad/sec   **b]** $\frac{88}{3\pi} \approx 9.3$ rev/sec

13. a] $\frac{200\pi}{3} \approx 209$ rad/min   **b]** $10\pi \approx 31.4$ m/min

5. a] $\dfrac{80}{3}$ cm/sec   **b]** 6.2 rad/sec

7. a] 38,000 rev (2 sig. digits)   **b]** 58 mi/hr

9. a] 80π cm/sec   **b]** $\dfrac{40\pi}{3}$ rad/sec $= \dfrac{20}{3}$ rev/sec

1. $v = 43.4\pi \approx 140$ cm/sec, $d = 434\pi \approx 1400$ cm

Chapter 1 Review Exercise (page 40)

1. a] 37.7°   **b]** 81.9°   **c]** 117°

3. a] 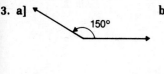   **b]** 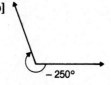   **c]** 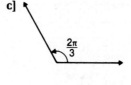   **d]**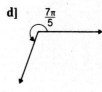

5. 340 cm²   **7.** $\theta < \beta < \alpha < \gamma$ **9.** 232.7 m²

11. 6.561 cm   **13.** 8   **15.** 19 : 17

Exercise 2.1 (page 47)

1. $\mathbf{R}^*$ **3.** $\mathbf{R}$ **5.** $\{x \mid -1 \le x \le 1\}$ **7.** $\mathbf{R}$

9. $\{x \mid x \ne 0\}$ **11.** $\{x \mid x \ge 0\}$ **13.** $\{x \mid x \ge -1, x \ne 0\}$ **15.** $-1; -1$

17. $-0.345; -\dfrac{\pi}{10}$ not in domain   **19.** $0.063; \dfrac{\sqrt{2}-\pi}{4}$ not in domain

21. $f(g(x)) = 1 + \dfrac{1}{\sqrt{x}}; D = \{x \mid x > 0\}$ **23.** $f(h(x)) = \dfrac{2 + 2x}{1 + 2x}; D = \{x \mid x \ne -\tfrac{1}{2}\}$

25. $g(h(x)) = \sqrt{1 + 2x}; D = \{x \mid x \ge -\tfrac{1}{2}\}$ **27.** $f(x) = 2x - 1; D = \{1, 2, 3, \ldots\}$

29. $f(x) = \dfrac{x}{|x|}; D = \{x \mid x \ne 0\}$ **31.** $\dfrac{1}{h(\sqrt{1 + h^2} + h)}$

33. $\sqrt{x + 1} - \sqrt{x}$ **35.** $\dfrac{1}{\sqrt{x}(x + t) + x\sqrt{x + t}}$

Exercise 2.2 (page 54)

1.

	sin	cos	tan	cot	sec	csc
$\dfrac{\pi}{6}$	$\dfrac{1}{2}$	$\dfrac{\sqrt{3}}{2}$	$\dfrac{1}{\sqrt{3}}$	$\sqrt{3}$	$\dfrac{2}{\sqrt{3}}$	2
$\dfrac{\pi}{4}$	$\dfrac{1}{\sqrt{2}}$	$\dfrac{1}{\sqrt{2}}$	1	1	$\sqrt{2}$	$\sqrt{2}$
$\dfrac{\pi}{3}$	$\dfrac{\sqrt{3}}{2}$	$\dfrac{1}{2}$	$\sqrt{3}$	$\dfrac{1}{\sqrt{3}}$	2	$\dfrac{2}{\sqrt{3}}$

3. a] 4/5   **b]** 3/5   **c]** 5/3

* $\mathbf{R}$ = set of real numbers.

5. a] $\dfrac{3\sqrt{5}}{7}$ **b]** $\dfrac{3\sqrt{5}}{7}$ **c]** $\dfrac{2\sqrt{5}}{15}$ **d]** 7/2

7. a] $\dfrac{\sqrt{5}}{3}$ **b]** $\dfrac{\sqrt{5}}{2}$ **c]** $\dfrac{\sqrt{5}}{3}$

9. a] 1.32 **b]** $1.32\sqrt{2} \approx 1.87$ **c]** $\dfrac{\sqrt{2}}{2} \approx 0.707$

11. a] $\dfrac{\pi}{4}$ rad **b]** $3\sqrt{2}$

13. a] $\sqrt{3}/3$ **b]** 4 **c]** 30° **15.** $\sqrt{13}, 2\sqrt{13}$

17.

	sin	cos	tan
α	$\dfrac{4}{2\sqrt{13}} = \dfrac{2}{\sqrt{13}}$	$\dfrac{6}{2\sqrt{13}} = \dfrac{3}{\sqrt{13}}$	$\dfrac{4}{6} = \dfrac{2}{3}$
β	$\dfrac{6}{2\sqrt{13}} = \dfrac{3}{\sqrt{13}}$	$\dfrac{4}{2\sqrt{13}} = \dfrac{2}{\sqrt{13}}$	$\dfrac{6}{4} = \dfrac{3}{2}$

19. a] 6.10 **b]** 0.387 **c]** 0.419 **21.** 168 m

Exercise 2.3 (page 59)

1. 0.4695 **3.** 0.3090 **5.** 1.2208 **7.** 0.6865

9. 0.9758 **11.** 0.3153 **13.** 0.4142 **15.** 1.0000

17. 1255.8; 0.027409 **19.** 14.1; 48.1; 92.6; 1260; 10,400

21. 0.2500; 0.2588 **23.** 1.7321; 1.1547 **25.** 2; 2 **27.** 1.99

29. 95.48 **31.** 9.94 **33.** 8.67 **35.** 1.17

37. 1.31 **39.** −0.33 **41.** −0.42 **43.** 2.94

45. 5.00 m

Exercise 2.4 (page 68)

1. $b \approx 6.1$, $c \approx 7.5$, $\beta = 55°$ **3.** $a \approx 37$, $c \approx 95$, $\beta = 67°$ **5.** $a \approx 10.9$, $b \approx 22.0$, $\alpha = 26°20'$

7. $c \approx 71$, $\alpha \approx 48°$, $\beta \approx 42°$ **9.** $b \approx 35.3$, $\alpha \approx 35°40'$, $\beta \approx 54°20'$

11. 13 **13.** 1600 (2 sig. digits) **15.** 5.3 cm **17.** 10.22 cm²

19. $|\overline{AE}| = a \cot \alpha - a$; $|\overline{AC}| = a \cot \alpha$; $|\overline{AB}| = a \csc \alpha$; $|\overline{FB}| = a \sec \alpha$

21. 77° **23.** 560 m (2 sig. digits) **25.** 68.31 cm² **27.** 2240 cm² (3 sig. digits)

29. 10,200 cm² for $n = 100$; 10,207 cm² for $n = 500$, area of circle is 10,207 cm²

31. 25.9 cm **33.** 2.47 cm² **35.** 487 m **37.** $62,000

39. 73.49 m; 20.21°

Exercise 2.5 (page 76)

1. a] 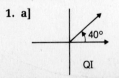 **b]** 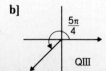 **c]** 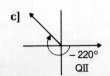 **d]**

3. a]

375°
QI

b]

−600°
QII

c]

8.47
QII

d]

3π

5. 260°

7. 2.19

9. 210°

11. No

13. Yes

15. No

17. −270°, 450°, 810°

19. $\left\{\alpha\middle|\alpha = -\dfrac{2\pi}{3} + k\cdot 2\pi\right\}$

21. $\left\{\alpha\middle|\alpha = \dfrac{\pi}{4} + k\cdot 2\pi\right\}$

23. $\left\{\alpha\middle|\alpha = (2k + 1)\pi\right\}$

Exercise 2.6 (page 84)

	$\sin\theta$	$\cos\theta$	$\tan\theta$	$\cot\theta$	$\sec\theta$	$\csc\theta$
1.	$-\dfrac{3}{5}$	$\dfrac{4}{5}$	$-\dfrac{3}{4}$	$-\dfrac{4}{3}$	$\dfrac{5}{4}$	$-\dfrac{5}{3}$
3.	$-\dfrac{2}{\sqrt{5}}$	$-\dfrac{1}{\sqrt{5}}$	2	$\dfrac{1}{2}$	$-\sqrt{5}$	$-\dfrac{\sqrt{5}}{2}$
5.	$\dfrac{\sqrt{3}}{\sqrt{5}}$	$\dfrac{\sqrt{2}}{\sqrt{5}}$	$\dfrac{\sqrt{3}}{\sqrt{2}}$	$\dfrac{\sqrt{2}}{\sqrt{3}}$	$\dfrac{\sqrt{5}}{\sqrt{2}}$	$\dfrac{\sqrt{5}}{\sqrt{3}}$

7. a] $\dfrac{\sqrt{3}}{2}$ **b]** $-\dfrac{\sqrt{3}}{2}$

9. a] -1 **b]** $\sqrt{2}$

11. a] Undefined **b]** -1

13. a] $\dfrac{\sqrt{3}}{2}$ **b]** $\dfrac{1}{2}$

15. a] $-1/2$ **b]** $-\dfrac{\sqrt{3}}{2}$

17. a] $\sqrt{3}$ **b]** -2

19.

	sin	cos	tan	cot	sec	csc
124°	+	−	−	−	−	+
−320°	+	+	+	+	+	+
3.04	+	−	−	−	−	+
−1.16	−	+	−	−	+	−

21. a] −1.483 **b]** 1.026

23. a] 0.746 **b]** −0.568

25. a] 0.653 **b]** 0.694

27. a] 0.223 **b]** −0.727

29. $\dfrac{2+\sqrt{3}}{2} \approx 1.866$

	sin	cos	tan	cot	sec	csc
35.	$\dfrac{4}{5} = 0.800$		$-\dfrac{4}{3} \approx -1.333$	$-\dfrac{3}{4} = -0.750$	$-\dfrac{5}{3} \approx -1.667$	$\dfrac{5}{4} = 1.250$
37.	$-\dfrac{4}{5} = -0.800$	$-\dfrac{3}{5} = -0.600$	$\dfrac{4}{3} \approx 1.333$		$-\dfrac{5}{3} \approx -1.667$	$-\dfrac{5}{4} = -1.250$
39.		$\dfrac{\sqrt{15}}{4} \approx 0.968$	$-\dfrac{1}{\sqrt{15}} \approx -0.258$	$-\sqrt{15} \approx -3.873$	$\dfrac{4}{\sqrt{15}} \approx 1.033$	$-4 = -4.000$

Exercise 2.7 (page 91)

1. $P(\cos 1, \sin 1) \approx (0.540, 0.841)$

3. $P\left(\dfrac{\sqrt{3}}{2}, \dfrac{1}{2}\right)$

5. $P(-\cos 1, -\sin 1) \approx (-0.540, -0.841)$

7. $P(0.866, -0.499)$

9. $P\left(\dfrac{\sqrt{2}}{2}, \dfrac{\sqrt{2}}{2}\right)$

11. $P(1, 0)$

13. $|\overline{OQ}| = \cos t; \ |\overline{PQ}| = \sin t$

15. $|\overline{OM}| = \sec t$

17. $|\overline{OS}| = \csc t; \ |\overline{OT}| = \sec t$

19.

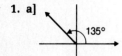

$P\left(-\dfrac{\sqrt{101}}{101}, \dfrac{-10\sqrt{101}}{101}\right)$

$R(0, -1)$
If Q is any point on arc $\overset{\frown}{PR}$ and t = the length of the arc $\overset{\frown}{AQ}$, then $\tan t > 10$.

21. $R(-\cos t, -\sin t)$

23. $\dfrac{\pi}{2} < t < \dfrac{3\pi}{2}$

25. $\dfrac{\pi}{2}, \dfrac{3\pi}{2}$

27. $\dfrac{\pi}{2}, \dfrac{3\pi}{2}$

Chapter 2 Review Exercise (page 94)

1. a]

b]

c]

d]

e]

f]

3. a] 1   **b]** $\dfrac{1}{\sqrt{3}}$ **c]** $-\dfrac{2}{\sqrt{3}}$ **d]** $-\dfrac{1}{2}$

    e] 0   **f]** 1   **g]** 1   **h]** -1

5. a] $-\dfrac{4}{5}$ **b]** $-\dfrac{5}{3}$ **c]** $\dfrac{3}{5}$ **d]** $\dfrac{4}{3}$ **e]** $\dfrac{5}{3}$ **f]** $\dfrac{4}{5}$

7. a] $270°$ **b]** $30°$ **c]** $135°$ **d]** $-45°$

9. a] $-\dfrac{\sqrt{3}}{2}$ **b]** 0   **c]** $\dfrac{1}{2}$ **d]** $\dfrac{\sqrt{3}}{2}$ **e]** $\sqrt{3}$ **f]** $\dfrac{2}{\sqrt{3}}$

11. a] 0.6820   **b]** -0.4877 **c]** 0.5407   **d]** 0.9004   **e]** 1.1897   **f]** 0.7771

13. a] 0.7880   **b]** 1.7646   **15. a]** 1   **b]** 1

17. a] True   **b]** False   **c]** True   **d]** False

19. $h \approx 3.74$, area ≈ 10.8 **21.** 22.7 cm   **23.** 13 cm^2 (2 sig. digits)

Exercise 3.1 (page 105)

1.

x	0	$\pi/20$	$\pi/10$	$3\pi/20$	$\pi/5$	$\pi/4$	$3\pi/10$	$7\pi/20$	$2\pi/5$	$9\pi/20$	$\pi/2$
cos x	1	0.99	0.95	0.89	0.81	0.71	0.59	0.45	0.31	0.16	0

3. $\sin\left(x + \dfrac{\pi}{2}\right) = \cos x$

5.

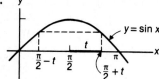

7. $-t,\ 2\pi - t,\ 3\pi + t$ **9.** $\dfrac{\pi}{2} - t,\ \dfrac{5\pi}{2} - t,\ \dfrac{7\pi}{2} + t$

11.

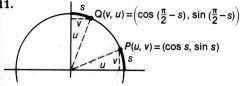

13.

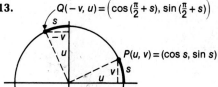

15.

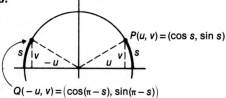

17. Valid   **19.** Valid   **21.** Not valid   **23.** Valid

25. $\dfrac{1}{\sqrt{2}}$ **27.** $\dfrac{-1}{\sqrt{2}}$ **29.** $\dfrac{\sqrt{3}}{2}$

Exercise 3.2 (page 111)

1.

x	π/4	3π/10	7π/20	2π/5	9π/20	π/2	11π/20	3π/5	13π/20	7π/10	3π/4
y	1	1.38	1.96	3.08	6.31	und.	−6.31	−3.08	−1.96	−1.38	−1

3.

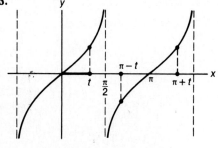

$\tan(\pi + t) = \tan t$
$\tan(\pi - t) = -\tan t$

5.

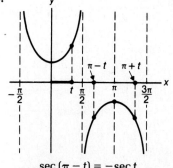

$\sec(\pi - t) = -\sec t$
$\sec(\pi + t) = -\sec t$

7. $\dfrac{5\pi}{2} - t,\ \dfrac{3\pi}{2} + t,\ -\dfrac{\pi}{2} + t$ **9.** $t,\ 2\pi + t,\ -\pi + t$

17. Valid   **19.** Not valid   **21.** Not valid   **23.** Valid

Exercise 3.3 (page 116)

1. $f^{-1}(x) = \frac{x-5}{3}$; Domain = **R**; $\frac{-5}{3}, \frac{-4}{3}, -3$ **3.** $f^{-1}(x) = \frac{2(x+4)}{5}$; Domain = **R**; $\frac{8}{5}, 2, 0$

5. $f^{-1}(x) = x^2$; Domain = $\{x \mid x \geq 0\}$; 0, 1, undef.

7. $f^{-1}(x) = \sqrt{x-1}$; Domain = $\{x \mid x \geq 1\}$; undef., 0, undef.

9. $f^{-1}(x) = \sqrt[3]{x}$; Domain = **R**; 0, 1, $-\sqrt[3]{4}$

11. $f^{-1}(x) = x - 3$

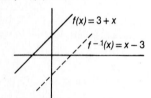

13. $f^{-1}(x) = \frac{2}{3}(x+3)$

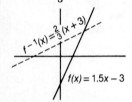

15. $f^{-1}(x) = x^2$ for $x \geq 0$

17. $f^{-1}(x) = \frac{x-4}{2}$ **19.** $f^{-1}(x) = \sqrt[3]{x+1}$

21. $g(x) = x^2 - 1$, $D(g) = \{x \mid x \geq 0\}$ **23.** $g(x) = x - 1$, $D(g) = \{x \mid x \geq 1\}$
$g^{-1}(x) = \sqrt{x+1}$, $D(g^{-1}) = \{x \mid x \geq -1\}$ $g^{-1}(x) = x + 1$, $D(g^{-1}) = \{x \mid x \geq 0\}$

Exercise 3.4 (page 126)

1. 0   **3.** $\frac{3\pi}{4}$ **5.** $\frac{\pi}{6}$ **7.** $-\frac{\pi}{3}$

9. $-\frac{\pi}{2}$ **11.** Undefined; $-\frac{2}{\sqrt{2}}$ not in domain   **13.** 0.39

15. 1.01   **17.** 2.02   **19.** 2.39   **21.** 0.55

23. −0.87

25. a]

x	−1.0	−0.8	−0.6	−0.4	−0.2	0	0.2	0.4	0.6	0.8	1.0
y	−1.57	−0.93	−0.64	−0.41	−0.20	0	0.20	0.41	0.64	0.93	1.57

27. $D = \{x \mid -1 \le x \le 1\}$
$R = \{y \mid -1 \le y \le 1\}$

29. $D = \mathbf{R}$
$R = \{y \mid 0 \le y \le \pi\}$

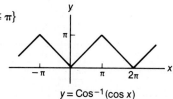

$y = \text{Cos}^{-1}(\cos x)$

31. $\dfrac{2}{7}$ **33.** $\dfrac{\pi}{3}$ **35.** $-\dfrac{1}{\sqrt{3}}$ **37.** 1

39. -1 **41.** $\dfrac{3}{7}$ **43.** $-\dfrac{2}{7}$ **45.** $-\dfrac{3}{5}$

47. 0.912   **49.** 0.973   **51.** Undefined   **53.** 0.954

55. 0.882   **57.** -0.46 **59.** No solution   **61.** -0.87

Exercise 3.5 (page 133)

1. $-\dfrac{\pi}{4}$ **3.** $\dfrac{\pi}{6}$ **5.** $\dfrac{\pi}{2}$ **7.** 0.636

9. 1.150   **11.** 1.467   **13.** $\dfrac{4}{3}$ **15.** $-\dfrac{7}{5}$

17. $-\dfrac{3}{5}$ **19.** $\dfrac{5}{13}$ **21.** $\dfrac{4}{7}$ **25.** No solution

27. -3.60 **29.** No solution

Exercise 3.6 (page 139)

1. a] $\dfrac{\pi}{6}$ **3. a]** $\dfrac{\pi}{2}$ **5. a]** $\dfrac{\pi}{4}$ **7. a]** π

  b] $\dfrac{2\pi}{3}$ **b]** π **b]** $-\dfrac{\pi}{4}$ **b]** $\dfrac{\pi}{2}$

9. a] $\dfrac{\pi}{6}$ **11. a]** $\dfrac{3}{4}$ **13. a]** $\dfrac{\sqrt{3}}{2}$ **15. a]** -1

  b] Undefined   **b]** $\dfrac{3}{\sqrt{5}}$ **b]** $-\dfrac{1}{2}$ **b]** Undefined

17. 2.748   **19.** 0.567   **21.** 2.618

23. a] 1.969   **25. a]** 1.289   **27. a]** Undefined   **29. a]** 0.629
  b] 0.234   **b]** 0.463   **b]** -1.047 **b]** 0.778

31. $\dfrac{7}{2}$ **33.** $\dfrac{12}{13}$ **35.** $-\dfrac{5}{13}$ **37.** 2

39. a] -1.01 **41. a]** 1.15   **43. a]** No solution   **45.** 0.576
  b] No solution   **b]** -3.01 **b]** -2.40

Chapter 3 Review Exercise (page 143)

1. $\dfrac{\pi}{4}$ **3.** $-\dfrac{\pi}{3}$ **5.** $\dfrac{2\pi}{3}$ **7.** 1

9. $-\dfrac{\sqrt{21}}{2}$ **11.** $\dfrac{1}{\sqrt{5}}$ **13.** $\dfrac{7}{2}$ **15.** $-\dfrac{\pi}{6}$

17. $-\dfrac{\sqrt{2}}{4}$ **19.** 0.35   **21.** 0.98   **23.** 0.60

25. −0.69   **27.** 0.43   **29.** −1.13

31. a]

x	0	0.2	0.4	0.6	0.8	1.0	1.2	1.4	1.6	1.8	2.0	2.2	2.4	2.6	2.8	3.0	3.2
sin x	0	0.20	0.39	0.56	0.72	0.84	0.93	0.99	1.00	0.97	0.91	0.81	0.68	0.52	0.33	0.14	−0.06
2 sin x	0	0.40	0.78	1.13	1.43	1.68	1.86	1.97	2.00	1.95	1.82	1.62	1.35	1.03	0.67	0.28	−0.12

33. $D = \{x \mid -1 \le x \le 1\}$
$R = \{y \mid -1 \le y \le 1\}$

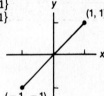

35. $D = \{x \mid -1 \le x \le 1\}$
$R = \{y \mid 0 \le y \le 1\}$

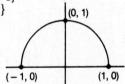

37. $\dfrac{3\pi}{4}$ **39.** $-\dfrac{\pi}{2}$ **41.** $\dfrac{1}{3}$ **43.** 0.86

45. 2.51   **47.** 0.31

Exercise 4.1 (page 149)

17. Identity   **19.** Not an identity

21. Identity, both sides are equal to $3(4x + 3)(1 - x)^{4/3}$ **23.** No solution

Exercise 4.3 (page 158)

1. Yes   **3.** No   **5.** No   **7.** No

9. Yes   **11.** No   **13.** No   **15.** No

17. No   **19.** Yes   **21.** Yes   **23.** No

25. No   **27.** Yes   **29.** Yes

Exercise 4.4 (page 164)

15. $\dfrac{\sqrt{2} - \sqrt{6}}{4} \approx -0.26$ **17.** $2 + \sqrt{3} \approx 3.73$ **19.** $-2 - \sqrt{3} \approx -3.73$ **21.** $2 + \sqrt{3} \approx 3.73$

23. $\dfrac{11}{3}$　　　　**25.** 3.3　　　　**27.** Yes　　　　**29.** No

31. No　　　　**33.** $\dfrac{\sqrt{2}+\sqrt{6}}{4} \approx 0.97$

Exercise 4.5 (page 166)

7. $\dfrac{1}{2}(\sin 8\theta - \sin 2\theta)$　　**9.** $\dfrac{1}{2}(\cos 2x - \cos 6x)$　　**11.** $\dfrac{1}{2}(\sin 6x + \sin 4x)$　　**13.** $2\cos x \sin 3x$

15. $-2\sin 4x \sin x$　　**17.** $\tan 2x$　　**19.** $\sin 3x$　　**21.** $2\sin x \cos x$

23. $\sin x \cos x + \dfrac{\sqrt{3}}{2}(\sin^2 x - \cos^2 x)$　　**25.** $-\dfrac{1}{\sqrt{2}}$　　**27.** $-\dfrac{1}{\sqrt{3}}$

31. b]

x	40	25	20	10	8	6	5	4	3.5	3.2	3.1	3.0	2.8	2.5	2.0	1.5	1.1	0.5
θ	2.14°	3.40°	4.22°	7.98°	9.59°	11.82°	13.19°	14.62°	15.26°	15.56°	15.64°	15.71°	15.80°	15.80°	15.26°	13.67°	11.36°	5.91°

c] 2.65 m

Exercise 4.6 (page 171)

25. a] $-\dfrac{120}{169}$　　**b]** $\dfrac{119}{169}$　　**c]** $-\dfrac{120}{119}$

27. a] 0.9507　　**b]** -0.3102　　**c]** -3.0652

29. a] $\dfrac{1}{4}$　　**b]** $\dfrac{\sqrt{3}}{2}$　　**c]** $-\dfrac{\sqrt{3}}{2}$

31. Identity　　**33.** Identity　　**35.** Identity　　**37.** Not an identity

39. Identity　　**43.** 1　　**45.** 1

Exercise 4.7 (page 176)

1. a] $\dfrac{1}{2}\sqrt{2+\sqrt{2}} \approx 0.9239$　　**b]** $\dfrac{1}{2}\sqrt{2+\sqrt{2}} \approx 0.9239$

c] $\dfrac{1}{2}\sqrt{2+\sqrt{3}} \approx 0.9659$　　**d]** $-\dfrac{1}{2}\sqrt{2-\sqrt{3}} \approx -0.2588$

3. a] $\dfrac{1}{2}\sqrt{2-\sqrt{3}} \approx 0.2588$　　**b]** $-\dfrac{1}{2}\sqrt{2-\sqrt{2}} \approx -0.3827$

c] $-\dfrac{1}{2}\sqrt{2+\sqrt{2}} \approx -0.9239$　　**d]** $2-\sqrt{3} \approx 0.2679$

5. a] $\dfrac{3}{\sqrt{13}}$　　**b]** $\dfrac{2}{\sqrt{13}}$　　**c]** $\dfrac{3}{2}$　　**d]** $\dfrac{\sqrt{13}}{2}$

7. $\cos\dfrac{\theta}{2} = -\dfrac{1}{2}\sqrt{2+\sqrt{3}}$; $\tan\dfrac{\theta}{2} = 2-\sqrt{3}$

9. a] $-\dfrac{5}{\sqrt{26}}$　　**b]** $\sqrt{\dfrac{\sqrt{26}+1}{2\sqrt{26}}}$　　**c]** $\dfrac{5}{13}$　　**d]** $-\dfrac{\sqrt{26}+1}{5}$

11. $-\dfrac{1}{8}$　　**13. a]** 0.2863　　**b]** 0.9581　　**c]** 0.2988

23. $-\dfrac{\sqrt{6}+\sqrt{2}}{4}$; $-\dfrac{\sqrt{2+\sqrt{3}}}{2}$　　**25. a]** $\dfrac{1}{\sqrt{5}}$　　**b]** $\sqrt{\dfrac{5+\sqrt{5}}{10}}$

Exercise 4.8 (page 180)

11. $-\sin x$ **13.** $-\cos x$ **15.** $-\cot x$ **17.** $\cos x$

19. $\dfrac{\tan x - 1}{1 + \tan x}$ **21.** $-\sec x$ **25.** $\tan 2\theta = -\dfrac{24}{7}$, $\sin \theta = \dfrac{4}{5}$, $\cos \theta = \dfrac{3}{5}$

27. $\tan 2\theta = \dfrac{24}{7}$, $\sin \theta = \dfrac{3}{5}$, $\cos \theta = \dfrac{4}{5}$ **29.** $\sec^2 x - \sec x \tan x$

31. $\dfrac{1}{2}(1 + \cos 2x)$ **33. a]** $\tan 3x$ **b]** $\cot \dfrac{x}{2}$

Chapter 4 Review Exercise (page 184)

27. Identity   **29.** Not an identity   **31.** Not an identity   **33.** $-\dfrac{4}{5}$

35. $\dfrac{1}{\sqrt{10}}$ **37.** $-\dfrac{56}{33}$ **39.** $\dfrac{119}{169}$ **41.** $-\dfrac{116}{845}$

43. $\dfrac{9}{25}$ **45.** $-\dfrac{5}{12}$ **47.** $-\dfrac{4}{5}$ **49.** $-\dfrac{3696}{4225}$

Exercise 5.1 (page 190)

1. 2   **3.** 1   **5.** $\dfrac{1}{2}$, -3

7. No real number solution   **9.** -3, 3   **11.** -1

13. $\{\sqrt{3} + 1\} \approx \{2.73\}$ **15.** $\{-5 - 3\sqrt{3}\} \approx \{-10.20\}$ **17.** $\left\{\dfrac{3 \pm \sqrt{17}}{4}\right\} \approx \{1.78, -0.28\}$

19. $\left\{\dfrac{-\sqrt{3} \pm \sqrt{31}}{2}\right\} \approx \{-3.65, 1.92\}$

Exercise 5.2 (page 194)

When general solutions are given, it is to be understood that the letter k ranges over the entire set of integers (zero, positive, and negative).

1. 120°, 240°   **3.** 30°, 210°   **5.** 90°   **7.** No solution

9. 135°, 315°   **11.** $\dfrac{\pi}{4}, \dfrac{3\pi}{4}$ **13.** $\dfrac{\pi}{4}, \dfrac{\pi}{2}, \dfrac{3\pi}{4}, \dfrac{3\pi}{2}$ **15.** $0, \dfrac{2\pi}{3}, \pi, \dfrac{5\pi}{3}, 2\pi$

17. $\left\{-\dfrac{\pi}{6}, \dfrac{5\pi}{6}\right\}$ **19.** $\left\{-\dfrac{\pi}{2}, \dfrac{\pi}{2}\right\}$ **21.** $\left\{-\dfrac{3\pi}{4}, -\dfrac{\pi}{2}, \dfrac{\pi}{4}, \dfrac{\pi}{2}\right\}$ **23.** No solution

25. $\{0.64, 5.64\}$ **27.** $\{1.85, 4.99\}$ **29.** $\{1.15, 4.29\}$

31. $\left\{x \mid x = \dfrac{2\pi}{3} + k \cdot 2\pi \ \ \text{or} \ \ x = \dfrac{4\pi}{3} + k \cdot 2\pi\right\}$ **33.** $\left\{x \mid x = k \cdot \pi \ \ \text{or} \ \ x = \pm\dfrac{\pi}{3} + k \cdot 2\pi\right\}$

35. $\left\{x \mid x = \pm\dfrac{\pi}{6} + k \cdot 2\pi\right\}$ **37.** $k \cdot 2\pi$

39. $\dfrac{3\pi}{2} + k \cdot 2\pi$ **41.** $\dfrac{3\pi}{2} + k \cdot 2\pi$ **43.** $k \cdot 2\pi$

45. a] $K = 25\pi \sec \theta$

    b] Domain $= \{\theta \mid 0 \le \theta \le \mathrm{Tan}^{-1} 6/5\}$, $\mathrm{Tan}^{-1} 6/5 \approx 0.876$

Exercise 5.3 (page 198)

When general solutions are given, it is to be understood that the letter k ranges over the entire set of integers (zero, positive, and negative).

1. $\{30°, 150°, 270°\}$ **3.** $\{180°\}$ **5.** $\{30°, 150°, 210°, 330°\}$ **7.** $\{60°, 300°\}$

9. $\left\{\dfrac{7\pi}{6}, \dfrac{11\pi}{6}\right\}$ **11.** $\{0.84, \pi, 5.44\}$ **13.** $\left\{\dfrac{3\pi}{4}, \dfrac{7\pi}{4}\right\}$ **15.** $\{0.97, 5.31\}$

17. $\left\{\dfrac{\pi}{2}, \dfrac{3\pi}{2}\right\}$ **19.** $\left\{\dfrac{3\pi}{2}\right\}$ **21.** $\left\{\dfrac{7\pi}{6}, \dfrac{11\pi}{6}\right\}$ **23.** No solution

25. $\{x\,|\,x = -30° + k \cdot 360°$ or $x = k \cdot 180°$ or $x = 210° + k \cdot 360°\}$

27. $\{x\,|\,x = \pm 60° + k \cdot 360°\}$

29. $\{x\,|\,x = 30° + k \cdot 180°$ or $x = 45° + k \cdot 180°\}$

Exercise 5.4 (page 202)

When general solutions are given, it is to be understood that the letter k ranges over the entire set of integers (zero, positive, and negative).

1. $\{0°, 90°, 360°\}$ **3.** $\{0°, 120°, 360°\}$ **5.** $\{120°, 300°\}$ **7.** $\{0°, 180°, 270°, 360°\}$

9. $\{30°\}$ **11.** $\left\{\dfrac{2\pi}{3}\right\}$ **13.** $\{0.64\}$ **15.** $\{1.57, 3.92\}$

17. No solution   **19.** $\{0, 1.57, 2.21, 4.71, 6.28\}$

21. $\left\{x\,|\,x = -\dfrac{\pi}{2} + k \cdot 2\pi$ or $x = \pi + k \cdot 2\pi\right\}$ **23.** $\left\{x\,|\,x = \dfrac{5\pi}{6} + k \cdot 2\pi\right\}$

Exercise 5.5 (page 206)

1. $\{30°, 150°\}$ **3.** $\{210°, 330°\}$ **5.** $\{30°, 90°, 150°, 270°\}$

7. $\{22.5°, 112.5°, 202.5°, 292.5°\}$ **9.** $\left\{\dfrac{\pi}{3}, \dfrac{5\pi}{3}\right\}$ **11.** $\left\{0, \dfrac{\pi}{2}, \pi, 2\pi\right\}$

13. $\left\{\dfrac{3\pi}{4}, \dfrac{7\pi}{4}\right\}$ **15.** $\left\{\dfrac{\pi}{4}, \dfrac{5\pi}{4}\right\}$ **17.** $\{0, 1.82, 3.14\}$ **19.** $\{0.37, 2.77\}$

21. $\{\pm 60°, \pm 90°\}$ **23.** $\{-30°, \pm 90°\}$ **25.** $\left\{\dfrac{\pi}{6}, \dfrac{5\pi}{6}\right\}$ **27.** No solution

29. $\left\{\dfrac{7\pi}{12}, \dfrac{\pi}{2}, \dfrac{11\pi}{12}\right\}$ **31.** $\left\{x\,|\,x = k\pi$ or $x = \pm \dfrac{2\pi}{3} + k \cdot 2\pi\right\}$

33. This is an identity; solution set consists of all allowable values of x.

35. Empty set   **37.** Empty set   **39.** $\left\{\dfrac{\pi}{4}, \dfrac{3\pi}{4}, \dfrac{5\pi}{4}, \dfrac{7\pi}{4}\right\}$

Exercise 5.6 (page 210)

1. 0   **3.** −0.74   **5.** 1.17   **7.** 1.76

9. 1.18   **11.** 1.11

Exercise 5.7 (page 214)

1. 1/2   **3.** −1/2   **5.** $-\sqrt{3}$ **7.** 1

9. No solution   **11.** 1/2   **13.** 0.5

15. $x = \frac{\pi}{6} + k \cdot 2\pi$ or $x = \frac{5\pi}{6} + k \cdot 2\pi$ **17.** $\frac{1}{\sqrt{2}}$ **19.** $x = \frac{\pi}{2} + k \cdot 2\pi$

21. $-\dfrac{1}{\sqrt{3}}$ **23.** 0.3   **25.** $\{0.97\}$ **27.** $\{\pm 0.92\}$

29. $\{x \mid 0.39 \le x \le 1\}$ **31.** $\{x \mid -1 \le x \le -0.8$ or $0.8 \le x \le 1\}$

33. a] $\{0.93\}$ **b]** $\{x \mid -1 \le x \le 0.93\}$

35. a] $\{0.4\}$ **b]** $\{x \mid x = 0.4 + k \cdot 2\pi$ or $x = 2.74 + k \cdot 2\pi\}$

Chapter 5 Review Exercise (page 217)

1. $\{45°, 315°\}$ **3.** $\{30°\}$ **5.** $\{60°, 120°\}$ **7.** $\{0°, 120°, 240°, 360°\}$

9. $\left\{\frac{\pi}{3}, \frac{2\pi}{3}, \frac{4\pi}{3}, \frac{5\pi}{3}\right\}$ **11.** $\left\{\frac{\pi}{3}, \frac{2\pi}{3}, \frac{4\pi}{3}, \frac{5\pi}{3}\right\}$ **13.** $\left\{\frac{3\pi}{4}\right\}$ **15.** $\{0, 2\pi\}$

17. $\{0.46\}$ **19.** $\{0.97, 2.18\}$ **21.** $\{0.34, 1.57, 2.80\}$ **23.** $\{\pm 0.8\}$

25. $\left\{x \mid 0 \le x \le \frac{\pi}{6}$ or $\frac{5\pi}{6} \le x \le 2\pi\right\}$ **27.** $\{x \mid 0.24 < x \le 1\}$

29. $\{x \mid -1 \le x \le -0.92$ or $0.92 \le x \le 1\}$

Exercise 6.1 (page 228)

For each problem we give (a) the fundamental interval denoted by FI, (b) period p, (c) frequency $f = 1/p$, and (d) amplitude A. If the fundamental cycle is upside-down, we indicate that by (U). The equivalent equations obtained by reduction formulas are used to determine the fundamental interval when listed.

Problem	Equivalent equation	FI	p	f	A
1.		$[0, 2\pi]$	2π	$\frac{1}{2\pi}$	2
3.		$[0, 2\pi]$	2π	$\frac{1}{2\pi}$	4 (U)
5.		$[0, 2\pi]$	2π	$\frac{1}{2\pi}$	$\frac{1}{2}$
7.		$\left[0, \frac{2\pi}{3}\right]$	$\frac{2\pi}{3}$	$\frac{3}{2\pi}$	1
9.		$\left[0, \frac{2\pi}{3}\right]$	$\frac{2\pi}{3}$	$\frac{3}{2\pi}$	2 (U)
11.	$y = 4 \sin\left(x - \frac{\pi}{4}\right)$	$\left[\frac{\pi}{4}, \frac{9\pi}{4}\right]$	2π	$\frac{1}{2\pi}$	4
13.		$[0, 4]$	4	$\frac{1}{4}$	3 (U)
15.	$y = -2 \cos \pi x$	$[0, 2]$	2	$\frac{1}{2}$	2 (U)
17.	$y = \cos x$	$[0, 2\pi]$	2π	$\frac{1}{2\pi}$	1
19.		$\left[-\frac{\pi}{4}, \frac{5\pi}{12}\right]$	$\frac{2\pi}{3}$	$\frac{3}{2\pi}$	2

Problem	Equivalent equation	FI	p	f	A
21.	$y = -4 \sin 3x$	$\left[0, \dfrac{2\pi}{3}\right]$	$\dfrac{2\pi}{3}$	$\dfrac{3}{2\pi}$	4 (U)
23.	$y = 2 \cos 2x$	$[0, \pi]$	π	$\dfrac{1}{\pi}$	2
25.		$\left[\dfrac{1}{3}, \dfrac{4}{3}\right]$	1	1	3 (U)
27.	$y = 3 \cos 2\pi x$	$[0, 1]$	1	1	3
29.	$y = -3 \sin 2\pi x$	$[0, 1]$	1	1	3 (U)
31.	$y = \cos x$	$[0, 2\pi]$	2π	$\dfrac{1}{2\pi}$	1
33.	$y = -4 \cos x$	$[0, 2\pi]$	2π	$\dfrac{1}{2\pi}$	4 (U)
35.	$y = 2 \cos \pi x$	$[0, 2]$	2	$\dfrac{1}{2}$	2

Exercise 6.2 (page 235)

For each problem we give (a) the fundamental interval, denoted by FI, (b) the location of vertical asymptotes, and (c) a point to determine the vertical scale. An upside-down fundamental cycle is indicated by (U). The letter k denotes any integer (zero, positive, or negative). If an equivalent function is used to determine the fundamental interval, the equivalent equation is listed.

Problem	Equivalent Equation	FI	Asymptotes	Point
1.		$\left(-\dfrac{\pi}{2}, \dfrac{\pi}{2}\right)$	$x = \dfrac{\pi}{2} + k\pi$	$\left(\dfrac{\pi}{4}, 3\right)$
3.		$(0, \pi)$	$x = k \cdot \pi$	$\left(\dfrac{\pi}{4}, -2\right)$ (U)
5.		$(-\pi, \pi)$	$x = \pi + k \cdot 2\pi$	$\left(\dfrac{\pi}{2}, -2\right)$ (U)
7.	$y = 4 \cot 2x$	$\left(0, \dfrac{\pi}{2}\right)$	$x = k \cdot \dfrac{\pi}{2}$	$\left(\dfrac{\pi}{8}, 4\right)$
9.	$y = -2 \cot \pi x$	$(0, 1)$	$x = k$	$\left(\dfrac{1}{4}, -2\right)$ (U)
11.	$y = -\sqrt{3} \cot \pi x$	$(0, 1)$	$x = k$	$\left(\dfrac{1}{4}, -\sqrt{3}\right)$ (U)
13.		$\left(-\dfrac{\pi}{2}, \dfrac{3\pi}{2}\right)$	$x = \dfrac{\pi}{2} + k\pi$	$(0, 3)$
15.		$\left(-\dfrac{\pi}{4}, \dfrac{3\pi}{4}\right)$	$x = \dfrac{\pi}{4} + k \cdot \dfrac{\pi}{2}$	$(0, -3)$ (U)
17.		$\left(0, \dfrac{2}{3}\right)$	$x = \dfrac{k}{3}$	$\left(\dfrac{1}{6}, 1\right)$
19.	$y = 3 \sec 2\pi x$	$\left(-\dfrac{1}{4}, \dfrac{3}{4}\right)$	$x = \dfrac{1}{4} + \dfrac{k}{2}$	$(0, 3)$
21.	$y = -2 \csc 2x$	$(0, \pi)$	$x = k \cdot \dfrac{\pi}{2}$	$\left(\dfrac{\pi}{4}, -2\right)$ (U)

Exercise 6.3 (page 238)

1.

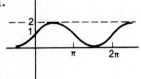

3.

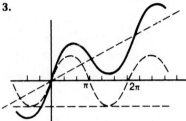

5.

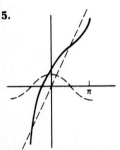

7.

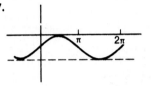

9.

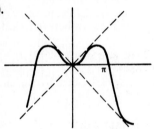

11.

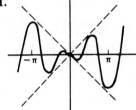

13.

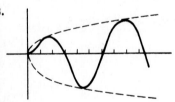

15.

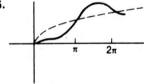

Exercise 6.4 (page 243)

For problems 1–19 we give a convenient form of the function (obtained by the use of identities) and describe the graph.

1. $y = 1 - \sin 2x$. Upside-down sine curve of period π oscillating about $y = 1$. $D = \mathbf{R}$, $R = [0, 2]$.

3. $y = 2 \sin x$, if $\cos x \neq 0$. Sine curve with holes, $A = 2$, fundamental interval $[0, 2\pi]$. $D = \{x \mid x \neq (2k - 1) \, \pi/2\}$, $R = (-2, 2)$.

5. $y = \sin 2x$, if $\sin x \neq 0$. Sine curve with holes, $A = 1$, fundamental interval $[0, \pi]$. $D = \{x \mid x \neq k\pi\}$, $R = [-1, 1]$.

7. $y = \sqrt{2} \sin (x + \pi/4)$. Sine curve, $A = \sqrt{2}$, fundamental interval $[-\pi/4, 7\pi/4]$. $D = \mathbf{R}$, $R = [-\sqrt{2}, \sqrt{2}]$.

9. $y = \sqrt{2} \cos(x + \pi/4)$. Cosine curve, $A = \sqrt{2}$, fundamental interval $[-\pi/4, 7\pi/4]$. $D = \mathbf{R}$, $R = [-\sqrt{2}, \sqrt{2}]$.

11. $y = 1$ if $\cos 2x \neq 0$. Horizontal line with holes. $D = \{x \mid x \neq (2k-1)\,\pi/4\}$, $R = \{1\}$.

13. $y = 2 \cos(2x + \pi/3)$. Cosine curve, $A = 2$, fundamental interval $[-\pi/6, 5\pi/6]$. $D = \mathbf{R}$, $R = [-2, 2]$.

15. $y = x$, $-1 \leq x \leq 1$. Line segment from $(-1, -1)$ to $(1, 1)$. $D = [-1, 1]$, $R = [-1, 1]$.

17. $y = 2 \cot 2x$. Cotangent curve, fundamental interval $(0, \pi/2)$, asymptotes: $k \cdot \pi/2$. $D = \{x \mid x \neq k\pi/2\}$, $R = \mathbf{R}$.

19. $y = x$, $D = \mathbf{R}$, $R = \mathbf{R}$.

21. LHS: $\sqrt{2} \sin(x + \pi/4)$; RHS: $2 \csc 2x$. No solutions.

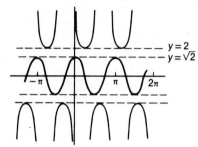

Chapter 6 Review Exercise (page 245)

For these problems, FI denotes the *fundamental interval*, p is the period, A is the amplitude, (U) indicates that the fundamental cycle is upside down, and k represents any integer.

1. FI: $[0, 2\pi]$, $p = 2\pi$, $A = 2$

3. FI: $\left[0, \dfrac{2\pi}{3}\right]$, $p = \dfrac{2\pi}{3}$, $A = \dfrac{1}{2}$

5. FI: $[0, 2\pi]$, $p = 2\pi$, $A = 2$ (U)

7. FI: $\left(-\dfrac{\pi}{4}, \dfrac{\pi}{4}\right)$, $p = \dfrac{\pi}{2}$, asymptotes: $\dfrac{\pi}{4} + k \cdot \dfrac{\pi}{2}$

9. FI: $\left(-\dfrac{\pi}{2}, \dfrac{\pi}{2}\right)$, $p = \pi$, asymptotes: $\dfrac{\pi}{2} + k\pi$. Basic tangent curve is translated 1 unit upward.

11. FI: $\left(0, \dfrac{\pi}{2}\right)$, $p = \dfrac{\pi}{2}$, asymptotes: $k \cdot \dfrac{\pi}{2}$

13.

15.

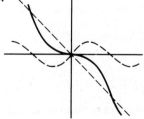

17. Equivalent equation is $y = \cot 2x$. FI: $\left(0, \dfrac{\pi}{2}\right)$, $p = \dfrac{\pi}{2}$, asymptotes: $k \cdot \dfrac{\pi}{2}$

19. FI: $\left[\dfrac{1}{3}, \dfrac{7}{3}\right]$, $p = 2$, $A = 1$

21. $y = \sqrt{2} \sin\left(x + \dfrac{\pi}{4}\right)$, FI: $\left[-\dfrac{\pi}{4}, \dfrac{7\pi}{4}\right]$, $p = 2\pi$, $A = \sqrt{2}$

23. $y = \sqrt{2} \sin x$, FI: $[0, 2\pi]$, $p = 2\pi$, $A = \sqrt{2}$

25. $y = x$, $-1 \leq x \leq 1$. Line segment from $(-1, -1)$ to $(1, 1)$

27.

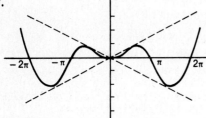

29. $y = \cos 4x$, FI: $\left[0, \dfrac{\pi}{2} \right]$, $p = \dfrac{\pi}{2}$, $A = 1$

Exercise 7.1 (page 253)

1. $b \approx 34$, $c \approx 35$, $\gamma = 80°$

3. $\beta = 21°$, $a \approx 64$, $b \approx 31$

5. $\gamma = 92°15'$, $b \approx 49.7$, $c \approx 60.4$

7. $\gamma = 80.79°$, $a \approx 4.734$, $b \approx 2.146$

9. $b \approx 57.5$ cm

11. Area ≈ 150 (2 sig. digits)

13. Area ≈ 20 (2 sig. digits)

15. 141 m

17. 141 m   **19.** 22 m

23. 23 m   **25.** 96 km/hr

27. 27 m   **35.** No

Exercise 7.2 (page 263)

1. $c \approx 48$, $\alpha \approx 31°$, $\beta \approx 106°$

3. $\alpha \approx 69°$, $\gamma \approx 28°$, $b \approx 90$

5. $a \approx 49$, $\beta \approx 30°$, $\gamma \approx 102°$

7. $\alpha \approx 58.7°$, $\beta \approx 31.3°$, $\gamma \approx 90.0°$

9. $\alpha \approx 46°$, $\beta \approx 30°$, $\gamma \approx 104°$

11. $a \approx 18.6$, $\beta \approx 91.0°$, $\gamma \approx 56.2°$

13. $\alpha \approx 33°20'$

15. $h \approx 2.83$, Area ≈ 7.56

17. 23.7

19. 85.5 km

21. 34.7°

27. 428 m

29. 7.6 cm, 11 cm (2 sig. digits)   **31. a]** 0.65

b] 200 (2 sig. digits)

33. Answers are rounded off to four decimal places.
   a] $\alpha \approx 41.4096°$, $\gamma \approx 82.8192°$

   b] $\alpha \approx 28.9550°$, $\gamma \approx 57.9100°$

   c] $\alpha \approx 33.5573°$, $\gamma \approx 67.1146°$

   d] $\alpha \approx 51.3178°$, $\gamma \approx 102.6356°$

37. a] If $m - n$ is an odd number, and m and n have no common factors, then the corresponding right-angle triple will have no common factors.

   b] If m and n have no common factors, then the corresponding double-angle triple will have no common factors.

Exercise 7.3 (page 272)

1. $\beta \approx 36°$, $\gamma \approx 91°$, $c \approx 110$ (2 sig. digits)

3. $\beta \approx 20°$, $\gamma \approx 36°$, $c \approx 2.5$

5. Two solutions: $c_1 \approx 63$, $\beta_1 \approx 35°$, $\gamma_1 \approx 120°$
$c_2 \approx 13$, $\beta_2 \approx 145°$, $\gamma_2 \approx 10°$

7. $\gamma \approx 23°$, $\alpha \approx 93°$, $a \approx 60$

9. $\alpha \approx 37°$, $\gamma \approx 26°$, $c \approx 41$

11. None   **13.** One

15. Two   **17.** $|\overline{BC}| \approx 5.5$, $|\overline{CD}| \approx 3.3$

19. a] 210 m (2 sig. digits)

21. $x = 36 - 12 \cos(10t) - 12\sqrt{4 - \sin^2(10t)}$

   b] 14°

Exercise 7.4 (page 276)

1. 879

3. 1010

5. 2110

7. 1190 m, Area ≈ 49300 m^2

9. 43200 m^2

11. $a \approx 6.89$ m, $b \approx 9.46$ m, $c \approx 7.46$ m

13. a] 1018.4 cm^2 (by factor of 4)
   b] 2291.4 cm^2 (by factor of 9)

17. Area $= \frac{1}{2}r^2 \left(\theta - \sin\theta\right)$, θ in radians

21. $36\pi \approx 113.097$

23. $72\sqrt{3}$

25. The sum of the areas of the three regions in Problem 24 is given by: Area $= 36n \sin(\pi/n)$. Evaluating for the given values of n gives 112.800, 113.023, 113.079, 113.097 (to 3 decimal places).

Exercise 7.5 (page 284)

1. 2.8 km, 32° west of north

5. 943 km, 5° west of south

7. 36 m, 17° east of north; 36 m, 17° west of south

9. 4.3 m, 34° east of north

11. a] $\sqrt{29} \approx 5.4$ **b]** $\sqrt{58} \approx 7.6$ **c]** 13

13. a] Same direction   **b]** Opposite direction   **c]** Perpendicular to each other

15. $(732, -217)$

Exercise 7.6 (page 288)

1. $3\vec{i} - 4\vec{j}$

3. $-4\vec{i} + 5\vec{j}$

5. $-9\vec{i} + 17\vec{j}$

7. $-4\vec{i} + 6\vec{j}$

9. $\sqrt{305} \approx 17.46$

11. θ for $\vec{u}$ is 30°; θ for $\vec{v}$ is 110°

13. $\vec{u} + \vec{v} \approx 0.21\vec{i} + 3.76\vec{j}$; magnitude ≈ 3.8 cm, direction 3° east of north

15. $487\vec{i} - 282\vec{j}$

17. $550\vec{i} - 1741\vec{j}$

19. 40° east of north

21. 2.0 km in the direction of 53° east of north

23. $\frac{3}{\sqrt{10}}\vec{i} + \frac{1}{\sqrt{10}}\vec{j}$ **25. a]** 60°15′

b] 60°15′

27. Sum of $2.50\vec{i} + 1.83\vec{j}$ and $0.50\vec{i} + 2.17\vec{j}$

29. b] $19\vec{i} + 22\vec{j}$

c] $17\vec{i} + 25\vec{j}$

31. 1052 km, 62° east of south

Chapter 7 Review Exercise (page 293)

1. $\beta = 59°$, $a \approx 6.7$, $c \approx 9.6$

3. $b \approx 87.1$, $\alpha \approx 22.8°$, $\gamma \approx 41.8°$

5. $\alpha \approx 78°$, $\beta \approx 42°$, $\gamma \approx 60°$

7. $\gamma = 20.2°$, $b \approx 90.3$, $c \approx 42.2$

9. $c \approx 52$, $\alpha \approx 92°$, $\beta \approx 33°$

11. $a \approx 54.3$, $\alpha \approx 59.3°$, $\beta \approx 30.7°$

13. One

15. Infinitely many

17. None

19. 30 (2 sig. digits)

21. 110 (2 sig. digits)

23. 5.1

25. 46.5°

27. a] $5\vec{i} + \vec{j}$

b] $\sqrt{26}$

29. $\vec{v} = -4\vec{i} - 4\sqrt{3}\vec{j}$

31. $\left(\dfrac{x_1 + x_2}{2}, \dfrac{y_1 + y_2}{2}\right)$

Exercise 8.1 (page 298)

1. a] P[3, 50°], [−3, 230°], [−3, −130°], [3, 410°]
   b] Q[4, −60°], [−4, 120°], [4, 300°], [4, 660°]
   c] T[2, 540°], [2, 180°], [2, −180°], [−2, 0°]

3. P_2 [3, 310°], Q_2 [4, 60°], T_2 [2, 180°]

5. $P_1\left[2, \dfrac{5\pi}{3}\right]$; $Q_1\left[3, \dfrac{\pi}{12}\right]$; $T_1\left[4, \dfrac{11\pi}{6}\right]$

7. a]
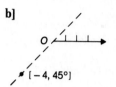

b]

c]
d]

9. a] $\left[3, \dfrac{4\pi}{3}\right]$ **b]** $\left[4, \dfrac{\pi}{4}\right]$ **c]** [2, π]   **d]** $\left[3, \dfrac{3\pi}{2}\right]$

Exercise 8.2 (page 302)

1.
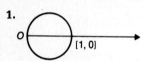

3.
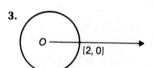

5.
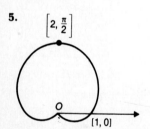

7.
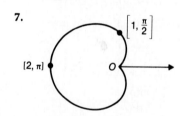

9.
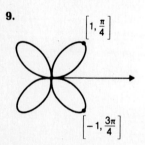

11.
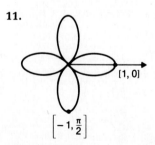

13.
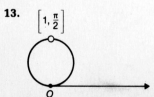

15.
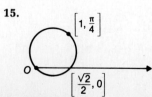

7.

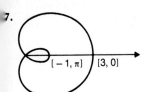

$[-1, \pi]$ $[3, 0]$

19.

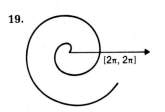

$[2\pi, 2\pi]$

Exercise 8.3 (page 305)

1. a] $[\sqrt{2}, 135°]$ **b]** $[2, 240°]$ **c]** $[5.09, 51.85°]$ **d]** $[2.89, 122.87°]$

3. a] $\left[3\sqrt{2}, \dfrac{3\pi}{4}\right]$ **b]** $[\sqrt{10}, 5.03]$ **c]** $[3.53, 0.48]$

5. a] $(0, 2)$ **b]** $\left(\dfrac{3\sqrt{2}}{2}, \dfrac{3\sqrt{2}}{2}\right)$ **c]** $(2.09, -0.81)$

7. a] Yes   **b]** No   **c]** Yes   **d]** No   **e]** No

11. $x^2 + y^2 - 2y = 0$ **13.** $y = (\tan\frac{4}{3})x$ or $y \approx 4.13x$

15. $y = 0.25x^2 - 1$ **17.** $x^2 + y^2 + 2x = 0$ **19.** $r = 1$, circle

21. $\theta = \mathrm{Tan}^{-1}\, 3 \approx 1.25$; line through the origin with slope 3

23. No, the origin is a point on $r = \sin\theta$ but not on $r\csc\theta = 1$ **25.** Yes

Chapter 8 Review Exercise (page 308)

1. a] $[1, 0]$ **b]** $[3, \pi]$ **c]** $\left[4\sqrt{2}, \dfrac{\pi}{4}\right]$ **d]** $\left[2\sqrt{2}, \dfrac{3\pi}{4}\right]$

e] $\left[2, \dfrac{7\pi}{6}\right]$ **f]** $\left[2, -\dfrac{\pi}{4}\right]$ **g]** $\left[4, \dfrac{\pi}{2}\right]$ **h]** $\left[3, \dfrac{3\pi}{2}\right]$

3. a] $(2, 2\sqrt{3})$ **b]** $(\sqrt{3}, -1)$ **c]** $(-4, 0)$

d] $\left(-\dfrac{1}{\sqrt{2}}, -\dfrac{1}{\sqrt{2}}\right)$ **e]** $\left(\dfrac{3}{\sqrt{2}}, \dfrac{3}{\sqrt{2}}\right)$

5. Graph is a circle of radius $\frac{1}{2}$ **7.** Graph is a circle of radius 1

9. Graph is a vertical line three units to the right of the origin

11. Graph is a spiral   **13.** $r^2 = 4$; circle with center at the origin and radius 2

15. $y^2 + 2x - 1 = 0$; parabola that opens to the left

Exercise 9.1 (page 314)

1. a] $-i$ **b]** -1 **c]** 1 **d]** i
e] i **f]** $-i$ **g]** 1 **h]** $-i$

3. a] 12 **b]** $12i$ **c]** -12 **d]** $-\frac{3}{4}i$
e] $\frac{3}{4}i$ **f]** $\frac{3}{4}$

5. a] 4 **b]** $-1 - 5i$ **c]** $\dfrac{4 - 3\sqrt{2}}{2} - \dfrac{2 + 3\sqrt{2}}{2}i$

9. a] $3 - 8i$ **b]** $2 + \sqrt{3}i$ **c]** $-4 + 4\sqrt{2}i$ **d]** 9
e] $\frac{1}{10} + \frac{3}{10}i$ **f]** $\dfrac{4}{17} + \dfrac{3\sqrt{2}}{17}i$

11. a] $2i; -\frac{1}{2}i$ **b]** $i; -3i$
c] $\dfrac{\sqrt{13} - 3}{2}i; -\dfrac{\sqrt{13} + 3}{2}i$ **d]** $i; -\frac{1}{2}i$

13. $x = -1, y = 2$ **15.** $x = -5, y = 13$ or $x = 2, y = 6$ **17.** Yes

19. a] No   **b]** No

Exercise 9.2 (page 318)

1. (3, 5)   **3.** (0, 4)   **5.** (0, 0)   **7.** (2, 1)

9. $-4i$ **11.** $-4 - 3i$

13.

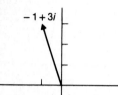

15.

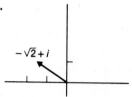

17.

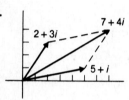

19.

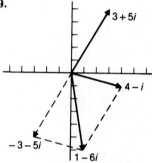

21. a] $3 - 4i$ **b]** $-3 + 4i$ **c]** $3 + 4i$ **d]** 3 **e]** $-4i$ **f]** 5

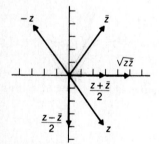

23. a] $-\dfrac{1}{2} + \dfrac{\sqrt{3}}{2}i$ **b]** $-\dfrac{1}{2} - \dfrac{\sqrt{3}}{2}i$ **c]** $-\dfrac{1}{2} - \dfrac{\sqrt{3}}{2}i$ **d]** 1

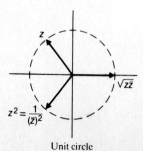

5. a] $x > 0$ and $y = 0$ **b]** $x = 0$ **c]** $x > 0$ and $y > 0$
    d] $x > 0$ **e]** $y < 0$

Exercise 9.3 (page 325)

3. a] $3(\cos 270° + i \sin 270°)$ **b]** $5(\cos 306.87° + i \sin 306.87°)$
    c] $\sqrt{2}\,(\cos 135° + i \sin 135°)$ **d]** $13(\cos 337.38° + i \sin 337.38°)$

5. a] $\dfrac{3\sqrt{2}}{2} + \dfrac{3\sqrt{2}}{2}i$ **b]** -5 **c]** $-\dfrac{1}{2} - \dfrac{\sqrt{3}}{2}i$

7. a] $4(\cos 315° + i \sin 315°)$ **b]** $3(\cos 120° + i \sin 120°)$
    c] $\cos \dfrac{\pi}{6} + i \sin \dfrac{\pi}{6}$

9. a] $\cos 45° + i \sin 45°$ **b]** $\dfrac{\sqrt{2}}{2} + \dfrac{\sqrt{2}}{2}i$

11. a] $2(\cos 120° + i \sin 120°)$ **b]** $-1 + \sqrt{3}\,i$

13. $18(\cos 180° + i \sin 180°) = -18$ **15.** $\dfrac{1}{6}\left[\cos(-30°) + i \sin(-30°)\right] = \dfrac{\sqrt{3}}{12} - \dfrac{1}{12}i$

17. a] $2\left[\cos(-30°) + i \sin(-30°)\right]$ **b]** $2\sqrt{2}\left[\cos(-135°) + i \sin(-135°)\right]$

19. a] $\dfrac{\sqrt{2}}{2}\left[\cos(-105°) + i \sin(-105°)\right]$ **b]** $\dfrac{\sqrt{2}}{2}(\cos 105° + i \sin 105°)$

23. a] $-4i$ **b]** -8

Exercise 9.4 (page 330)

1. a] $\cos 150° + i \sin 150° = -\dfrac{\sqrt{3}}{2} + \dfrac{1}{2}i$ **b]** $16\left[\cos(-180°) + i \sin(-180°)\right] = -16$
    c] $\cos 240° + i \sin 240° = -\dfrac{1}{2} - \dfrac{\sqrt{3}}{2}i$

3. a] $8\left[\cos(-90°) + i \sin(-90°)\right] = -8i$ **b]** $\cos 180° + i \sin 180° = -1$

5. a] $16(\cos 0° + i \sin 0°) = 16$ **b]** $16(\cos 240° + i \sin 240°) = -8 - 8\sqrt{3}i$
    c] $\dfrac{\sqrt{2}}{4}(\cos 225° + i \sin 225°) = -\dfrac{1}{4} - \dfrac{1}{4}i$

7. a] $256(\cos 180° + i \sin 180°) = -256$ **b]** $8(\cos 90° + i \sin 90°) = 8i$

9. $-5 + i$ **11.** $4 + 8\sqrt{3}i$ **13. a]** $1 + 3i$ **b]** $-9 - 3i$

15. $\sin 3\theta = 3 \sin \theta \cos^2 \theta - \sin^3 \theta = 3 \sin \theta - 4 \sin^3 \theta$;
$\cos 3\theta = \cos^3 \theta - 3 \sin^2 \theta \cos \theta = 4 \cos^3 \theta - 3 \cos \theta$

Exercise 9.5 (page 335)

1. $1; -\dfrac{1}{2} + \dfrac{\sqrt{3}}{2}i; -\dfrac{1}{2} - \dfrac{\sqrt{3}}{2}i$

3. $1.12 - 0.24i; 0.57 + 0.99i; -0.77 + 0.85i; -1.05 - 0.47i; 0.12 - 1.14i$

5. $\dfrac{\sqrt{3}}{2} + \dfrac{1}{2}i; i; -\dfrac{\sqrt{3}}{2} + \dfrac{1}{2}i; -\dfrac{\sqrt{3}}{2} - \dfrac{1}{2}i; -i; \dfrac{\sqrt{3}}{2} - \dfrac{1}{2}i$

7. $2.36 + 0.31i; -0.31 + 2.36i; -2.36 - 0.31i; 0.31 - 2.36i$

9. $\tfrac{1}{2}(\sqrt{3} + i); \tfrac{1}{2}(-\sqrt{3} + i); -i$ **11.** $2 + i; 1 - i$ **13.** $i; -1 - i$

15. $-1; \dfrac{\sqrt{2}}{2}(-1 + i); \dfrac{\sqrt{2}}{2}(1 - i)$ **17.** $2 - i; -2 + i$ **19.** $2 + 3i; -2 - 3i$

Chapter 9 Review Exercise (page 337)

1. $-2 + 2i$ **3.** $-7 - 24i$ **5.** $-\frac{1}{2}i$ **7.** $-\frac{\sqrt{3}+1}{32} + \frac{\sqrt{3}-1}{32}i$

9. $51.25 + 47.50i$ **11.** $2 - 2i$ **13.** 24   **15.** $-3 + 2\sqrt{3}i$

17. $0.98 - 0.17i; -0.34 + 0.94i; -0.64 - 0.77i$ **19.** $i; -i; -\frac{\sqrt{2}}{2} + \frac{\sqrt{2}}{2}i; \frac{\sqrt{2}}{2} - \frac{\sqrt{2}}{2}i$

Exercise 10.1 (page 343)

1. 4.729   **3.** 0.062   **5.** 7.389   **7.** 0.072
9. 3   **11.** 45   **13.** 54   **15.** 2048
17.

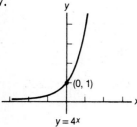

$y = 4^x$

19.

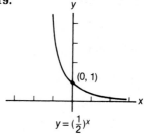

$y = (\frac{1}{2})^x$

21. a] 3   **b]** 3.8   **c]** $4 - \frac{\sqrt{5}}{5} \approx 3.553$ **d]** $4 - 5^{0.24} \approx 2.529$

23. a] 48,000   **b]** 51,000   **c]** 70,000   **25.** No

Exercise 10.2 (page 349)

1. 5   **3.** $\frac{5}{2}$ **5.** 2   **7.** -2

9. -1 **11.** Undefined   **13.** Undefined   **15.** $\frac{5}{2}$

17. 2.5743   **19.** 0.6309   **21.** 1.7223   **23.** 0.4307

25. $\log_3\left(\frac{25}{4}\right)$ **27.** 81   **29.** 2   **31.** 1

33. 2   **35.** True   **37.** Meaningless   **39.** True

41. $\{x \mid x > -1\}$ **43.** $\{x \mid x < 0\}$ **45.** $\{x \mid x < 0 \quad \text{or} \quad x > 4\}$

Exercise 10.3 (page 356)

1. 1.6094   **3.** 0.2718   **5.** 2.2912   **7.** Undefined
9. -0.4666 **11.** 0.7730   **13.** 1.43   **15.** -0.42

17. $\frac{1}{2}$ **19.** 4.6274   **21.** 0.4809   **23.** -0.1335

25. No solution   **27.** 42.2561   **29.** 1.2019   **31.** 0.9423

33. $\{x \mid x > 1\}$

35.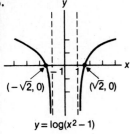

$y = \log(x^2 - 1)$

$(-\sqrt{2}, 0)$ $(\sqrt{2}, 0)$

37.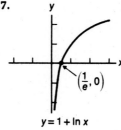

$\left(\frac{1}{e}, 0\right)$

$y = 1 + \ln x$

39. 2.34

Chapter 10 Review Exercise (page 358)

1. 0.903   **3.** 3.135   **5.** 10.751   **7.** 0.520

9. 1.292   **11.** 0.166   **13.** Undefined   **15.** $\frac{7}{2}$

17. $\dfrac{3}{\log e} \approx 6.908$ **19.** $e^{10} \approx 22026$ **21.** $\dfrac{\ln 10}{3 + \ln 10} \approx 0.434$ **23.** $1 + \dfrac{\ln 4}{\ln 3} \approx 2.262$

25. No solution

27.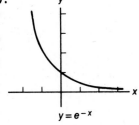

$y = e^{-x}$

29.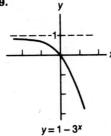

$y = 1 - 3^x$

31. $\dfrac{1 + \sqrt{1 + 4e}}{2} \approx 2.223$

Index

Index

TRIGONOMETRIC FUNCTIONS

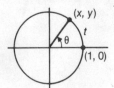 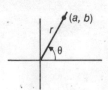

$$\sin t = \sin \theta = y = \frac{b}{r} \qquad \cos t = \cos \theta = x = \frac{a}{r}$$

$$\tan t = \tan \theta = \frac{y}{x} = \frac{b}{a} \qquad \cot t = \cot \theta = \frac{x}{y} = \frac{a}{b}$$

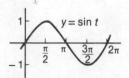 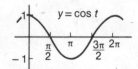 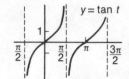

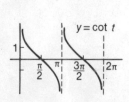 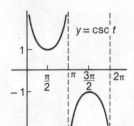 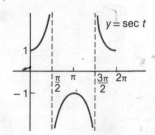

FORMULAS FROM TRIGONOMETRY

Angle Measures

$$180° = \pi \text{ radians} \qquad x \text{ deg} = \left(x \cdot \frac{\pi}{180}\right) \text{rad} \qquad x \text{ rad} = \left(x \cdot \frac{180}{\pi}\right) \text{deg}$$

Triangles

Law of sines

$$\frac{\sin \alpha}{a} = \frac{\sin \beta}{b} = \frac{\sin \gamma}{c}$$

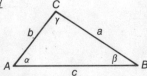

Law of cosines

$$a^2 = b^2 + c^2 - 2bc \overset{\cos}{\cancel{\sin}} \alpha$$

$$b^2 = a^2 + c^2 - 2ac \cancel{\sin} \beta$$

$$c^2 = a^2 + b^2 - 2ab \cancel{\sin} \gamma$$

area: $A = \dfrac{1}{2} ab \sin \gamma = \dfrac{1}{2} bc \sin \alpha = \dfrac{1}{2} ac \sin \beta$

$$A = \sqrt{s\,(s-a)(s-b)(s-c)} \qquad \text{where } s = \frac{a+b+c}{2}$$